Metallic Glasses and Their Composites

Updated 2nd Edition

D.V. Louzguine (Luzgin)

WPI Advanced Institute for Materials Research,

Tohoku University, Japan

Published by **Materials Research Forum LLC**
Millersville, PA 17551, USA

Published as part of the book series
Materials Research Foundations
Volume 85 (2021)
ISSN 2471-8890 (Print)
ISSN 2471-8904 (Online)

Print ISBN 978-1-64490-100-7
ePDF ISBN 978-1-64490-101-4

Distributed worldwide by

Materials Research Forum LLC
105 Springdale Lane
Millersville, PA 17551
USA
http://www.mrforum.com

Manufactured in the United States of America
10 9 8 7 6 5 4 3 2 1

Table of Contents

Preface

The formation process of both conventional metallic glasses and bulk metallic glasses through melt solidification using the glass-transition phenomenon (either by kinetic freezing or by a kind of phase transition) is still not fully understood. Two general mechanisms of glass formation related to stability of the supercooled liquid are presented here: the limiting of nucleation and the limiting of growth in relation to the formation of crystals and quasicrystals. Since the properties of metallic glasses, especially mechanical, are defined by their atomic structure, the research conducted on the structure of glassy alloys is emphasized as a particularly important topic in modern physical metallurgy.

Due to the absence of a crystalline lattice and dislocations, a unique deformation mechanism can be observed in bulk glassy alloys, which allows them to exhibit high strength, extreme hardness, good wear resistance and large elastic deformation. Though, metastable, but long-lived at room temperature, glassy alloys nevertheless devitrify/crystallize, upon heating. Devitrification term includes both formation of the supercooled liquid and crystallization upon heating. In some cases it is used to produce crystal/glassy composites with enhanced mechanical properties. Magnetic materials make up another extremely practically important and quickly developing application field of metallic glassy alloys.

Due to the excellent structural, functional, magnetic, chemical and biological properties metallic glasses are suitable for different applications, including sportive goods, cases, wire cords, elastic hinges, springs, diaphragms, membranes, knife blades, electromagnetic wave shields, optical mirrors, power inductors, Coriolis flow meters, biological implants, etc. Especially, metallic glasses are well applicable for high added value components where their relatively high material cost is not the limiting factor, such as: ferromagnetic cores of mini transformers, micro-geared motor parts in microelectromechanical devices, pressure sensors, orthopaedic screws, precision surgical instruments and so on. Future development of metallic glasses is likely going to be connected with dual-phase composite/hybrid materials utilizing attractive properties of both: the metallic glassy phase and an additional one.

Nomenclature

Abbreviations:

3D – three-dimensional

3DAP – 3D atom probe

AFM – atomic force microscopy

AXS – anomalous X-ray scattering

BMG – bulk metallic glass

BCC – body centered cubic (lattice)

CHT – continuous–heating-transformation (diagram)

DOS – density of states (total)

DMA – dynamic mechanical analysis

DSC – differential scanning calorimetry

EDX – energy dispersive X-ray (analysis)

ETM – early transition metal

EXAFS – extended X-ray absorption fine structure

FCC – face centered cubic (lattice)

GFA – glass-forming ability

HAADF – high-angle annular dark-field (image)

HCP – hexagonal close-packed (lattice)

LTM – late transition metal

MD – molecular dynamics

MRO – medium range order

PDOS – partial density of states

RE – rate earth (metal)

SAED – selected-area electron diffraction

SEM – scanning electron microscope

STM – scanning tunneling microscopy

STEM – scanning transmission electron microscopy

STZ – shear transformation zone

TM – transition metal

TEM – transmission electron microscopy

(HR)TEM – high-resolution transmission electron microscopy

TTT – time–temperature-transformation (diagram)

VFTH – Vogel-Fulcher-Tammann-Hesse

XAFS – X-ray absorption fine structure

XANES – X-ray absorption near edge structure

XRD – X-ray diffraction

Latin symbols:

AP – area under the peak

APR – ratio of the first to second peak areas (AP1/AP2)

a – lattice parameter (crystal)

a_q – quasilattice constant (quasicrystal)

B_s – magnetic flux density

Bi – Biot number

C_p – specific heat capacity

c – concentration

D – diffusion coefficient

D_c – critical diameter (for glass formation)

D_f – liquid fragility (thermodynamic)

D_{hkl} – coherent scattering area size (X-ray)

d – grain size

d_p – penetration depth

E – Young's modulus

E' – Storage modulus

E'' – Loss modulus

f – frequency

f_m – X-ray atomic scattering factor

G – shear modulus

H – enthalpy

H^* – magnetic field strength

H_c – coercive force (field)

HV – Vickers microhardness

h – heat transfer coefficient at the interface

K – bulk modulus

K_{Ic} – fracture toughness

k – thermal conductivity

L – diffusion length

M_s – saturation magnetization

m – liquid fragility (kinetic)

n – Avrami exponent

P – pressure

P_a – power absorbed per unit volume

$PDF(R)$ – pair distribution function

Q – scattering vector

Q^{-1} – internal friction

Q_a – apparent activation energy for crystallization

Q_g – activation energy for growth

Q_f – heat flow

$Q_i(Q)$ - interference function

Q_n – activation energy for nucleation

Q_v – activation energy of viscous flow

R – distance

R – gas constant

R_D – liquid fragitity parameter

$RDF(R)$ – radial distribution function

r – radius of a crystal nucleus

S – entropy

S_C – configurational entropy

S_d – entropy of configurons

$S(Q)$ - structure factor

T – temperature

T_c – Curie temperature

T_g – glass transition temperature

T_K – Kauzmann temperature

T_l – liquidus temperature

T_m – melting (solidus) temperature

T_{rg} – reduced glass transition temperature

T_x – crystallization temperature

ΔT_x – supercooled liquid region

t – time

V – volume

V_m – molar volume

*when the symbols are similar the difference is clear from the context

Greek symbols:

α – thermal diffusivity

β – thermal expansion coefficient

β_i – initial cooling rate

γ – parameter indicating GFA

δ – parameter indicating GFA

δ – phase lag between cycling stress and strain

ε – strain

ε' – strain rate

η – viscosity

η_s – von-Mises shear strain

θ – diffraction angle

θ_A – Arrhenius crossover temperature

θ_D – Debye temperature

λ – wavelength (X-ray)

λ_l – thermal conductivity of a liquid

λ_t – topological instability criterion

λ_s – magnetostriction

μ – magnetic permeability

ν – Poisson's ratio

ρ – mass density

$\rho(R)$ – radial number density

σ – stress

σ_y – yield strength

$\sigma_{0.2}$ – proof strength (usually equal to σ_y)

σ_e – electrical conductivity

τ – relaxation time

$\tau_{1,2}$ – time constants

ω – angular frequency

CHAPTER 1

Formation of Metallic Glasses and Their General Physical Properties

Metallic glasses and bulk metallic glasses (BMGs), in particular, are usually produced by solidification of a melt while amorphous alloys can be produced by various other techniques including physical and chemical vapor deposition, electrodeposition, mechanical attrition, ion implantation, sintering and others. The formation mechanisms of bulk metallic glasses and their general physical properties are discussed in the present chapter.

Contents

1.1   Formation of metallic glasses

The first metallic glassy alloy (or metallic glass) [1], namely Au-Si one, was produced by rapid solidification of the melt in 1960 [2]. Ternary Au-Si-Ge [3] alloys followed by other system alloys [4,5] developed later, demonstrated a better glass-forming ability (GFA). They can be produced in the form of thin ribbons of about 1-10 mm in width and 20-50 μm in thickness (Fig. 1.1).

Fig. 1.1. A metallic glassy ribbon sample produced by the melt-spinning technique.

In general, metallic glasses can be produced using various processing methods. Amorphous metallic materials were produced since the middle of the past Century by thin film deposition [6]. However, the films and powders produced by a solid state reaction using mechanical attrition, such as ball milling [7] or by severe plastic deformation [8] and later compacted using thermal (hot pressing in a WC die [9]), microwave (MW) [10,11] (metallic powders absorb microwave radiation) or spark plasma sintering (SPS) (Fig. 1.2a) [12] are usually called amorphous. Nevertheless, Ni-based bulk metallic glassy (BMG) samples with a size of 20 mm and nearly 100% relative density (Fig. 1.3a) were fabricated by spark plasma sintering of the gas-atomized $Ni_{52.5}Nb_{10}Zr_{15}Ti_{15}Pt_{7.5}$ (all alloy compositions are given in nominal atomic percents (at.%)) glassy powders (Fig. 1.3b) (here and elsewhere the chemical composition is given in nominal at.%) [13]. when cases sintering is performed within the supercooled liquid their properties are similar to those formed by casting. The initial ingots of metallic glasses are usually prepared either by arc-melting or by induction melting of elemental metals having more than 99.9 mass% purity in an argon atmosphere (Fig. 1.3 (c)).

Composite materials based on a polymer and a metallic glass, for example polyethylene terephthalate (PET about 1% by mass) and $Al_{85}Y_8Ni_5Co_2$ metallic glass, were also produced by mechanical alloying and consequent spark plasma sintering (Fig. 1.2b). It was found that composite sample ($Al_{85}Y_8Ni_5Co_2$ / PET) have a better thermal conductivity in comparison with the pure PET samples [14].

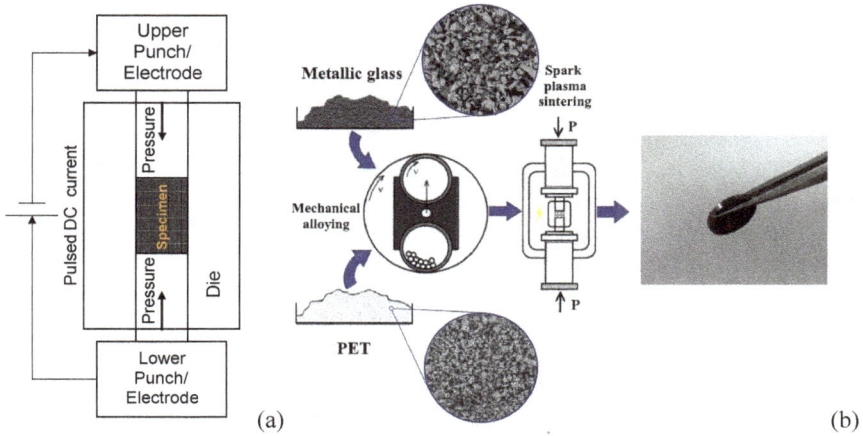

Fig. 1.2. (a) Schematic representation of the spark plasma sintering technique. (b) A scheme of the metallic glass-composite samples manufacturing using a mechanical alloying technique and subsequent spark plasma sintering. Reproduced from [14] with permission of MDPI.

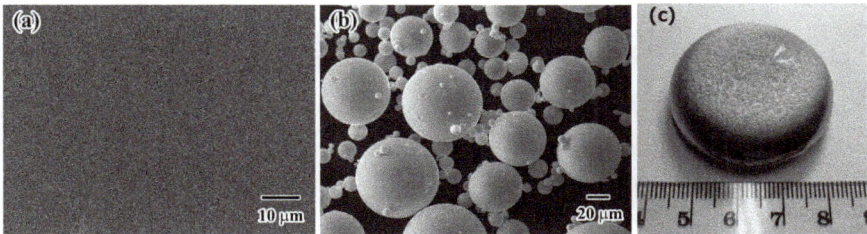

Fig. 1.3. SEM micrograph of the cross section of the sintered $Ni_{52.5}Nb_{10}Zr_{15}Ti_{15}Pt_{7.5}$ specimen obtained at a sintering temperature of 773 K (a). The micrograph of the original powder (b) is also shown for comparison. Reproduced from Ref. [13] with permission of American Institute of Physics. A photograph of an alloy ingot prepared by arc-melting later used for casting or atomization (c).

The application of pulsed voltage induces various phenomena caused by electrical and thermal effects, providing advantages which could not be realized using conventional sintering processes. In the SPS process, pulse electric current directly flows through the powder material being sintered, and a high heating efficiency is achieved. These phenomena can be summarized as follows [13]: electrical breakdown of a surface oxide film and removal of the contaminated layer on the particle surface by spark generation and sputtering effect; destruction of the surface oxide film and necks formation between the powder particles; Joule heating at the neck, and enhanced migration of an atom by the temperature difference between the neck and the particle core. Therefore, the sintering can be carried out in a shorter time than during conventional processes. Thus, the SPS process (Fig. 1.2a) or MW sintering process (Fig. 1.4) [10] can be applied to powdered materials that require suppressing crystallization and grain growth. Al-based bulk glassy samples of high relative density were obtained by warm extrusion of the atomized amorphous powders [15] because they have rather low glass-forming ability owing to a high density of so-called quenched-in nuclei in some glasses. Electrodeposition (electroplating) from a solution can also be used for producing metallic glasses [16].

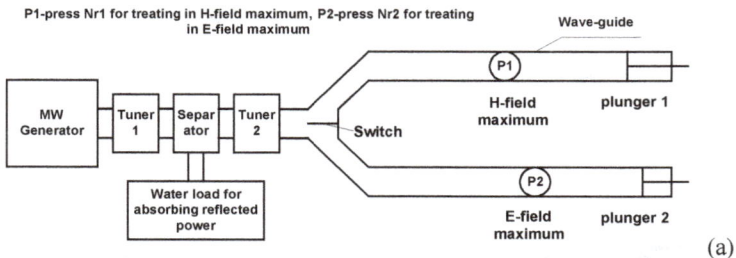

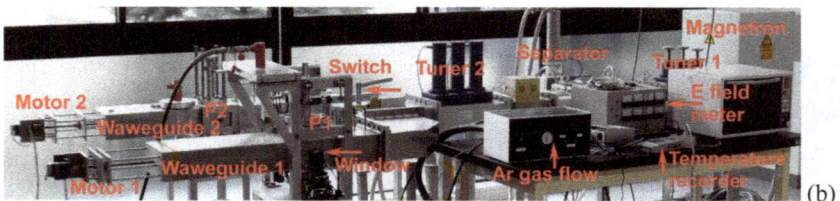

Fig. 1.4. A scheme (a) and a photograph (b) of a custom made microwave sintering apparatus consisting of a 5 kW power microwave generator (915 MHz), two tuners, a separator for absorbing the reflected power, two MW waveguides with standing waves tuned by plungers 1 and 2 used to treat powdered samples either in electrical or magnetic field maxima and press them in alumina dies (P1 and P2). Each MW waveguide and press has a small quartz window to control the sample temperature through an optical waveguide connected to an infrared pyrometer.

Bulk metals are opaque to electromagnetic radiation and microwaves in particular. However, microwave absorption takes place in a thin skin layer and the penetration depth (d_p) is:

$$d_p = \frac{1}{\sqrt{\pi f \mu \sigma}} \tag{1.1}$$

where f is frequency, μ is the magnetic permeability, and σ is the electrical conductivity. At 2.45 GHz frequency typical penetration depth in metallic materials is about several microns. Although it is about 1.6 times thicker if 915 MHz radiation is used (see Fig. 1.4) bulk metals are still not heated. The materials with a lower electrical conductivity like metallic glasses or intermetallic compounds have a thicker skin layer compared to pure metals. The heating rate (dT/dt) depends upon density (ρ), specific heat capacity (C_p) and power absorbed per unit volume (P_a):

$$\frac{dT}{dt} = \frac{P_a}{\rho C_p} \tag{1.2}$$

As shown in Fig. 1.4 there is possible to produce a standing wave in a waveguide and heat up the sample either in the magnetic (H) or electrical (E) field. The most of non-ferrous metallic glasses were heated in H-field maximum owing to eddy current heating mechanism, whereas no heating was observed in E-field maximum. Ferromagnetic metallic glasses were heated effectively in both fields: in H-field owing to both magnetic loss and eddy current heating mechanism and in E-field owing to the magnetic flux losses produced by the alternating E-field.

Much more productive technique for making BMG samples is solidification from a liquid phase by Cu-mold casting, for example, using a quartz crucible through a nozzle (Fig. 1.5 (a)). Among other alternative ways of casting one can enumerate tilt casting when the melt flows to the mould under gravity and suction casting into the mould by negative pressure difference in the chamber and inside the mold (Fig. 1.5 (b)) directly from the arc-melting furnace. Water cooling technique is another useful way of making BMG samples. Here a sample is molten in the glass pipe under Ar atmosphere and then cooled in water (Fig. 1.5 (c)).

Additive manufacturing (AM) is a novel materials manufacturing technique. It implies layer by layer addition and sintering of powder feedstock using a computer controlled laser beam. It is used to produce complex shaped functional metallic components for aerospace, automotive and biomedical industries. Laser sintering, melting and deposition are presently three most used AM processes and even bulk metallic glassy samples were produced by this method [17,18]. This procedure was claimed to drastically increase the fracture toughness of Fe-based metallic glassy composites [19]. Also, large scale 3D metallic glassy samples of complex geometry were produced by the Laser Foil Printing technology either by foil-welding by a continuous wave fiber laser or by foil-cutting by a pulsed laser [20].

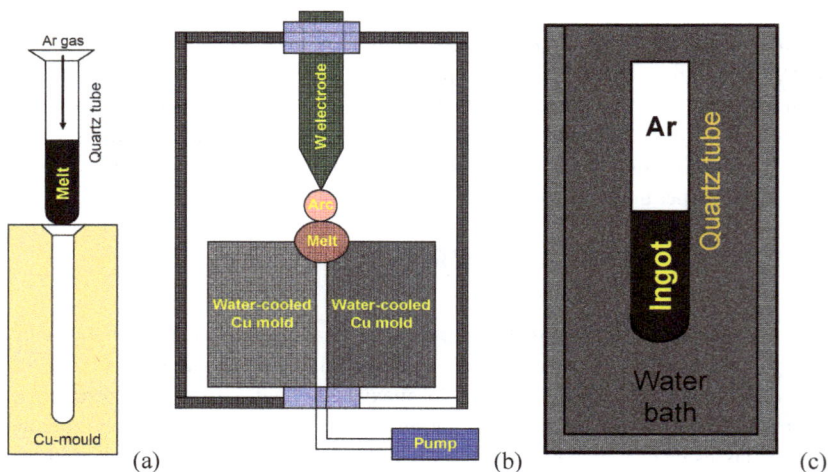

Fig. 1.5. Schematic representation of a bulk glassy sample Ar-gas injection casting technique (a), suction casting technique (b) and water cooling process of a BMG glass-forming melt encapsulated in a quartz tube under an Ar atmosphere (c).

After general saturation of the research activities in the field of conventional crystalline alloys since the end of the past century [21] bulk metallic glasses and glassy-crystal dual phase materials, owing to their good mechanical, chemical, magnetic and other properties, are at the research frontier among metallic materials [22]. These materials having a non-crystalline structure exhibit high strength [15], good wear [23] and corrosion resistance [24]. Bulk glassy alloys having an extraordinary high GFA [25] among metallic alloys have been widely produced after the breakthrough achieved in the end of 1980s and beginning of 1990s. They can be produced at the cooling rates of the order of 100, 10, 1 K/s and even less (for example, the famous $Pd_{40}Cu_{30}Ni_{10}P_{20}$ alloy [26]), which is much lower than those of 10^4-10^6 K/s required for vitrification of marginal glass-formers by using a rapid solidification technique.

The photographs of the cast BMG alloy samples produced by using Ar gas pressure casting (typical sample size samples up to 10 mm) and gravity casting (typical size from 5 to 25 mm) are shown in Fig. 1.6 (a,b). Larger size samples are usually produced by water cooling in quartz or steel tubes. One should be noted that $Au_{49}Cu_{26.9}Ag_{5.5}Pd_{2.3}Si_{16.3}$ bulk metallic glassy rods have a thin crystalline surface layer which consists of the Au-based solid solution grains embedded in the amorphous phase [27] responsible for gold-yellow color of the samples (Fig. 1.6 (c)). The lattice parameter of the Au solid solution equals to 0.401 nm, which corresponds to about 20 at.% Cu dissolved. The atomic sizes of Ag and Pd atoms are similar to that of Au and do not change the lattice parameter significantly while Si is insoluble in Au. FCC Au nucleated heterogeneously from the surface crystalline layer of Cu mold.

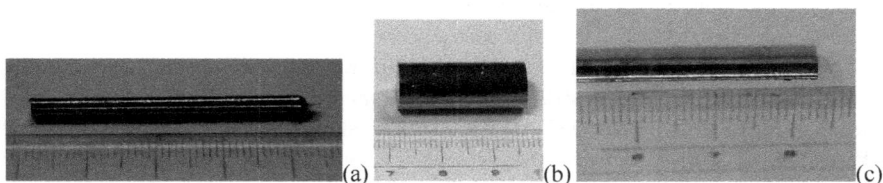

Fig. 1.6. Two photographs of the cast Zr-based BMG alloy samples produced by the Ar gas pressure casting (a) and the gravity casting (b) techniques. The ingots have a usual metallic grey color. A photograph of an Au-based metallic glass (c). As it will be discussed in a later chapter gold color here comes from Au precipitated on the surface of the BMG sample.

Bulk metallic glasses of 1-2 mm diameter were first produced in the noble metals-based metallic systems [28]. A large number of bulk metallic glasses also called bulk metallic glassy alloys defined as 3-dimentional massive glassy articles with a size of not less than 1 mm (by another definition not less than 10 mm) in every spatial dimension have been produced during the last 20 years in the thickness range of 10^0-10^2 mm by using various casting processes [29,30,31,32].

Rapid Joule heating processing allows homogeneous, volumetric and controllable rapid heating of a bulk metallic glass at rates up to 10^6 K/s to the supercooled liquid region. It is used for quick shaping of bulk metallic glasses in the supercooled liquid region [33] or even production of metal-crystal composites [34]. Rapid thermoplastic forming of the liquid into complex net shapes is implemented under rheological conditions used in molding of plastics.

Bulk metallic glasses can be roughly classified into the metal-metal group glasses (the majority of BMGs like Zr-Cu-Al, for example) and metal-metalloid glasses (Fe-B-C-P with other elements, Pd-Ni-Cu-P, etc...). Although well-known noble-metals- [28] and lanthanide-metals- [35] based bulk metallic glasses are not particularly attractive for engineering applications due to their high cost and low oxidation resistance, respectively, Ti [36], Zr [32,37] and Mg-based [38] bulk metallic glasses discovered later on are quite promising engineering materials. BMG alloys with exceptionally large critical size were obtained in Ag-[39], Au-[40], Ca- [41,42], Cu-[43], Co-[44], Fe-[45], La-[46], Mg-[47], Nd-[48], Ni-[49], Pd-[50], Pt- [51], Ti-Zr- [52] and Zr-based [30], alloy systems. Large size ingots of Ti-Zr-Ni-Be, Zr-Ti-Ni-Be [53] and Zr-Cu-Ag-Al-Be [54] BMGs are produced because of Be addition. The addition of Be suppresses crystallization of Zr and Ti alloys as it forms deep eutectics with Zr/Ti and because there are no stable intermetallic compounds in the Zr,Ti-rich area. Moreover, Be-rich intermetallic compounds are mostly incongruently melting, and thus, difficult to form from the melt as well as in case of Mg-Y system [55]. B, P and C are well-known additions stimulating formation of the glassy phase in Fe-based BMG alloys. Multicomponent Fe-based BMGs were obtained in various metal-metalloid multicomponent alloy systems including Fe–(Al,Ga)–(P,C,B,Si) [56], Fe-Cr-Mo-Nb-C-B, Fe-Cr-Mo-Ta-C-B [57], Fe–Cr–Mo–P–C–B [58] and other systems. Cr-based BMGs [59] with high corrosion resistance and Co-based [44] BMGs were produced by substituting Fe for Cr and Co, respectively. Typical values of critical diameter of the ingot (D_c) and typical T_g values are shown in Table 1.1.

Multicomponent equiatomic bulk glassy alloys without a major component, such as $Ti_{20}Zr_{20}Hf_{20}Cu_{20}Ni_{20}$ [60], $Pd_{20}Pt_{20}Cu_{20}Ni_{20}P_{20}$ [61], Er-Gd-Y-Al-Co [62] and other system alloys [63] were also produced by analogy with the multiprincipal element alloys which are also called high-entropy alloys. Although the equiatomic compositions possess high configurational entropy the lowest liquidus temperatures enhancing the glass-forming ability are rarely found in the center of multicomponent phase diagrams.

Although Al-based metallic glasses are rather marginal glass-formers the addition of Co [64] improved the glass-forming ability of the Al-RE (RE-rare earth metal) and ternary Al-RE-TM (TM-transition metals) [65] alloys (critical ingot size reached 0.5 mm and even larger) and increased their mechanical strength. Mischmetal (a natural mixture of RE metals) can also be added to substitute RE metals in Al-based [66] and other alloys thus reducing the total cost. Quaternary $Al_{85}Y_8Ni_5Co_2$ alloy shows one of the widest supercooled liquid region on heating among Al-based metallic glasses [64]. The GFA of this alloy was further improved by the addition of Ca [67]. Based on these principles Al-based BMGs with a critical size of 1 mm were also produced [68].

Table 1.1. Maximum critical ingot diameter (D_c) and typical glass-transition temperature values in different alloy systems.

Alloys system	D_c (mm)	T_g (K)
Zr-Cu-TM-Al	30	650-750
Pd-Cu-Ni-P	80	550-600
Ti-Cu-Sn-TM	7	600-700
Mg-Cu-RE	25	400-450
Ca-Mg-Al-TM	10	400-500
La-TM-Al	20	400-450
Fe-Co-TM-ME	15	800-850

RE denotes rare-earth metal(s), ME denotes metalloid(s), TM denotes other transition metal(s)

Bulk metallic glass-metallic crystal [69] -ceramics [70], -polymer [71,72] composites were also produced by various methods including casting (both in-situ and ex-situ composites) and sintering (ex-situ composites). A secondary phase in the glassy matrix, for example crystalline one act as active barriers for shear band propagation, enabling multiple shear deformation and preventing sudden brittle rupture.

The ex-situ $Mg_{58}Cu_{28.5}Gd_{11}Ag_{2.5}$ bulk metallic glass composites dispersion strengthened by spherical Ti particles pre-mixed in the melt with different volume fractions and particle sizes were produced by injection casting [73]. The Ti powder particles absorb the deformation energy and branch the primary shear bands into multiple shear bands. They decrease the stress concentration in front of a shear band and enhance plasticity. A high-quality interface was found between Ti particles and the amorphous matrix. For a given volume fraction of Ti of 40% it is found that smaller particles improve the compression plasticity from 12 % plastic strain (at the particle size of 90 μm) up to 25 % strain (at the size of 50 μm). Higher volume fraction of Ti leads to even larger compression plasticity. This subject will be also discussed in Chapter 5.

1.2   General physical properties

From Fig. 1.6 one can admit good surface quality (shiny surfaces) of bulk metallic glasses suitable for near-net-shape processing. Their mass density is somewhat lower than that in the crystalline state. The specific heat capacity (C_p) of metallic glasses at room temperature is close to 3R but slightly higher than that the competing crystalline phase(s) [74]. The C_p of a metallic glass measured by using differential scanning calorimetry under continuous heating conditions increases gradually with increasing temperature. In the glass-transition region the slope of the plot changes quickly and the specific heat reaches a maximum in the supercooled liquid state (Fig. 1.7). The specific heat remains virtually constant in the supercooled liquid state, and decreases to a lower value (similar to that of the glassy state) on crystallization. Visible decrease in C_p starting at about 480 K is connected with heat release connected with the process of structural relaxation. Fast DSC measurements allowed measuring heat capacities of the metastable phases (supercooled liquid and metastable solids) [75]. The glass transition and structural relaxation processes will be discussed in the subsequent Chapters.

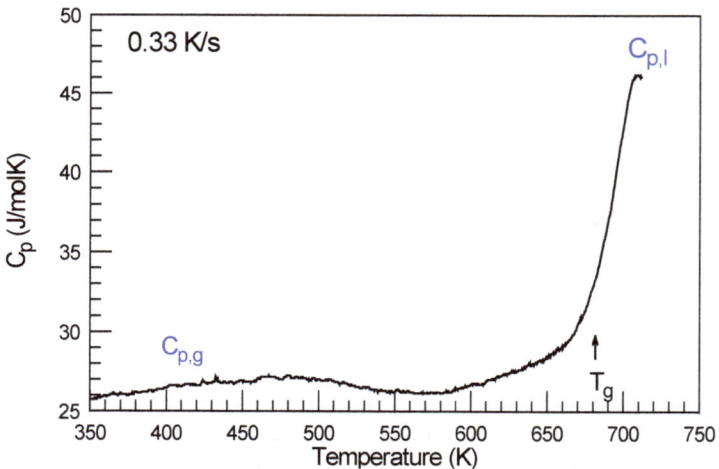

Fig. 1.7. C_p of the $Cu_{44}Ag_{15}Zr_{36}Ti_5$ metallic glass and liquid phases measured on heating at 0.33 K/s. $C_{p,g}$ and $C_{p,l}$ are C_p of the glassy and liquid state, respectively.

Metallic Glasses and Their Composites – 2nd Edition   Materials Research Forum LLC
Materials Research Foundations **85** (2021)   https://doi.org/10.21741/9781644901014

A non-zero residual entropy of the $Ni_{0.333}Zr_{0.667}$ metallic glass at 0 K was found to be 2.7 $\pm$ 2.1 J mol^{-1} K^{-1} is significantly lower than the residual entropy of oxide glasses [76]. This indicates a much higher degree of order in metallic glasses in comparison with glassy oxides.

Owing to their non-crystalline structure, bulk metallic glasses exhibit an electrical resistivity (1-3 $\mu\Omega\cdot$m), which is higher than that of the corresponding crystalline alloys. Although some intermetallic compounds, like FeAl or Fe_3Al, also have high resistivity [77], in general, it is lower than that of metallic glasses.

The thermal conductivity and diffusivity of BMGs and supercooled liquids of bulk metallic glass formers has been measured together with electrical conductivity for Au-[78], Zr-[79] and Pd-based [80] BMGs and the results are shown in Table 1.2. The thermal diffusivity, thermal and electrical conductivity of the metallic glasses and amorphous solid were weakly temperature-dependent, with slightly positive temperature coefficients.

Table 1.2. Room temperature specific heat capacity (C_p), thermal diffusivity (α), thermal (λ) and electrical conductivity (σ_e) of bulk metallic glasses (at 300 K).

Alloy	α (mm^2/s)	C_p (J/K·mol)	λ (W/m·K)	σ_e ($\mu\Omega\cdot$m)$^{-1}$	Ref.
$Au_{49}Cu_{26.9}Ag_{5.5}Pd_{2.3}Si_{16.3}$	5.0	23.8	15	0.68	[78]
$Pd_{40}Ni_{10}Cu_{30}P_{20}$	1.6	24	5.1	0.54	[74,80]
$Zr_{55}Al_{10}Ni_5Cu_{30}$	2.2	25	5.0	0.53	[79]

Many metallic glasses exhibit moderately negative temperature dependence of the electrical resistivity (Fig. 1.8) though still exhibit metallic type bonding. However, it was shown that the temperature coefficient of resistivity in the $Pd_{40}Ni_{40-x}Cu_xP_{20}$ glassy alloys changes from negative to positive with increase in Cu content [80]. A similar change from the negative to the positive value was reported to occur as the P content is reduced in the Pd-Ni-P alloy system [81].

Fig. 1.9 shows the temperature dependence of the relative electrical resistivity (ρ) of a Cu-Zr-Al-Mn metallic glassy alloy normalized by the value at room temperature (ρ_0) together with the DSC plot. The electrical resistivity decreases with temperature. As one can see there are some variations in the resistivity curve related to structural relaxation below T_g, a clear change of the slope of ρ/ρ_0 at T_g and a rapid decrease in ρ at T_x. There is about 13% difference in the electrical resistivity between the supercooled liquid and crystalline state. After the initial crystallization reflected in the exothermic peak at 495 °C the temperature dependence of resistivity becomes positive which is typical for crystalline alloys.

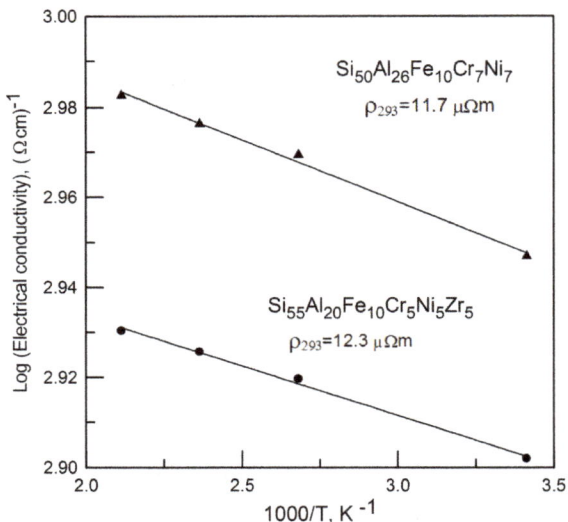

Fig. 1.8. Inverse temperature dependence of the logarithmic electrical conductivity of the Si-Al-TM (TM-transition metals) alloys.

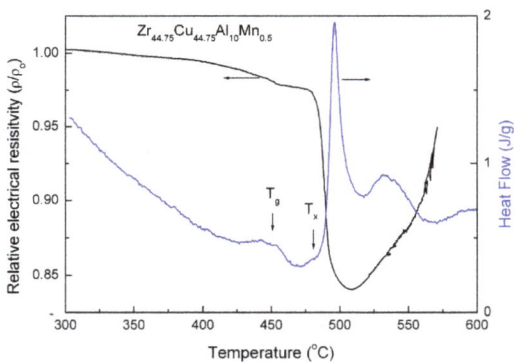

Fig. 1.9. Temperature dependence of the relative electrical reistivity of a $Zr_{44.75}Cu_{44.75}Al_{10}Mn_{0.5}$ metallic glassy alloy together with the DSC plot. Courtesy of A. Bazlov.

As shown in Table 1.3 for Cu-Zr-Al-Mn alloys the sign of temperature coefficient of electrical resistivity (TCR) changes even in the same alloy as a function of temperature as a result of structural relaxation [82]. The addition of a very small amount of manganese to the $Zr_{45}Cu_{45}Al_{10}$ alloy also doubles its strain sensitivity coefficient (SSC) which reaches 5.1 at 2 at.% Mn. The values of TCR are also quite small in these alloys.

The Ziman theory was found to explain the negative temperature dependence of the electrical resistivity of some liquid and amorphous alloys [83]. A negative thermal coefficient of resistivity is found to occur if the Fermi sphere diameter $(2k_F)$ to the wave number corresponding to the first peak K_p is close to unity, otherwise it is found to be positive. The electronic structure and properties of other metallic glasses were also studied [84].

Atomic diffusion in metallic glasses and supercooled liquids has been studied for many years and typical values of the diffusion coefficient for Zr, Al and Ni in a Zr-based BMG [85] in the temperature range of 580-680 K are shown in Fig. 1.10. Please, note a large difference in the diffusion coefficients between large size Zr, medium size Al and small size Ni atom. The diffusion measurements of Zr-Ti-Cu-Ni-Be BMGs demonstrate that two distinct processes contribute to long-range transport in the supercooled liquid state: single-atom hopping and collective motion which is the dominant process. A change in the slope of diffusion coefficient at T_g is connected with a transition to the liquid state [86]. However, it was shown that the atomic transport mechanism of the $Pd_{40}Cu_{30}Ni_{10}P_{20}$ metallic glass is thermally activated hopping in a wide temperature range even above the glass transition temperature [87]. Also, unlike diffusion in crystals, hopping in the metallic glassy alloys was suggested to be a highly collective process, involving many atoms.

Diffusion is significantly enhanced in the shear bands. Fast shear band diffusion was found in Pd-Ni-P BMGs (the diffusion coefficients (D) are of the order of 10^{-16} m²/s). The Pd-Ni-P with 1 at.% Co in some bands exhibits diffusion coefficients of the order of 10^{-14} m²/s) [88].

Table 1.3 - TCR of the Cu-Zr-Al-Mn system alloys in different state and SSC at room temperature. Reproduced from Ref. [82] with permission of Elsevier.

Alloy	State	Temperature interval, K	TCR, K^{-1}	SSC
$Zr_{45}Cu_{45}Al_{10}$	As cast	293 – 423	$-3,7 \cdot 10^{-5}$	2.6
		423 – 553	$2,1 \cdot 10^{-5}$	
	673 K, 5 min	293 – 623	$-1,5 \cdot 10^{-4}$	
	673 K, 10 min		$-6,3 \cdot 10^{-5}$	
	673 K, 20 min		$-3,7 \cdot 10^{-5}$	
$Zr_{44.75}Cu_{44.75}Al_{10}Mn_{0.5}$	As cast	293 – 423	$-2,5 \cdot 10^{-5}$	3.7
		423 – 553	$1,2 \cdot 10^{-5}$	
	673 K, 5 min	293 – 623	$-6,7 \cdot 10^{-5}$	
	673 K, 10 min		$-6.0 \cdot 10^{-5}$	
	673 K, 20 min		$-9,8 \cdot 10^{-5}$	
$Zr_{44.5}Cu_{44.5}Al_{10}Mn_1$	As cast	293 – 423	$-7,4 \cdot 10^{-5}$	4.2
		423 – 553	$2,4 \cdot 10^{-5}$	
	673 K, 5 min	293 – 623	$-6,7 \ 10^{-5}$	
	673 K, 10 min		$-6,6 \ 10^{-5}$	
	673 K, 20 min		$-7,4 \ 10^{-5}$	
$Zr_{44}Cu_{44}Al_{10}Mn_2$	As cast	293 – 423	$-8,5 \ 10^{-5}$	5.1
		423 – 553	$3,09 \cdot 10^{-5}$	
	673 K, 5 min	293 – 623	$-1,1 \cdot 10^{-4}$	
	673 K, 10 min		$-4,5 \ 10^{-5}$	
	673 K, 20 min		$-6,3 \ 10^{-5}$	

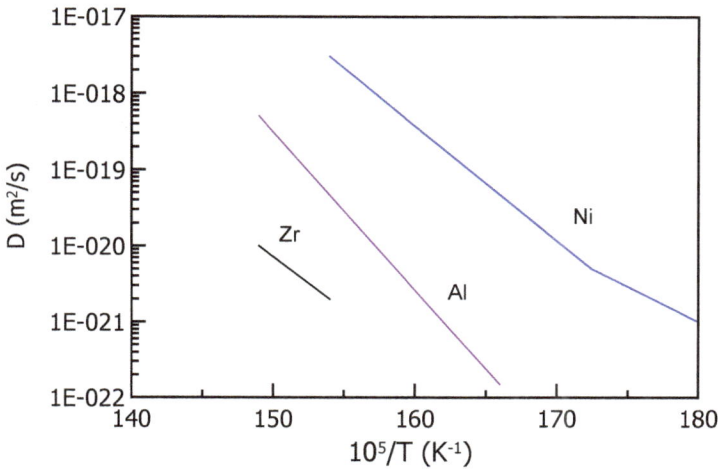

Fig. 1.10. A schematic temperature dependence of tracer diffusion coefficients in Vitreloy 4. The data points are obtained from Ref. [85]. There is a change in the slope of ln(D) as a function of 1/T near the glass-transition temperature visible for Ni.

Equilibrium viscosity (η) of the melts increases several orders in magnitude on cooling and reaches 10^{11}-10^{12} Pa·s at T_g on cooling [89]. By one of definitions η should be identical to 10^{12} Pa·s at T_g. However, this temperature not always corresponds to thermally manifested T_g (Fig. 1.11). Viscosity of metallic glasses at room temperature is very high (no visible flow is observed) and diffusion coefficients below T_g are also low, though significantly higher than those in crystals. It suggested that the glass-forming ability parameters/indicators of the metallic glasses can be deduced from their viscosity [90]. Nonequilibrium viscosity of metallic glasses was also measured below T_g [91] but it is significantly strain rate dependent.

The optical properties of the $Pd_{77.5}Cu_6Si_{16.5}$ BMG were characterized by spectral ellipsometry, and conduction band energy calculations. Refractive index of the metallic glass alloy is lower than that in a crystallized state. The conduction band energy increases from 1.78 ± 0.06 eV in the amorphous material to 2.32 ± 0.03 eV in the bulk crystalline specimen [92]. As other metals metallic glasses are usually wet with water with the wetting angle ranging from 70 to 100 ° [93].

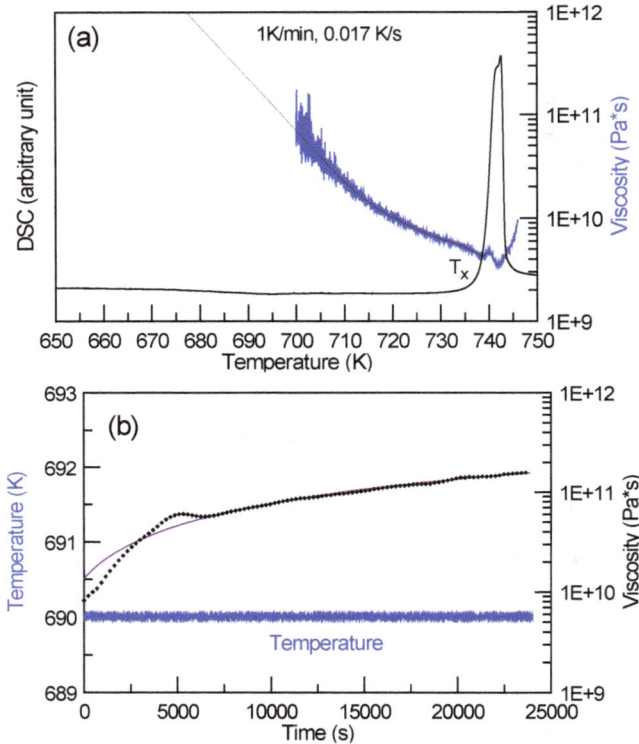

Fig. 1.11. (a) Viscosity of a $Zr_{48}Cu_{36}Al_8Ag_8$ bulk glassy sample as a function of temperature derived from the length change at the heating rate of 1 K/min (0.017 K/s). DSC trace is given for comparison (black curve). (b) Viscosity as a function of time at 690 K (diamonds). Violet line is fitting with exponential decay function. Please, note that the heating rate before this measurement was 10 K/min (0.17 K/s) and 690 K nearly corresponds to the beginning of the glass-transition in such a case. Reproduced from [94] with permission of Springer.

References

[1] A. L. Greer, New horizons for glass formation and stability, Nat. Mater., 14 (2015) 542–546. https://doi.org/10.1038/nmat4292

[2] W. Klement, R. H. Willens and P. Duwez, Non-crystalline structure in solidified gold–silicon alloys, Nature, 187 (1960) 869-873. https://doi.org/10.1038/187869b0

[3] H. S. Chen, D. Turnbull, Thermal evidence of a glass transition in gold-silicon alloy, Applied Physics Letters, 10 (1967) 284-286. https://doi.org/10.1063/1.1754813

[4] A. Inoue, K. Ohtera, K. Kita and T. Masumoto, New amorphous Mg-Ce-Ni alloys with high strength and good ductility, Japan. J. Appl. Phys., 27 (1988) L2248. https://doi.org/10.1143/JJAP.27.L2248

[5] V.V. Molokanov and V.N. Chebotnikov, Glass forming ability, structure and properties of Ti and Zr-intermetallic compound based alloys, Key Engineering Materials, 40-41 (1990) 319-332. https://doi.org/10.4028/www.scientific.net/KEM.40-41.319

[6] W. Buckel, R. Hilsch, Einfluß der Kondensation bei tiefen temperaturen auf den elektrischen Widerstand und die supraleitung für verschiedene Metalle, Z. Physik, 138 (1954) 109-115. https://doi.org/10.1007/BF01337903

[7] A. W. Weeber and H. Bakker, Amorphization by ball milling. A review, Physica B, 153 (1988) 93-102. https://doi.org/10.1016/0921-4526(88)90038-5

[8] R. Z. Valiev, Structure and mechanical properties of ultrafine-grained metals, Materials Science and Engineering A, 234-236 (1997) 59-65. https://doi.org/10.1016/S0921-5093(97)00183-4

[9] A. Inoue and H. M. Kimura, High strength Al- and Mg-based alloys with nanocrystalline or nanoquasicrystalline phase, J. Metast. Nanocr. Mater., 9 (2001) 41-56. https://doi.org/10.4028/www.scientific.net/JMNM.9.41

[10] D. V. Louzguine-Luzgin, G. Xie, S. Li, A. Inoue, N. Yoshikawa, K. Mashiko, S. Taniguchi and M. Sato, Microwave-induced heating and sintering of metallic glasses, Proceedings of 14th International Symposium on Metastable and Nano-Materials (ISMANAM-2007), Journal of Alloys and Compounds, 483 (2009) 78-81. https://doi.org/10.1016/j.jallcom.2008.07.158

[11] S. Li, G.Q. Xie, D. V. Louzguine-Luzgin, M. Sato and A. Inoue, Microwave-induced sintering of Cu-based metallic glass matrix composites in a single-mode 915-MHz applicator, Metallurgical and Materials Transactions A, 42A (2011) 1463-1467. https://doi.org/10.1007/s11661-010-0346-8

[12] G. Xie, W. Zhang, D. V. Louzguine-Luzgin, H. M. Kimura and A. Inoue, Fabrication of porous Zr-Cu-Al-Ni bulk metallic glass by spark plasma sintering process, Scripta Mater, 55 (2006) 687. https://doi.org/10.1016/j.scriptamat.2006.06.034

[13] G. Xie, D. V. Louzguine-Luzgin, H. M. Kimura and A. Inoue, Nearly full density $Ni_{52.5}Nb_{10}Zr_{15}Ti_{15}Pt_{7.5}$ bulk metallic glass obtained by spark plasma sintering of gas atomized powders, Applied Physics Letters, 90 (2007) 241902. https://doi.org/10.1063/1.2748102

[14] V. Zadorozhnyy, M. Churyukanova, A. Stepashkin, M. Zadorozhnyy, A. Sharma, D. Moskovski, J.Q. Wang, E. Shabanova, S. Ketov, D.V. Louzguine-Luzgin and S. Kaloshkin, Structure and Thermal Properties of an Al-Based Metallic Glass-Polymer Composite, Metals, 8 (2018) 1037. https://doi.org/10.3390/met8121037

[15] A. Inoue and H. M. Kimura, Fabrications and mechanical properties of bulk amorphous, nanocrystalline, nanoquasicrystalline alloys in aluminum-based system, J. Light Met., 1 (2001) 31-41. https://doi.org/10.1016/S1471-5317(00)00004-3

[16] T. Yamasaki, P. Schlossmacher, K. Ehrlich and Y. Ogino, Formation of amorphous electrodeposited Ni-W alloys and their nanocrystallization, Nanostructured Materials, 10 (1998) 375-380. https://doi.org/10.1016/S0965-9773(98)00078-6

[17] Z. Mahbooba, L. Thorsson, M. Unosson, P. Skoglun, H. West, T. Horn, C. Rock, E. Vogli, O. Harrysson, Additive manufacturing of an iron-based bulk metallic glass larger than thecritical casting thickness, Applied Materials Today, 11 (2018) 264–269. https://doi.org/10.1016/j.apmt.2018.02.011

[18] P. Bordeenithikasem, M. Stolpe, A. Elsen, D.C. Hofmann, Glass forming ability, flexural strength, and wear properties of additively manufactured Zr-based bulk metallic glasses produced through laser powder bed fusion, Additive Manufacturing, 21 (2018) 312-317. https://doi.org/10.1016/j.addma.2018.03.023

[19] N. Li, J. Zhang, W. Xing, D. Ouyang, L. Liu, 3D printing of Fe-based bulk metallic glass composites with combined high strength and fracture toughness, Materials & Design, 143 (2018) 285-296. https://doi.org/10.1016/j.matdes.2018.01.061

[20] Y. Shen, Y. Li, C. Chen, H.L. Tsai, 3D printing of large, complex metallic glass structures, Materials and Design, 117 (2017) 213–222. https://doi.org/10.1016/j.matdes.2016.12.087

[21] K. Lu, The future of metals, Science, 328 (2010) 319-320. https://doi.org/10.1126/science.1185866

[22] D.V. Louzguine-Luzgin, and A. Inoue, Bulk metallic glasses. formation, structure, properties, and applications, Handbook of Magnetic Materials, Edited by K.H.J. Buschow, Elsevier 21 (2013) 131. https://doi.org/10.1016/B978-0-444-59593-5.00003-9

[23] D.V. Louzguine-Luzgin, H.K. Nguyen, K. Nakajima, S.V. Ketov, and A.S. Trifonov, A study of the nanoscale and atomic-scale wear resistance of metallic glasses, Materials Letters, 185 (2016) 54–58. https://doi.org/10.1016/j.matlet.2016.08.035

[24] S. Pang, T. Zhang, K. Asami and A. Inoue, Bulk glassy Ni(Co–)Nb–Ti–Zr alloys with high corrosion resistance and high strength, Materials Science and Engineering A, 375 (2004) 368-371. https://doi.org/10.1016/j.msea.2003.10.152

[25] A. Inoue, and A. Takeuchi, Recent progress in bulk glassy alloys, Materials Trans., 43 (2002) 1892-1906. https://doi.org/10.2320/matertrans.43.1892

[26] A. Inoue, N. Nishiyama and T. Matsuda, Preparation of bulk glassy $Pd_{40}Ni_{10}Cu_{30}P_{20}$ alloy of 40 mm in diameter by water quenching, Mater. Trans. JIM, 37 (1996) 181. https://doi.org/10.2320/matertrans1989.37.181

[27] S. V. Ketov, N. Chen, A. Caron, A. Inoue, and D. V. Louzguine-Luzgin, Structural features and high quasi-static strain rate sensitivity of $Au_{49}Cu_{26.9}Ag_{5.5}Pd_{2.3}Si_{16.3}$ bulk metallic glass, Applied Physics Letters, 101 (2012) 241905. https://doi.org/10.1063/1.4770072

[28] H. S. Chen, Thermodynamic considerations on the formation and stability of metallic glasses, Acta Metall. 22 (1974) 1505. https://doi.org/10.1016/0001-6160(74)90112-6

[29] H. W. Kui, A. L. Greer and D. Turnbull, Formation of bulk metallic glass by fluxing, Appl. Phys. Lett., 45 (1984) 615-619. https://doi.org/10.1063/1.95330

[30] A. Inoue, High strength bulk amorphous alloys with low critical cooling rates, Mater. Trans. JIM, 36 (1995) 866. https://doi.org/10.2320/matertrans1989.36.866

[31] W. L. Johnson, Bulk glass-forming metallic alloys: science and technology, MRS Bull., 24 (1999) 42. https://doi.org/10.1557/S0883769400053252

[32] A. Inoue, Stabilization of metallic supercooled liquid and bulk amorphous alloys, Acta Mater., 48 (2000) 279-306. https://doi.org/10.1016/S1359-6454(99)00300-6

[33] W. L. Johnson, G. Kaltenboeck, M. D. Demetriou, J. P. Schramm, X. Liu, K. Samwer, C. P. Kim, Beating crystallization in glass-forming metals by millisecond heating and processing. Science 332, (2011) 828–833. https://doi.org/10.1126/science.1201362

[34] I. V. Okulov, I. V. Soldatov, M. F. Sarmanova, I. Kaban, T. Gemming, K. Edström & J. Eckert, Flash Joule heating for ductilization of metallic glasses, Nat. Commun. 6 (2015) 7932. https://doi.org/10.1038/ncomms8932

[35] W. Zhang, F. Jia and A. Inoue, Formation and properties of new La-based bulk glassy alloys with diameters up to centimeter order, Materials Transactions, 48 (2007) 68-73. https://doi.org/10.2320/matertrans.48.68

[36] T. Zhang and A. Inoue, Thermal and mechanical properties of Ti-Ni-Cu-Sn amorphous alloys with a wide supercooled liquid region before crystallization, Materials Transactions, JIM, 39 (1998) 1001-1006. https://doi.org/10.2320/matertrans1989.39.1001

[37] K. Jin, J.F. Löffler, Bulk metallic glass formation in Zr–Cu–Fe–Al alloys, Appl. Phys. Lett., 86 (2005) 1–3. https://doi.org/10.1063/1.1948513

[38] A. Inoue, A. Kato, T. Zhang, S. Kim, and T. Masumoto, Mg-Cu-Y amorphous alloys with high mechanical strengths produced by metallic mold casting method, Mater. Transactions JIM, 32 (1991) 609-616. https://doi.org/10.2320/matertrans1989.32.609

[39] K. J. Laws, K. F. Shamlaye, M. Ferry, Synthesis of Ag-based bulk metallic glass in the Ag–Mg–Ca–[Cu] alloy system, Journal of Alloys and Compounds, 513 (2012) 10–13. https://doi.org/10.1016/j.jallcom.2011.10.097

[40] J. Schroers, B. Lohwongwatana, W. L. Johnson and A. Peker, Gold based bulk metallic glass, Appl. Phys. Lett., 87 (2005) 061912. https://doi.org/10.1063/1.2008374

[41] E.S. Park, D.H. Kim, Formation of Ca–Mg–Zn bulk glassy alloy by casting into cone-shaped copper mold, J. Mater. Res., 19 (2004) 685–688. https://doi.org/10.1557/jmr.2004.19.3.685

[42] O. Senkov, D. Miracle, V. Keppens et al., Development and characterization of low-density Ca-based bulk metallic glasses. An overview, Metall and Mater. Trans. A, 39 (2008) 1888. https://doi.org/10.1007/s11661-007-9334-z

[43] D.H. Xu, G. Duan, and W.L. Johnson, Unusual glass-forming ability of bulk amorphous alloys based on ordinary metal copper, Phys. Rev. Lett., 92 (2004) 245504. https://doi.org/10.1103/PhysRevLett.92.245504

[44] T. Zhang, Q. Yang, J. I. Yun, L. I. Ran, S. J. Pang, J. F. Wang and T. Xu, Centimeter-scale-diameter Co-based bulk metallic glasses with fracture strength exceeding 5000 MPa, Chinese Science Bulletin, 56 (2011) 3972-3977. https://doi.org/10.1007/s11434-011-4765-8

[45] K. Amiya and A. Inoue, Fe-(Cr, Mo)-(C, B)-Tm bulk metallic glasses with high strength and high glass-forming ability, Rev. Adv. Mater. Sci., 18 (2008) 27-29.

[46] Q.K. Jiang, G.Q. Zhang, L. Yang, X.D. Wang, K. Saksl, H. Franz, R. Wunderlich, H. Fecht, and J.Z. Jiang, La-based bulk metallic glasses with critical diameter up to 30 mm, Acta Mater., 55 (2007) 4409–4418. https://doi.org/10.1016/j.actamat.2007.04.021

[47] K. Amiya and A. Inoue, Preparation of bulk glassy $Mg_{65}Y_{10}Cu_{15}Ag_5Pd_5$ alloy of 12 mm in diameter by water quenching, Materials Transactions, 42 (2001) 543-545. https://doi.org/10.2320/matertrans.42.543

[48] A. Inoue and T. Zhang, Thermal stability and glass-forming ability of amorphous Nd-Al-TM (TM = Fe, Co, Ni or Cu) alloys, Mater. Sci. Eng. A, 226–228 (1997) 393–396. https://doi.org/10.1016/S0921-5093(97)80050-0

[49] Y. Zeng, N. Nishiyama, T. Yamamoto, and A. Inoue, Ni-rich bulk metallic glasses with high glass-forming ability and good metallic properties, Mater. Trans. JIM, 50 (2009) 2441–2445. https://doi.org/10.2320/matertrans.MRA2008453

[50] N. Nishiyama, K. Takenaka, H. Miura, N. Saidoh, Y. Zeng, and A. Inoue, The world's biggest glassy alloy ever made, Intermetallics, 30 (2012) 19–24. https://doi.org/10.1016/j.intermet.2012.03.020

[51] J. Schroers, and W.L. Johnson, Ductile bulk metallic glass, Phys. Rev. Lett., 93 (2004) 255506. https://doi.org/10.1103/PhysRevLett.93.255506

[52] M.Q. Tang, H.F. Zhang, Z.W. Zhu, H.M. Fu, A.M. Wang, H. Li, and Z.Q. Hu, TiZr-base bulk metallic glass with over 50 mm in diameter, J. Mater. Sci. Technol., 26 (2010) 481–486. https://doi.org/10.1016/S1005-0302(10)60077-1

[53] A. Peker and W. L. Johnson, A highly processable metallic glass: $Zr_{41.2}Ti_{13.8}Cu_{12.5}Ni_{10.0}Be_{22.5}$, Appl. Phys. Lett., 63 (1993) 2342. https://doi.org/10.1063/1.110520

[54] H.B. Lou, X.D. Wang, F. Xu, S.Q. Ding, Q.P. Cao, K. Hono J.Z. Jiang, 73 mm-diameter bulk metallic glass rod by copper mold casting, Appl. Phys. Lett., 99 (2011) 051910. https://doi.org/10.1063/1.3621862

[55] W. F. Gale, and T. C. Totemeier, editors. Smithells Metals Reference Book 8-th Edition, Elsevier Butterworth-Heinemann Ltd., Oxford UK (2004) p. 11-1.

[56] A. Inoue, Y. Shinohara, J.S. Gook, Thermal and magnetic properties of bulk Fe-Based Glassy alloys prepared by copper mold casting, Mater Trans. JIM, 36 (1995) 1427-1430. https://doi.org/10.2320/matertrans1989.36.1427

[57] S. J. Pang, T. Zhang, K. Asami, A. Inoue, New Fe-Cr-Mo-(Nb, Ta)-C-B glassy alloys with high glass-forming ability and good corrosion resistance, Mater. Trans. JIM, 42 (2001) 376-383. https://doi.org/10.2320/matertrans.42.376

[58] X. J. Gu, S. J. Poon, G. J. Shiflet, M. Widom, Ductility improvement of amorphous steels Roles of shear modulus and electronic structure, Acta Materialia, 56 (2008) 88–94. https://doi.org/10.1016/j.actamat.2007.09.011

[59] T. Xu, S.J. Pang, H.F. Li, T. Zhang, Corrosion resistant Cr-based bulk metallic glasses with high strength and hardness, Journal of Non-Crystalline Solids, 410 (2015) 20-25. https://doi.org/10.1016/j.jnoncrysol.2014.12.006

[60] L. Ma, L. Wang, T. Zhang, A. Inoue, Bulk glass formation of Ti-Zr-Hf-Cu-M (M= Fe, Co, Ni) alloys, Mater. Trans., 43 (2002) 277–280. https://doi.org/10.2320/matertrans.43.277

[61] A. Takeuchi, N. Chen, T. Wada, Y. Yokoyama, H. Kato A.Inoue, J.W.Yeh, $Pd_{20}Pt_{20}Cu_{20}Ni_{20}P_{20}$ high-entropy alloy as a bulk metallic glass in the centimeter, Intermetallics, 19 (2011) 1546–1554. https://doi.org/10.1016/j.intermet.2011.05.030

[62] J. Kim, H.S. Oh, J. Kim, H.J. Chang, E.S. Park, Utilization of high entropy alloy characteristics in Er-Gd-Y-Al-Co high entropy bulk metallic glass, Acta Materialia, 155 (2018) 350-361. https://doi.org/10.1016/j.actamat.2018.06.024

[63] A. Inoue, Z. Wang, D.V. Louzguine-Luzgin, Y. Han, F.L. Kong, E. Shalaan, F. Al-Marzouki, Effect of high-order multicomponent on formation and properties of Zr-based bulk glassy alloys, J. Alloys Compd., 638 (2015) 197–203. https://doi.org/10.1016/j.jallcom.2015.03.078

[64] A. Inoue, N. Matsumoto, and T. Masumoto, Al–Ni–Y–Co amorphous alloys with high mechanical strengths, wide supercooled liquid region and large glass-forming capacity, Mater. Trans. JIM, 31 (1990) 493-495. https://doi.org/10.2320/matertrans1989.31.493

[65] A. Inoue, K. Ohtera, A. P. Tsai, T. Masumoto, Aluminum-based amorphous alloys with tensile strength above 980 MPa (100 kg/mm^2), Jpn. J. Appl. Phys., 27 (1988) 280-282. https://doi.org/10.1143/JJAP.27.L479

[66] D. V. Louzguine and A. Inoue, Proceedings of the ISMANAM 2001 International Symposium on Metastable, Mechanically Alloyed and Nanocrystalline Materials, Materials Science Forum, 386-388 (2002) 117.

[67] J.Q. Wang, Y.H. Liu, S. Imhoff, N. Chen, D.V. Louzguine-Luzgin, A. Takeuchi, M.W. Chen, H. Kato, J.H. Perepezko, and A. Inoue, Enhance the thermal stability and glass forming ability of Al-based metallic glass by Ca minor-alloying, Intermetallics, 29 (2012) 35-40. https://doi.org/10.1016/j.intermet.2012.04.009

[68] B.J. Yang, J.H. Yao, J. Zhang, H.W. Yang, J.Q. Wang and E. Ma, Al-rich bulk metallic glasses with plasticity and ultrahigh specific strength, Scripta Materialia, 61 (2009) 423–426. https://doi.org/10.1016/j.scriptamat.2009.04.035

[69] S. V. Madge, P. Sharma, D. V. Louzguine-Luzgin, A. L. Greer and A. Inoue, New La-based glass–crystal ex situ composites with enhanced toughness, Scripta Materialia, 62 (2010) 210-213. https://doi.org/10.1016/j.scriptamat.2009.10.029

[70] G. Xie, D. V. Louzguine-Luzgin and A. Inoue, Characterization of interface between the particles in NiNbZrTiPt metallic glassy matrix composite containing SiC fabricated by spark plasma sintering, Journal of Alloys and Compounds, 483 (2009) 239-242. https://doi.org/10.1016/j.jallcom.2008.07.226

[71] S. Li, D. V. Louzguine-Luzgin, G. Xie, M. Sato and A. Inoue, Development of novel metallic glass/polymer composite materials by microwave heating in a separated H-field, Materials Letters, 64 (2010) 235-238. https://doi.org/10.1016/j.matlet.2009.10.017

[72] M.Yu. Zadorozhnyy, D.I. Chukov, M.N. Churyukanova, M.V. Gorshenkov, V.Yu. Zadorozhnyy, A.A. Stepashkin, A.A. Tsarkov, D.V. Louzguine-Luzgin and S.D. Kaloshkin, Investigation of contact surfaces between polymer matrix and metallic glasses in composite materials based on high-density polyethylene, Materials and Design, 92 (2016) 306–312. https://doi.org/10.1016/j.matdes.2015.12.031

[73] J.S.C. Jang, J.B. Li, S.L. Lee, Y.S. Chang, S.R. Jian, J.C. Huang and T.G. Nieh, Prominent plasticity of Mg-based bulk metallic glass composites by ex-situ spherical Ti particles, Intermetallics, 30 (2012) 25-29. https://doi.org/10.1016/j.intermet.2012.03.038

[74] X. Hu, T.B. Tan, Y. Li. G. Wilde, J.H. Perepezko, The glass transition of $Pd_{40}Ni_{10}Cu_{30}P_{20}$ studied by temperature-modulated calorimetry, Journal of Non-Crystalline Solids, 260 (1999) 228-234. https://doi.org/10.1016/S0022-3093(99)00579-7

[75] J.E.K. Schawe, S. Pogatscher, J.F. Löffler, Thermodynamics of polymorphism in a bulk metallic glass: Heat capacity measurements by fast differential scanning calorimetry, Thermochimica Acta, 685 (2020) 178518. https://doi.org/10.1016/j.tca.2020.178518

[76] K.S. Gavrichev, L.N. Golushina, V.E. Gorbunov, A.I. Zaitsev, N.E. Zaitseva, B.M. Mogutnov, V.V. Molokanov, A.V. Khoroshilov, The absolute entropy of $Ni_{0.667}Zr_{0.333}$ and $Ni_{0.333}Zr_{0.667}$ amorphous alloys, J. Phys. Condens. Matter, 16 (2004) 1995-2002. https://doi.org/10.1088/0953-8984/16/12/008

[77] Y. Terada, K. Ohkubo, T. Mohri and T. Suzuki, Thermal conductivity of intermetallic compounds with metallic bonding, Materials Transactions, 43 (2002) 3167-3176. https://doi.org/10.2320/matertrans.43.3167

[78] V. V. Pryadun, D. V. Louzguine-Luzgin, L. V. Shvanskaya and A. N. Vasiliev, Thermoelectric properties of Au-based metallic glass at low temperatures, JETP Letters, 101 (2015) 465–468. https://doi.org/10.1134/S0021364015070127

[79] M. Yamasaki, S. Kagao, Y. Kawamura, Thermal diffusivity and conductivity of $Zr_{55}Al_{10}Ni_5Cu_{30}$ bulk metallic glass, Scripta Materialia, 53 (2005) 63-67. https://doi.org/10.1016/j.scriptamat.2005.03.021

[80] H. M. Kimura, A. Inoue, N. Nishiyama, K. Sasamori, O. Haruyama, and T. Masumoto, Thermal, mechanical, and physical properties of supercooled liquid in Pd–Cu–Ni–P amorphous alloy, Sci. Rep. RITU A43 (1997) 101–106.

[81] B.Y. Boucher, Influence of phosphorus on the electrical properties of Pd-Ni-P amorphous alloys, J. Non-Cryst. Solids, 7 (1972) 277–284. https://doi.org/10.1016/0022-3093(72)90028-2

[82] A.I. Bazlov, M.S. Parhomenko, O.I. Mamzurina, D.Yu. Karpenkov, I. Serhiienko, A. S Prosviryakov, E. N Zanaeva, D.V. Louzguine-Luzgin, Effect of manganese addition on thermal and electrical properties of $Zr_{45}Cu_{45}Al_{10}$ metallic glass, Journal of Non-Crystalline Solids, 542 (2020) 120103. https://doi.org/10.1016/j.jnoncrysol.2020.120103

[83] J. M. Ziman, A theory of the electrical properties of liquid metals. I The monovalent metals, Phil. Mag., 6 (1961) 1013-1034. https://doi.org/10.1080/14786436108243361

[84] U. Mizutani, Electronic structure of metallic glasses, Progress in Materials Science, 28 (1983) 97–228. https://doi.org/10.1016/0079-6425(83)90001-4

[85] F. Faupel, W. Frank, M.-P. Macht, H. Mehrer, V. Naundorf, K. Rätzke, H. R. Schober, S. K. Sharma, and H. Teichler, Diffusion in metallic glasses and supercooled melts, Rev. Mod. Phys. 75 (2003) 237-278. https://doi.org/10.1103/RevModPhys.75.237

[86] X. P. Tang, U. Geyer, R. Busch, W. L. Johnson, Y. Wu, Diffusion mechanisms in metallic supercooled liquids and glasses, Nature 402, (1999) 160–162. https://doi.org/10.1038/45996

[87] V. Zollmer, K. Ratzke, F. Faupel, A. Rehmet and U. Geyer, Evidence of diffusion via collective hopping in metallic supercooled liquids and glasses, Physical Review B, 65, (2002) 220201. https://doi.org/10.1103/PhysRevB.65.220201

[88] R. Hubek, M. Seleznev, I. Binkowski, S.V. Divinski, G. Wilde, Intrinsic heterogeneity of shear banding: Hints from diffusion and relaxation measurements of Co micro-alloyed PdNiP-based glass, Journal of Applied Physics, 127 (2020) 115109. https://doi.org/10.1063/1.5142162

[89] R. Busch, The thermophysical properties of bulk metallic glass-forming liquids, JOM, 52 (2000) 39-42. https://doi.org/10.1007/s11837-000-0160-7

[90] A. Takeuchi and A. Inoue, Calculations of dominant factors of glassforming ability for metallic glasses from viscosity, Mater. Sci. Eng. A, 375–377 (2004) 449–454. https://doi.org/10.1016/j.msea.2003.10.199

[91] O.P. Bobrov, V.A. Khonik, S.A. Lyakhov, K. Csach, K. Kitagawa, H. Neuhäuser, Shear viscosity of bulk and ribbon glassy $Pd_{40}Cu_{30}Ni_{10}P_{20}$ well below and mear the glass transition, J. Appl. Phys., 100 (2006) 033518. https://doi.org/10.1063/1.2226984

[92] F. Pinnock, J. Ketkaew, A. C. Martin, JanSchroers, André D.Taylor, Measured optical constants of $Pd_{77.5}Cu_6Si_{16.5}$ bulk metallic glass, L. McMillon-Brown, P. Bordeenithikasem, Optical Materials: X, 1 (2019) 100012. https://doi.org/10.1016/j.omx.2019.100012

[93] P. Yiu, W. Diyatmika, N. Bönninghoff, B.-Z. Lai, J.P. Chu, Thin film metallic glasses: Properties, applications and future, Journal of Applied Physics, 127 (2020) 030901. https://doi.org/10.1063/1.5122884

[94] D. V. Louzguine-Luzgin, I. Seki, T. Wada, and A. Inoue, Structural relaxation, glass transition, viscous formability, and crystallization of Zr-Cu-Based bulk metallic glasses on heating, Metallurgical and Materials Transactions A, 43 (2012) 2642-2648. https://doi.org/10.1007/s11661-011-1005-4

CHAPTER 2

Glass-Transition Process

Although, the glass-transition phenomenon in metallic glasses has been studied extensively there are still considerable uncertainties in this field of materials science. There are some gaps in obtaining a clear picture of the glass transition. Two general mechanisms of glass formation by limiting crystal nucleation and by limiting their growth will be discussed in the present Chapter.

Contents

2.1   Crystal nucleation suppression mechanism

2.1.1   Glass-transition and glass-forming ability

Liquids retain their volume as crystals but flow under the action of gravity and constantly undergo changes in their atomic structure. It is easy to cool a liquid below its liquidus temperature (T_l) owing to kinetic reasons related to difficulties in a crystal nucleation and growth at low temperature, and if the crystal nucleation is suppressed, then it will vitrify forming a glassy phase [1,2]. On the other hand, significant superheating of a solid above its melting temperature and a liquid above its boiling temperature is less likely because the diffusion processes in solids close to the melting temperature, and especially in liquids, are very fast providing no such kinetic limitations to the phase transformation, except for the thermodynamic energy barrier for nucleation of the high-temperature liquid or gaseous phase, respectively.

From the melt metallic glasses (glassy alloys) were first formed by using a rapid solidification technique at a very high cooling rate of 10^6 K/s [3]. For a long time Pd-Cu-Si and Pd-Ni-P were known to be the best metallic glass formers. The bulk glassy alloys possess three common features summarized by Inoue [4], i.e., belong to multicomponent systems (three or more components, in general), have significant atomic size ratios above 12% (but below some critical value) and exhibit negative heats of mixing among the constituent elements. All of them represent indispensable conditions leading to good glass-forming ability (GFA) [1] and relatively high thermal stability against crystallization [5]. The role of size ratio, strong chemical bonding, and valence-electron concentration was discussed in Ref. [6]. The first principle (the number of components) leads to the formation of dense packed atomic structures, "deep" ternary and quaternary eutectics by elemental "confusion" on solidification.

Good correlation was found between the alloy system complexity (the number of alloying elements) and critical size of the glassy sample (D_c) [7]. The glass-forming ability is shown to increase with increasing the number of alloying elements. Statistical analysis shows that the difference in D_c among binary, ternary and quaternary alloys is meaningful. In ternary alloys the GFA is maximized in case of the $A_{70}B_{20}C_{10}$, $A_{65}B_{25}C_{10}$, $A_{65}B_{20}C_{15}$, $A_{56}B_{32}C_{12}$, $A_{55}B_{28}C_{17}$, $A_{44}B_{43}C_{13}$ and $A_{44}B_{38}C_{18}$ alloy compositions (A, B and C symbols represent the constituent elements). Quaternary and quinary bulk glass-forming alloys with a large critical diameter, in general, have a larger supercooled liquid region than those of ternary alloys, so that the addition of a fourth alloying element not only enhances GFA but also stabilizes the supercooled liquid region. Bulk glassy alloys were reported to be obtained not only in the ternary but even in some binary systems like Ni-Nb, Ca-Al, Pd-Si, Cu-(Zr or Hf) [8,9]. However, addition of the third element significantly enhances the glass-forming ability.

Machine learning procedure was also used to design new metallic glasses [10]. Based on a large database machine learning was also used to model the critical casting diameter (D_c), and the supercooled liquid region. The model was used to evaluate 2.7 million possible adjustments to the existing alloys. Several predicted compositions outperform the original alloys in terms of the supercooled liquid region [11]. A computational method was designed for the prediction of possible glass-forming compositions and calculation of their critical cooling rates [12]. The obtained results were tested for Pd-Ni-P, Cu-Mg-Ca, and Cu-Zr-Ti alloy systems. A comparison with experimental results for the quaternary Ti-Zr-Cu-Ni system shows a promising overlap of calculation and experiment.

Role of the second principle (atomic size ratio) has been rationalized by Egami and Waseda [13] who investigated the minimum atomic concentration necessary to destabilize terminal solid solution phases. Later, Miracle and Senkov developed the topological criterion for glass formation [14]. However, the concept of atomic size has some deficiencies. In metallic glasses all bounds are considered to be non-directional, though the existence of some degree of covalent or even partly ionic bonding in the intermetallic compounds (and thus in metallic glasses) have been discussed for years [15]. The atomic size of an element derived from the bond length with different elements varies owing to different bond character, charge transfer, etc... Significant bond shortening was observed in case of Au-Al [16] and other Al-bearing alloys.

The third principle (large negative mixing enthalpy) is responsible for the formation of eutectics and dense packing of atoms of different kind. The importance of an efficient atomic packing for the formation of metallic glasses was shown [17,18]. One should note that even, pure metals, for example, Ni, Fe can be made amorphous at a high enough cooling rate estimated at 10^8-10^9 K/s [19] which are, however, not stable at room temperature unless separated into the nanometer scale spheres [20].

As has already been mentioned (see Chapter 1) alloys having a low GFA can be prepared in an amorphous state by condensation from a vapor phase or by electrodeposition in a solution without using the molten state. Some alloys can be produced by a solid state reaction using mechanical attrition or by severe plastic deformation. However, melt casting is a much more productive method.

The specific volume (V) of a liquid phase decreases faster (while its density (ρ) increases faster) with temperature than that of a competing crystal. According to several studies, the linearity of $d\rho/dT$ slope maintains not only in the equilibrium liquid but in the supercooled liquid region as well [21,22]. Fig. 2.1 shows a typical specific volume versus temperature (T) diagram for a pure substance. The glass-transition region is related to the area of change in the volumetric thermal expansion coefficient from a relatively high value characteristic of a liquid phase to a lower value for a solid glassy phase as shown in Fig. 2.1. Such a change of the thermal expansion coefficient on glass-transition was clearly demonstrated in many works, see for example, Ref. [23].

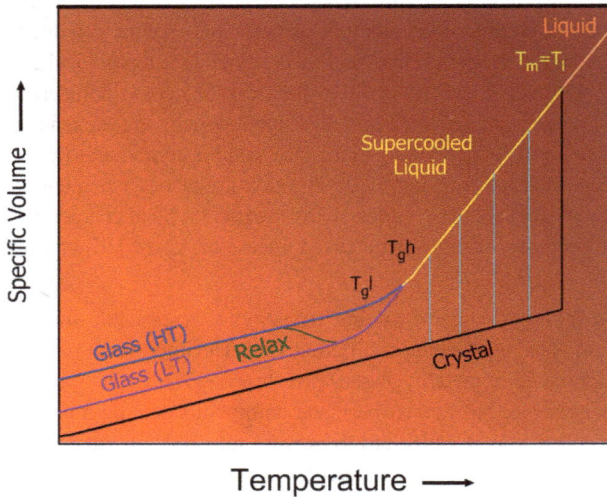

Fig. 2.1. Schematic specific volume versus temperature diagram for a pure substance without a solidification temperature interval. Black line indicates crystallization of liquid at T_l (an ideal case at infinitely slow cooling). In reality crystallization always takes place at some supercooling as shown by light blue lines. Higher supercooling is achieved at a higher cooling rate. T_g^l and T_g^h illustrate T_g obtained at a low and high cooling rate, respectively. The behavior is opposite to the first order phase transitions during crystallization when faster cooling induces higher supercooling.

By analogy with the well-known Kauzmann's entropy crisis [24] illustrated in Fig. 2.2 [25] one can suggest that a liquid metal should not have a lower volume (or a higher density), at a given temperature, than its crystalline counterpart provided that it contracts upon solidification and there are no changes in the chemical bond character. Face-centered cubic (FCC) and hexagonal close packed (HCP) lattices are the densest packing structures for crystalline pure metals, and thus, a liquid or a glassy monoatomic metallic glassy phase should not become denser than the corresponding crystal.

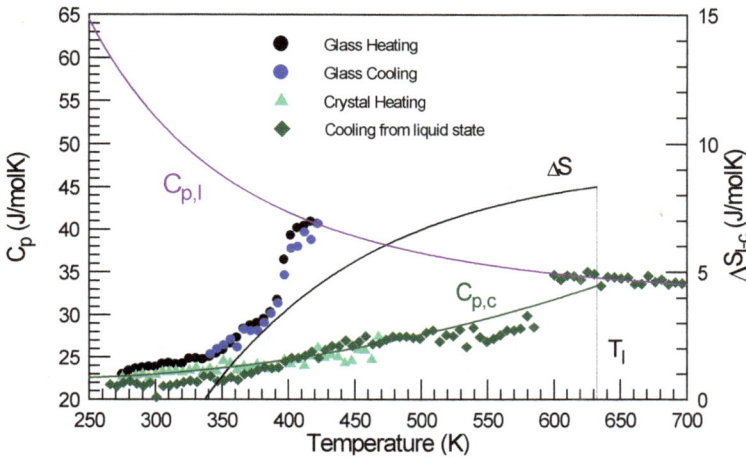

Fig. 2.2. C_p of the $Au_{49}Cu_{26.9}Ag_{5.5}Pd_{2.3}Si_{16.3}$ alloy in liquid, glassy and crystalline state as a function of temperature measured in a step scan mode. The Kauzmann temperature is 340 K. Reproduced from [25] with permission of Elsevier.

The glass-transition phenomenon is also characterized by the specific heat capacity (C_p) change as one can observe in Fig. 2.2. The glass-transition phenomenon in the $Au_{49}Cu_{26.9}Ag_{5.5}Pd_{2.3}Si_{16.3}$ metallic glass studied using the step-scan calorimetry method on heating showed the existence of two different slopes within the glass-transition region indicating two glass-transition processes one starting at a low temperature around 340 K and the other at a high temperature around 380 K. This phenomenon is likely related to the different diffusion coefficients of the alloying elements in this alloy. An arbitrary glass-transition (T_g) point is defined as a temperature at which viscosity of a liquid reaches 10^{12} Pa·s on cooling, while the transition takes place in a temperature interval. For some materials this value belongs to the glass-transition region defined by C_p measurement, but it is not the case for other materials [26]. T_g is high (up to 900 K) for Ferrous-metals based BMGs but for the Ca-Li-based BMGs it is close to room temperature being as low as 308 K for the $Ca_{65}Li_{14.54}Mg_{12.46}Zn_8$ one [27]. It is also found that the compressibility of glass-forming liquids and the difference in compressibility between the liquid and glassy phases decreases towards T_g [28].

There is also a concept of fictive glass transition temperature which is the temperature at which a property when extrapolated for the glassy phase intersects that of the equilibrium liquid phase [29].

Although, the glass-transition phenomenon in metallic glasses has been studied extensively there are still considerable uncertainties in this field of materials science [30]. Several theories are used to describe glass transition [31,2] though none is really comprehensive so far. The glass transition looks like a second-order phase transformation with continuity of material volume and entropy, and discontinuity of their derivatives which are therefore used in practice to detect the transformation. However, there are some reasons to suggest that the glass transition is a first-order melting transition in the $Pt_{80-x}Cu_xP_{20}$ bulk metallic glass-forming alloys [32]. It is suggested that at low Cu content the extrapolated Kauzmann temperature becomes indistinguishable from the glass transition temperature determined by DSC. For the alloys with $x < 17$, the liquid configurational enthalpy displays a drop with temperature. The entropy drop for this first-order liquid/glass transition is approximately two-thirds of the entropy of fusion of the crystallized eutectic alloy. However, the alloys with $x = 14\text{-}16$ are still bulk glass formers owing to high T_{rg}.

It is also considered that melting of an amorphous material occurs when broken bonds form a percolation cluster that changes macroscopically the rigidity of system. Below glass-liquid transition temperature T_g the amount of broken atomic bonds (configurons) can be calculated as a function of temperature using an equivalent two-level system that consists of unbroken and broken bonds [33]. As atoms in metals are more flexibly attracted to each other the metallic bond is more collective in nature than ionic or covalent. Nevertheless, effectively the elementary act of gradual debonding on increase of temperature can be schematically considered as due to bond breaking though in metals it is difficult to exactly localize the configuron to a certain orientation.

The glass-transition process also has similarities with a diffusionless phase transformation which takes place within a temperature interval provided that not all but a certain volume of the matrix phase is transformed at a definite temperature between the beginning (T_{bg}) and finish (T_{fg}) glass-transition temperatures. The diffusionless character of this transformation implies that the atoms are shifted to the distances smaller than the interatomic distances. One can suppose that the supercooled liquid regions are gradually transformed to the glassy regions upon cooling. The diffusionless transformation can be a natural source of the excess volume and stresses [34]. The transition at T_g can be related to a change of the liquid Gibbs free energy, which is the driving force of the glass transition. In this approach the classical Gibbs free energy change for a crystal formation can be modified to account for the enthalpy saving which allowed decent description of liquid-liquid and glass transitions [35].

The reduced glass-transition temperature $T_{rg}=T_g/T_l$ (T_l is the liquidus temperature, the temperature at which first crystals precipitate from the liquid on cooling) [36] is one of the most widely known physical-mathematical parameters indicating GFA, derived assuming homogeneous nucleation. As it will be described in detail in Chapter 4 upon crystallization homogeneous nucleation rate depends on the activation energy for transferring atoms across the surface (Q_N) of the nucleus and the nucleation barrier (ΔG^*) the free energy required to form a nucleus of critical size while growth rate depends upon the activation energy for growth (Q_g) and the thermodynamic driving force ΔG. Data in the literature often contain non-equilibrium T_l and non Kauzmann T_g values determined at different heating rates. However, the equilibrium T_l values obtained from the corresponding phase diagrams should be used. T_g increases with increasing cooling rate during solidification from the liquid state [37] and alloys solidified at different cooling rate have different fictive T_g. One should also note that the glass-transition does not occur at a discrete temperature and T_g is defined as an inflection point either on heating (Fig. 2.3(a)) upon devitrification or on cooling during vitrification (see Fig. 2.1).

It has been shown that the width of the supercooled liquid region (ΔT_x) on heating of the glassy sample defined as T_x-T_g where T_x is the onset crystallization temperature (Fig. 2.3(a)) as an indicator of the stability of the supercooled liquid against devitrification also correlates quite well with the glass-forming ability [38]. However, the crystallization temperature T_x is not purely intrinsic as it depends not only on the heating/cooling rate (β) but also on the nucleation reaction type, the as-solidified structure of the glassy alloy, its purity, existence of heterogeneous nucleation sites [39], and thus, cannot be regarded as purely intrinsic parameter. As one can see in Fig. 2.3(b) crystallization starts even at isothermal conditions well below T_x. The same speculation is applied to the γ parameter $\gamma=T_x/(T_g+T_l)$ [40], which takes into account both T_{rg} and ΔT_x criteria as well as the γ_m parameter [41]. Other parameters which take into account both T_{rg} and the fragility index m (see 2.1.2) of the supercooled liquid were introduced [42,43]. However, the T_{rg} concept works well when the alloys of similar fragility are treated as shown in Table 2.1 [44] and plotted in Fig. 2.4 (a). The critical diameter is used for a cylindrical sample as an indicator of the GFA but for a plate-shaped sample it is its thickness.

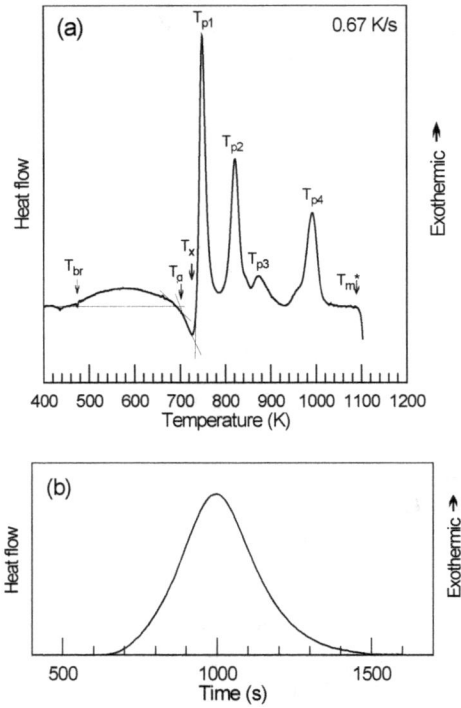

Fig. 2.3. A typical DSC (a) and differential isothermal calorimetry (b) traces of some bulk metallic glasses. T_{br}-temperature indicates the beginning of structural relaxation. An incubation period of about 650 s can be clearly seen in (b).

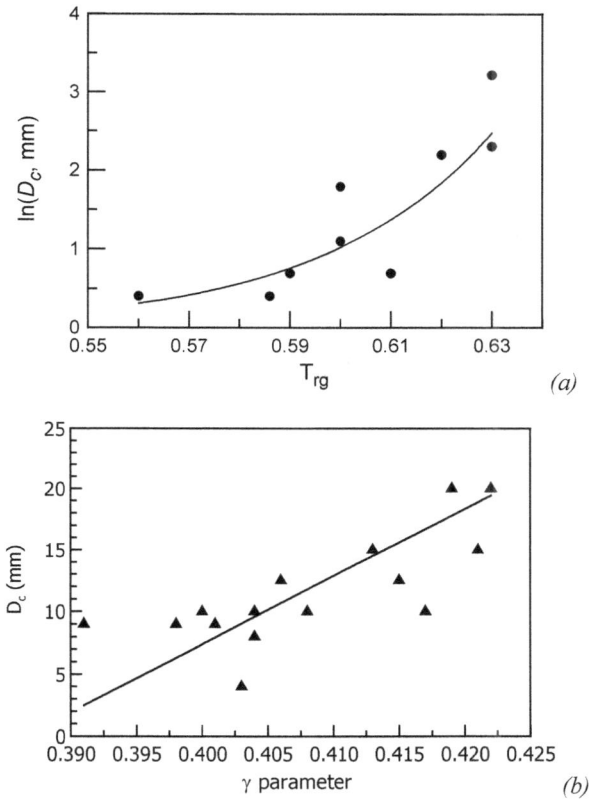

Fig. 2.4. (a) ln(D_{cr}) of the Cu-Zr-based alloys presented in Table 2.1 as a function of T_{rg}. (b) D_{cr} of the Zr-Cu-Fe-Al alloys presented in Table 2.2 as a function of γ parameter.

The parameters for Zr-Cu-Fe-Al BMGs are presented in Table 2.2 [45]. In this system there is no correlation with T_{rg} but better correlation of D_c with γ parameter. From Table 2.2, one can also see that ΔT_x increases with addition of Fe (below 7.5%) but decreases at higher Fe content. Large ΔT_x values are found at 5~7.5 at.% Fe. ΔT_x and γ values decrease when Al content increases from 10 at.% to 12.5 at.% or decreased to 7.5 at.%. At 10 at.% Al γ parameter and D_c decrease with Fe content. Good glass-formers are found at about 5~7.5 at.% Fe and 10 at.% Al.

Table 2.1. The T_{rg} and γ parameters, the critical diameter (D_c) and its natural logarithm of the Cu-Zr-based alloys.

BMG	T_g/T_l	γ	D_c, mm	$\ln(D_{cr}, mm)$
$Cu_{55}Zr_{45}$	0.59	0.399	1.5	0.41
$Cu_{50}Zr_{50}$	0.59	0.380	2.0	0.69
$Cu_{45}Zr_{45}Ag_{10}$	0.60	0.412	6.0	1.79
$Cu_{35}Zr_{45}Ag_{20}$	0.56	0.404	1.5	0.41
$Cu_{55}Zr_{30}Ti_{10}Ag_5$	0.60	0.397	2.0	1.1
$Cu_{45}Zr_{45}Al_{5}Ag_5$	0.61	0.416	9.0	2.2
$Cu_{44}Ag_{15}Zr_{36}Ti_5$	0.62	0.425	10.0	2.3
$Cu_{36}Zr_{48}Al_8Ag_8$	0.63	0.404	25.0	3.2

Table 2.2. The T_{rg} and γ parameters, as well as the critical glassy ingot diameter (D_c) of the Zr-Cu-Fe-Al BMGs. The data are from Ref. [45].

BMG	T_{rg}	γ	D_c (mm)
Base alloy $Zr_{60}Cu_{30}Al_{10}$	0.549	0.404	8
$Zr_{60}Cu_{27.5}Fe_{2.5}Al_{10}$	0.552	0.408	10
$Zr_{60}Cu_{25}Fe_5Al_{10}$	0.557	0.419	20
$Zr_{60}Cu_{22.5}Fe_{7.5}Al_{10}$	0.554	0.413	15
$Zr_{60}Cu_{20}Fe_{10}Al_{10}$	0.559	0.404	10
$Zr_{60}Cu_{17.5}Fe_{12.5}Al_{10}$	0.561	0.398	~9
$Zr_{60}Cu_{15}Fe_{15}Al_{10}$	0.551	0.401	~9
$Zr_{60}Cu_{22.5}Fe_5Al_{12.5}$	0.555	0.415	~12.5
$Zr_{62.5}Cu_{22.5}Fe_5Al_{10}$	0.548	0.422	20
$Zr_{60}Cu_{27.5}Fe_5Al_{7.5}$	0.556	0.406	~12.5
$Zr_{62.5}Cu_{20}Fe_{7.5}Al_{10}$	0.559	0.421	15
$Zr_{60}Cu_{20}Fe_{7.5}Al_{12.5}$	0.531	0.417	10
$Zr_{60}Cu_{20}Fe_{12.5}Al_{7.5}$	0.538	0.391	~9
$Zr_{60}Cu_{22.5}Fe_{10}Al_{7.5}$	0.536	0.4	10
$Zr_{55}Cu_{25}Fe_{10}Al_{10}$	0.559	0.403	~4

However, there are alloy systems like La-Al-Ag-Cu and La-Al-Ni-Cu ones, in which quite poor correlation between $\ln(D_c)$ and the parameters used to evaluate GFA is achieved (Fig. 2.5). The characteristic temperatures, T_{rg} and γ parameters, the critical glassy ingot diameter (D_c) and its natural logarithm of the La-Al-Ni-Cu BMGs are shown in Table 2.3. Poor statistical correlation with the coefficient of determination of $R^2 < 0.5$ is observed with $\ln(D_c)$ for both T_{rg} and γ parameters.

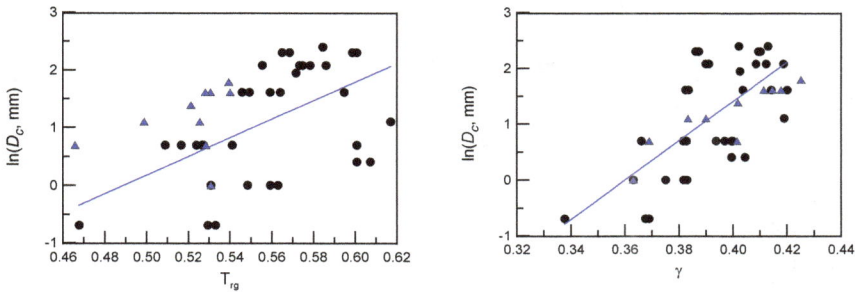

Fig. 2.5. $\ln(D_{cr})$ of the La-Al-Ag-Cu (triangles) and La-Al-Ni-Cu (circles) alloys presented in Table 2.3 as a function of T_{rg} and γ parameter.

Table 2.3. The characteristic temperatures, T_{rg} and γ parameters, the critical glassy ingot diameter (D_c) and its natural logarithm of the La-based BMGs. "Ref" indicates the literature source from references 46,47,48,49,50,51.

Alloy	D_c	T_g, K	T_x, K	T_l, K	ΔT_x, K	T_{rg}	γ	$\ln(D_c)$	Ref
$La_{65}Al_{10}Cu_{20}Ag_5$	5	380	458	716	78	0.531	0.418	1.609	46
$La_{62.5}Al_{12.5}Cu_{20}Ag_5$	6	389	472	721	83	0.540	0.425	1.792	46
$La_{60}Al_{15}Cu_{20}Ag_5$	5	401	481	759	80	0.528	0.415	1.609	46
$La_{55}Al_{15}Cu_{20}Ag_{10}$	2	416	483	787	67	0.529	0.401	0.693	46
$La_{55}Al_{17.5}Cu_{20}Ag_{7.5}$	3	425	498	852	73	0.499	0.390	1.099	46
$La_{55}Al_{20}Cu_{20}Ag_5$	4	429	503	823	74	0.521	0.402	1.386	46
$La_{55}Al_{25}Cu_{15}Ag_5$	3	452	503	860	51	0.526	0.383	1.099	46
$La_{62.5}Al_{12.5}Cu_{20}Ag_5$	6	389	472	721	83	0.540	0.425	1.792	47
$La_{60}Al_{15}Cu_{20}Ag_5$	5	410	481	759	71	0.540	0.411	1.609	47

$La_{55}Al_{15}Cu_{20}Ag_{10}$	2	416	483	893	67	0.466	0.369	0.693	47
$La_{55}Al_{17.5}Cu_{20}Ag_{7.5}$	3	425	498	852	73	0.499	0.390	1.099	47
$La_{55}Al_{20}Cu_{20}Ag_{5}$	4	429	503	823	74	0.521	0.402	1.386	47
$La_{70}Al_{12.4}Cu_{8.8}Ni_{8.8}$	1	403	422	759	19	0.531	0.363	0.000	48
$La_{70}Al_{14}Cu_{8}Ni_{8}$	0.5	404	429	763	25	0.529	0.368	-0.693	48
$La_{66}Al_{14}Cu_{10}Ni_{10}$	1.5	405	431	674	26	0.601	0.399	0.405	48
$La_{68}Al_{14}Cu_{9}Ni_{9}$	1	405	431	724	26	0.559	0.382	0.000	48
$La_{66}Al_{14}Cu_{10}Ni_{10}$	2	405	431	674	26	0.601	0.399	0.693	48
$La_{68}Al_{13.2}Cu_{9.4}Ni_{9.4}$	1	407	431	742	24	0.549	0.375	0.000	48
$La_{70}Al_{14}Cu_{8}Ni_{8}$	0.5	410	435	769	25	0.533	0.369	-0.693	49
$La_{68}Al_{14}Cu_{9}Ni_{9}$	1	411	437	730	26	0.563	0.383	0.000	49
$La_{64}Al_{14}Cu_{11}Ni_{11}$	8	411	439	715	28	0.575	0.390	2.079	48
$La_{66}Al_{14}Cu_{10}Ni_{10}$	1.5	413	442	680	29	0.607	0.404	0.405	49
$La_{64.6}Al_{14.6}Cu_{10.4}Ni_{10.4}$	5	414	448	696	34	0.595	0.404	1.609	48
$La_{64}Al_{14}Cu_{11}Ni_{11}$	8	417	445	721	28	0.578	0.391	2.079	49
$La_{62}Al_{14}Cu_{12}Ni_{12}$	10	417	446	738	29	0.565	0.386	2.303	48
$La_{63.1}Al_{15.2}Cu_{10.85}Ni_{10.85}$	10	420	459	699	39	0.601	0.410	2.303	48
$La_{59}Al_{14}Cu_{13.5}Ni_{13.5}$	5	422	457	773	35	0.546	0.382	1.609	48
$La_{62}Al_{15.7}Cu_{11.15}Ni_{11.15}$	11	422	460	722	38	0.584	0.402	2.398	48
$La_{62}Al_{14}Cu_{12}Ni_{12}$	10	423	452	744	29	0.569	0.387	2.303	49
$La_{61.4}Al_{15.9}Cu_{11.35}Ni_{11.35}$	11	426	477	729	51	0.584	0.413	2.398	48
$La_{60.5}Al_{16.3}Cu_{11.6}Ni_{11.6}$	8	426	471	727	45	0.586	0.408	2.079	48
$La_{59.6}Al_{16.6}Cu_{11.9}Ni_{11.9}$	8	426	482	743	56	0.573	0.412	2.079	48
$La_{57}Al_{14}Cu_{14.5}Ni_{14.5}$	2	427	474	815	47	0.524	0.382	0.693	48
$La_{59}Al_{14}Cu_{13.5}Ni_{13.5}$	5	428	463	779	35	0.549	0.384	1.609	49
$La_{58.6}Al_{17}Cu_{12.2}Ni_{12.2}$	5	431	495	764	64	0.564	0.414	1.609	48
$La_{57}Al_{14}Cu_{14.5}Ni_{14.5}$	2	433	480	821	47	0.527	0.383	0.693	49
$La_{57.6}Al_{17.5}Cu_{12.45}Ni_{12.45}$	8	435	510	783	75	0.556	0.419	2.079	48
$La_{50.2}Al_{20.5}Cu_{14.65}Ni_{14.65}$	0.5	435	461	930	26	0.468	0.338	-0.693	48

$La_{55.4}Al_{18.4}Cu_{13.1}Ni_{13.1}$	2	436	508	844	72	0.517	0.397	0.693	48
$La_{56.5}Al_{17.9}Cu_{12.8}Ni_{12.8}$	2	440	501	813	61	0.541	0.400	0.693	48
$La_{65}Al_{13}Ni_{10}Cu_{12}$	10	397	434	663	37	0.599	0.409	2.303	50
$La_{66}Al_{14}Ni_{10}Cu_{10}$	3	398	437	645	39	0.617	0.419	1.099	50
$La_{62}Al_{14}Ni_{12}Cu_{12}$	7	403	446	705	43	0.572	0.403	1.946	50
$La_{55}Al_{25}Ni_{5}Cu_{15}$	2	456	495	896	39	0.509	0.366	0.693	51
$La_{55}Al_{25}Ni_{10}Cu_{10}$	5	467	547	835	80	0.559	0.420	1.609	51
$La_{55}Al_{25}Ni_{15}Cu_{5}$	2	474	541	900	67	0.527	0.394	0.693	51

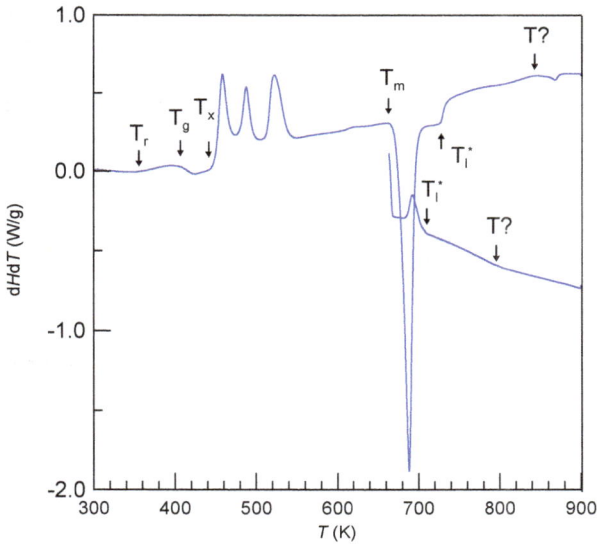

Fig. 2.6. DSC plot of a $La_{62}Al_{14}Cu_{12}Ni_{12}$ bulk metallic glassy alloy obtained at the heating and cooluing rate of 0.33 K/s. One can see the beginning of structural relaxation, glass transition, crystallization temperatures, as well as approximate melting and liquidus temperatures marked with an asterisk.

The source of poor correlation observed for these plots might be in part connected with difficulty of determination of the liquidus temperature. It is rare case for the scientists to accurately determine T_l. Usually a single DSC run is used to determine the characteristic temperatures. As the heating and cooling rates are not infinitely low, T_m and T_l are influenced by overheating and supercooling/undercooling phenomena. For example, the approximate T_l^* values of a $La_{62}Al_{14}Cu_{12}Ni_{12}$ bulk metallic glassy alloy measured at 0.33 K/s on heating and on cooling are 714 and 705 K, respectively while T_m is approximately 667 K on both cooling and heating (Fig. 2.6). At the same time there are changes in the slope of the curves above T_l^* values (marked at T?) which may be related to a high-temperature primary phase. In a reference paper 705 K is mentioned as T_l^* [50].

The intrinsic α parameter equal to $(T_e-T_g)/(T_l-T_g)$ [52] (here T_e is the eutectic temperature) or $(T_s-T_g)/(T_l-T_g)$ in a more general form, takes into account the difference T_l-T_s, and thus, $\alpha \cdot T_g/T_l$ criterion has better correspondence with the glass-forming ability of the alloy than T_g/T_l if T_s (solidus temperature) is significantly lower than T_l. One should note that T_l should be properly measured and extrapolated to $\beta \rightarrow 0$. For example, the degree of supercooling in nearly eutectic $Cu_{36}Zr_{48}Al_8Ag_8$ alloy varies with the cooling rate when measured with a differential thermal analysis (DTA) technique (Fig. 2.7) and if extrapolated to $\beta=0$ T_l should be about 1093.5 K [44]. T_l is also often measured on heating as the peak maximum temperature on heating. However, such temperature obtained at the heating rate of 20 K/min of 1115 K is significantly higher. It was also shown that the liquid structure and the glass-forming ability, especially in Al alloys containing rare earth metals, depend on superheating of the melt [53]. It is connected with crystal-like clusters existing even above the liquidus temperature [54].

Another purely intrinsic parameter, ΔT_l [55] indicates lowering of the melting temperature compared to the hypothetical ideal solution. Although, in general, compositions of metallic glasses follow deep eutectics [56], Ni-Si alloys containing from 45 to 58 at.% Si [57] with a drastically reduced liquidus temperature and Au-Al alloys [16] could be treated as good glass-formers, which they are not mainly due to their low atomic size ratio. The liquidus temperature of some Au-Al system alloys is as low as about 798 K, which is significantly lower than those for the pure elements. It indicates that this thermodynamically stable liquid is significantly under-cooled compared to that of pure elements. The most common mechanism of deep eutectic formation is due to the enthalpy and size effects. However, the difference in the effective size between Al and Au is small, 4% at most. Therefore the deep depression of the liquidus in Al-Au on the Au-rich side most likely originates from the instability of crystalline phases on the Au-rich side. The local structure of the $AlAu_4$ phase, with icosahedral clusters and strong local distortions, must be very similar to that of the liquid phase. The example of Al-Au also indicates that if the heat of formation is too strongly negative it is harmful to glass formation, because it leads to formation of stable intermetallic compounds. Formation of strong local chemical clusters stabilizing crystals may also be detrimental to glass formation.

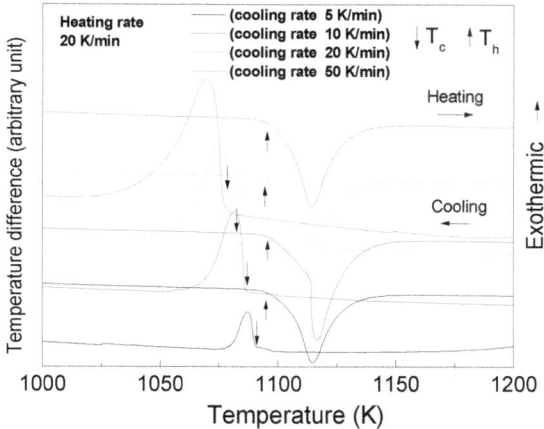

Fig. 2.7. DTA curves of the $Cu_{36}Zr_{48}Al_8Ag_8$ alloy heated up at 20 K/min and cooled down at different rate. Arrows indicate the points of beginning of melting (nonequilibrium solidus temperature) on heating and solidification on cooling (nonequilibrium liquidus temperature).

Notwithstanding on the formation of a deep eutectic and low stability of oP8 PdSi compound, $Si_{52}Pd_{48}$ alloy also did not produce a glassy phase, though Pd-Si alloys are good glass formers and even bulk glass-formers when fluxed [58,59]. It indicates that glass-formation is hampered in metalloid-rich areas likely owing to unfavorable atomic size ratios. The number of glassy alloys with $R_B/R_A<1$ (solute to solvent atomic size ratio) is three times larger than with $R_B/R_A>1$.

Although eutectic alloys have the lowest T_l values the best glass-forming compositions are not found at the equilibrium eutectic point but somewhat shifted usually towards a more refractory phase [60], while T_g is not significantly different in the observed range. This is most likely due to the shift of the eutectic point with supercooling at high enough cooling rate as casting conditions of bulk glassy samples are far from the equilibrium conditions. This may be a result of deep supercooling or existence of the competing crystalline phases in the system. An influence of the kinetic and thermodynamic parameters was also shown [61,62].

Another important intrinsic parameter is thermal conductivity of a molten alloy (λ_l) or its thermal diffusivity. A reasonably good correlation between the initial cooling rate (β_i) for four different alloys ($Cu_{60}Zr_{30}Ti_{10}$, $Zr_{65}Al_{7.5}Ni_{10}Pd_{17.5}$, $Ni_{53}Nb_{20}Ti_{10}Zr_8Co_6Cu_3$ and $Au_{49}Ag_{5.5}Pd_{2.3}Cu_{26.9}Si_{16.3}$) and λ_l values for pure metals was shown [63]. The cooling rates obtained for the $Zr_{55}Cu_{30}Ni_5Al_{10}$, $Ti_{41.5}Zr_{2.5}Hf_5Cu_{42.5}Ni_{7.5}Si_1$, $Pd_{40}Ni_{10}Cu_{30}P_{20}$, and $Ni_{59.5}Nb_{33.6}Sn_{6.9}$ bulk glass-forming alloys using the cooling curves (Fig. 2.8) were also found to scale with the thermal conductivities of these glassy alloys (Table. 2.4, Fig. 2.9) [64]. Thus, contrary to the melt spinning procedure when thermal gradient within the liquid is low the thermal conductivity is among the important factors influencing the GFA upon casting of bulk glassy alloys.

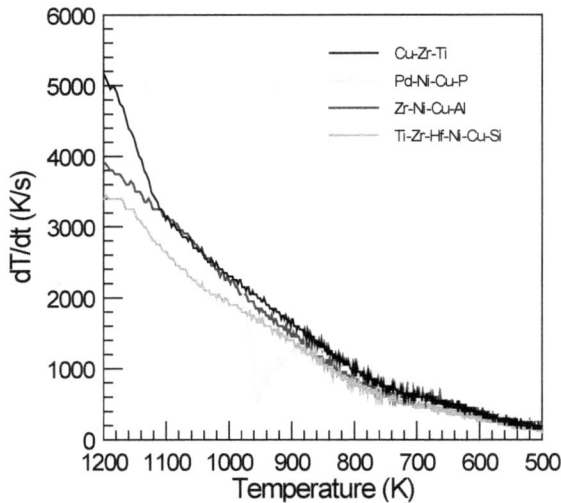

Fig. 2.8. The derivative dT/dt as a function of temperature for the $Zr_{55}Cu_{30}Ni_5Al_{10}$, $Ti_{41.5}Zr_{2.5}Hf_5Cu_{42.5}Ni_{7.5}Si_1$, $Pd_{40}Ni_{10}Cu_{30}P_{20}$, and $Ni_{59.5}Nb_{33.6}Sn_{6.9}$ bulk glass-forming alloys. Reproduced from [64] with permission of Cambridge University Press.

Table 2.4. The cooling rate (β) of the studied alloys as a function of temperature and their thermal conductivities (λg) at room temperate in the glassy state. Reproduced from [64] with permission of Cambridge University Press.

System alloy	Cu-Zr-Ti	Zr-Cu-Ni-Al	Ti-Zr-Hf-Cu-Ni-Si	Pd-Ni-Cu-P
β, K/s at 1173 K	4800	3700	3500	-
β, K/s at 1073 K	2850	2900	2400	-
β, K/s at 973 K	2150	2050	1800	1850
β, K/s at 873 K	1500	1350	1200	1200
β, K/s at 773 K	900	750	650	900
β, K/s at 723 K	650	650	500	700
$\sum(\beta_i/T_i)/N$, s^{-1} (1173-723 K; N=6)	2.95	2.64	2.31	-
$\sum(\beta_i/T_i)/N$, s^{-1} (973-723 K; N=4)	2.20	2.03	1.75	2.00
λ_g, W/mK	9	5	4.5	7

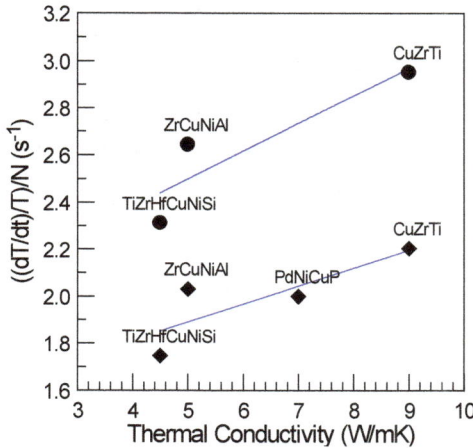

Fig. 2.9. Average cooling rate $\sum(\beta_i/T_i)/N$ of the $Zr_{55}Cu_{30}Ni_5Al_{10}$, $Ti_{41.5}Zr_{2.5}Hf_5Cu_{42.5}Ni_{7.5}Si_1$, $Pd_{40}Ni_{10}Cu_{30}P_{20}$, and $Ni_{59.5}Nb_{33.6}Sn_{6.9}$ bulk glass-forming alloys divided by temperature as a function of the thermal conductivities of these alloys. Reproduced from [64] with permission of Cambridge University Press.

One should also point out the importance of the melting/liquidus temperature of the alloy. The high liquidus temperature alloy being cast will cause a higher melt-mould temperature gradient, and thus, a higher initial cooling rate. The nose of the TTT diagram of such an alloy is also likely located at higher temperature. On the other hand, as expected the casting temperature was found to have little effect on the cooling rate of a single alloy with constant liquidus temperature (Fig. 2.10) [65].

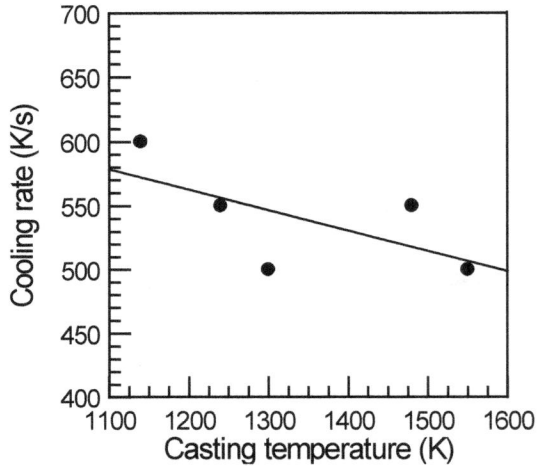

Fig. 2.10. The cooling rate of the bulk $Cu_{44}Ag_{15}Zr_{36}Ti_5$ sample (3 mm diameter) at T_g as a function of casting temperature. Reproduced from [65] with permission of Cambridge University Press.

The thermal diffusivity ($\lambda/C_p\rho_l$) accounts for the specific volume heat capacity C_p and density ρ_l of the alloy being cast are also useful characteristics. These parameters differ within one order of magnitude (in general 2-3 times) for different molten metals.

Other factors influencing the GFA of alloys are related to the electronic structure [66]. It is shown that, the addition of 1 at. % of Cu or Pd to Al-Y-Ni-Co alloy [67] or Cu to Al-Y-Fe alloys [68] drastically reduces their GFA and causes precipitation of the primary α-Al particles while non of the other parameters such as the average atomic size or overall mixing enthalpy are affected. Replacing Ni in the $Al_{85}Ni_9La_6$ alloy with Cu also decreases its thermal stability [69]. It is known that Al forms readily intermetallic compounds with Cu as well as with Fe, Ni or Y. Thus, the effect cannot be connected with possible repulsive interatomic interactions. A small amount of Si affects crystallization of the $Al_{86}Ni_2Co_{5.8}Gd_{5.7}$ alloy [70]. A comparison study of the GFAs of (Si,Ge)-Ni-Nd alloys [71] and later work [72] also showed that the principles for achieving good GFA known so far are important necessary conditions but not sufficient.

Stability of the supercooled liquid or metallic glass against crystallization is another very important parameter. For example, the alloys which form congruently melting intermetallic compounds like Mg-Yb-Cu [73] have lower GFA compared to alloys with incongruently melting compounds like Mg-Y-Cu even though the liquidus and eutectic temperature for the Mg-Yb alloys are lower than those for the Mg-Y alloys.

Packing density for non-crystalline structures, as a geometrical factor influencing GFA, has been verified using a hard spheres model [74]. An importance of topology [75] and the efficient atomic packing for the formation of metallic glasses was shown [18,76]. Actually the formation of dense structures in glassy alloys derives from a low density difference between the bulk glassy alloys and the crystallized counterparts [77]. Geometrical aspects of the atomic packing have been considered in binary metallic glasses. It was found that those having the solute-to-solvent radius ratios ranging from 0.60 to 1.45, produce structure-forming, solute-centered clusters with coordination numbers varied from 7 to 20. Also a strong preference was found for special radius ratios, which give efficient local atomic packing in the first coordination shell [78]. There is also a theory of the atomic level stresses in solid solutions and the stress criteria for the instability of the solid solution [79]. However, local distortions may exist even in crystalline pure metals and are quite common in equilibrium intermetallic compounds. The atomic size difference is found even in cI58 α-Mn between atoms of the same sort [80,81].

Fig. 2.11 shows the number of metallic glassy alloys as a function of the atomic size ratio with at least three clear maxima at 0.7, 0.8 and 1.25, a small number of alloys with atomic size ratios from 0.9-1.1 and absence of alloys having R_B/R_A exceeding 1.45 [72]. Similar data were also obtained earlier [76,78,82]. Also, surprisingly low atomic size ratio of 1.14 is found for the Ni-Nb system in which a bulk glass-former was produced [83].

It was shown that the electronegativity difference between the constituent elements is an important factor influencing glass formation and the temperature interval of the supercooled liquid region of the glass-forming alloys. The supercooled liquid region in the $Al_{85}RE_8Ni_5Co_2$ alloys was found to increase almost linearly with an increase in the electronegativity of the constituent RE metal [84]. Following this concept the width of the supercooled liquid region in Al-based metallic glasses the $Al_{85}Y_6Ni_5Co_2Zr_2$ and $Al_{84}Y_6Ni_4Co_2Sc_4$ glassy alloys with the largest supercooled liquid region known so far of about 50 K were created [85]. Nearly the same results have been obtained for other Al alloys [86]. High crystallization temperatures of the $Al_{85}Y_6Ni_5Co_2Zr_2$ and $Al_{84}Y_6Ni_4Co_2Sc_4$ glassy alloys (compared to the $Al_{85}Y_8Ni_5Co_2$ one) lead to a large supercooled liquid regions of 43 and 48 K, respectively [85].

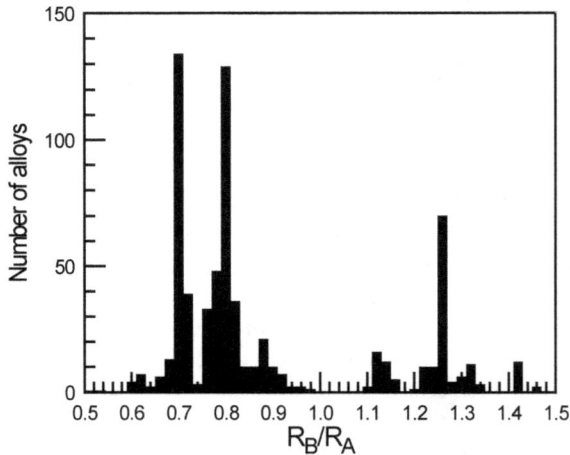

Fig. 2.11. The number of metallic glassy alloys produced by melt spinning as a function of the atomic size ratio of solute to solvent atom (R_B/R_A). Reproduced from [72] with permission of Springer.

Thus, upon homogeneous nucleation intrinsic parameters and factors indicating high GFA include T_{rg}, αT_{rg}, δ, λ_t, $\lambda_t/C_p\rho_l$[87], efficient cluster packing as well as a topological criterion λ_t [13,79, 88], ΔT_l[54], σ [89] liquid fragility [42,43,90] which have been summarized and separated for intrinsic and extrinsic factors [91]. ΔT_x and γ can be considered only as nearly intrinsic, because T_x is dependent on other factors such as heterogeneous vs. homogeneous nucleation [92]. The influence of extrinsic factors can also explain why semi-empirical γ parameter better describes the experimental data within ternary systems than T_{rg} because γ is sensitive to both extrinsic and intrinsic factors.

Nevertheless, there is difficult to create a unique parameter indicating the glass-forming ability because some factors such as T_{rg} deal with the dynamics of liquid (through its viscosity) towards glass transition while others like T_x indicate stability of the liquid versus crystallization: thermodynamic factors in terms of the driving force and kinetic factors in terms of nucleation and growth mechanisms. Moreover, T_x is measured on heating of a glass from room temperature while the high-temperature part, where the nose of the TTT diagram can be located, remains untreated.

The ideal equivolume glass-transition temperature (temperature at which volumes of liquid and solid phase become equal) was also derived from the thermal expansion data for corresponding liquids and solids [23]. The analysis of the thermal expansion curves allowed to derive a new parameter [93]

$$\delta = \alpha_l \cdot \rho_l (T_m - 298)/\Delta\rho_{s-l} \tag{2.1}$$

where α_l is the volumetric thermal expansion coefficient of the liquid alloy, ρ_m is the density of a liquid measured at a certain temperature, usually close to the melting temperature (T_m) and $\Delta\rho_{s-l}$ is the density difference between the solid and liquid phases. The influence of thermal expansion coefficient of a solid α_s is much smaller than that of the liquid and is excluded for convenience. The δ criterion indicates how fast a metal/alloy can reach the glass transition temperature, while it does not provide information on the stability of the resulted glass against crystallization [94]. One should also note that as predicted [95] the smallest density changes on crystallization of Cu-Zr glasses [96] were found at compositions close to those of the alloys with the best glass-forming ability. Molecular dynamics (MD) simulation results of the liquid viscosity [97] and direct viscosity measurements [98] indicate strong liquid character of the best glass-forming liquids. $Cu_{64}Zr_{36}$ alloy, one of the best glass-formers in the binary Cu-Zr system, is characterized by a pronounced icosahedral short range order in the liquid and glassy states.

By using the equivolume (isochoric) criterion the ideal glass-transition temperature (T_g) of Ni was found to be at about 1000-1200 K [99]. Bu using MD simulation the coordination numbers in the first coordination shell at 1800 and 1000 K were found to be 11.23 and 11.86, respectively. The former value is typical for a liquid metal [100] while the later one likely represents a glass. An empirical criterion for the glass transition based on the reduced radial distribution function or pair distribution function ($PDF(R)$) described by: $\mathfrak{R}=PDF(R)_{min}/PDF(R)_{max}$ where "max" and "min" indicate the first maximum and minimum of $PDF(R)$, respectively, [101] also indicated the glass-transition temperature of about 1200 K.

Thus, the ideal T_{rg} of pure Ni may be as high as ~0.6-0.7, which is an extremely high value for a pure metal. In general, bulk glassy alloys are formed at the compositions with high T_g/T_l ratio exceeding approximately 0.6 [36]. However, as has been mentioned earlier, poor GFA of pure metals even if they have high T_{rg} is related to their large fragility which will be discussed in the following part.

It is also interesting to note that T_g predicted by the point of equal volume for liquid and crystalline Fe of about 1200 K [93] is very close to that found in the MD simulation. The results of MD simulation of liquid Fe preformed by LAMMPS package [102] at constant pressure with an embedded atom potential [103] are shown in Fig. 2.12. Boiling was suppressed by using the periodic boundary conditions. The same T_g temperature of 1200 K was found by monitoring the dependence of the size of the largest cluster formed by mutually penetrating and contacting icosahedrons [104].

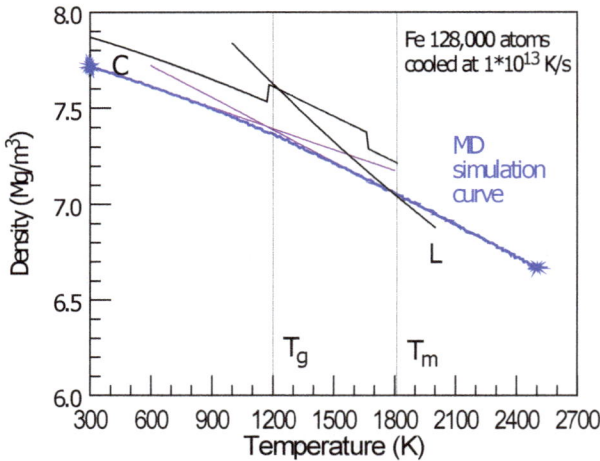

Fig. 2.12. MD simulation of liquid Fe preformed by LAMMPS package at nearly constant pressure with an embedded atom potential: density as a function of temperature (blue curve). Black curves represent the experimental values for liquid (L) and crystalline (C) iron.

As can be seen from Table 2.5, metals, which were found to vitrify easily in the experiment [20], i.e. Ni, Fe, Co, V have a relatively large δ parameter and a large activation energy for the temperature dependence of viscosity (Q_v) in the equilibrium liquid state. Most of the data is from Ref. [22].

Table 2.5. Physical properties of several elements in the liquid state. T_m and T_l are the melting and liquidus temperatures, respectively, which are equal for pure metals; Q_v is the activation energy for temperature dependence of viscosity, dV/V_l is the volume change upon melting, and the state: crystalline (Crst) or amorphous (Am) obtained upon rapid solidification in Ref. [20]. Dash character indicates that the parameter is not known.

El.	Lattice at room T and at melting	$T_m=T_l$, K	Q_v, kJ/mol	dV/V_l, %	δ	State
Ti	HCP, BCC	1958	-	1.9	0.99	Crst
Zr	HCP, BCC	2123	-	3.7	0.99	Crst
V	BCC	2185	-	2.1	1.33	Am
Fe	BCC, FCC*, BCC	1809	41.4	3.5	1.59	Am
Co	HCP, FCC	1766	44.4	3.5	1.46	Am
Ni	FCC	1727	50.2	5.5	1.71	Am
Pd	FCC	1825	-	6	1.25	Crst
Pt	FCC	2042	-	6.7	1.94	Crst
Cu	FCC	1358	30.5	4	0.88	Crst
Ag	FCC	1234	22.2	4	0.73	-
Au	FCC	1336	15.9	5	0.8	Crst

*intermediate state

2.1.2 Liquid viscosity and fragility

The plot scaling liquid viscosity by the reduced temperature at ambient pressure introduced by Laughlin-Uhlmann [105] and extended by Angell [106] shows the temperature dependence of an equilibrium liquids' viscosity (η) normalized by T_g (Fig. 2.13). Not all of the substances reach the equilibrium viscosity of 10^{12} Pa·s at the glass-transition temperature monitored by the heat capacity variation, and, in general this value ranges from about 10^{10} to 10^{12} Pa·s [107]. The viscosity is proportional to the relaxation time [108]. Moreover, the relation of $\eta = \sigma/3\dot{\varepsilon}$, where σ is the flow stress and $\dot{\varepsilon}$ is the strain rate, is used to calculate viscosity from the strain rate.

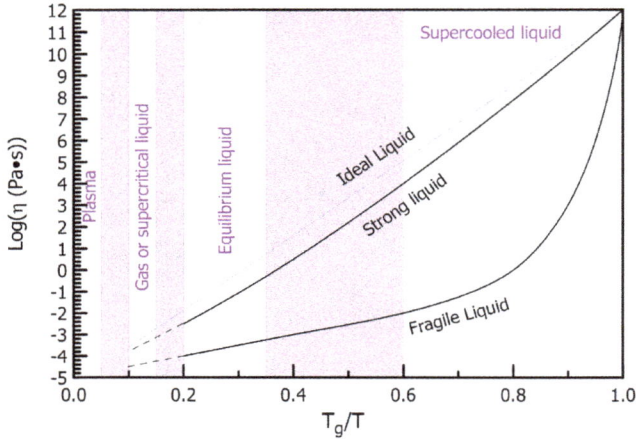

Fig. 2.13. A schematic plot of the viscosity as a function of temperature for strong and fragile liquids together with the corresponding phase regions. The "ideal liquid" shows the Arrhenius behavior while none of the known substances follows this law in the entire range from T_g to boiling temperature. Filled areas represent the temperature regions of liquidus, boiling, and plasma forming temperature depending on the alloy composition.

This plot differentiates "strong" and "fragile" liquids [109,110] which exhibit different degrees of deviation from the Arrhenius law:

$$\eta = \eta_0 \exp(-Q_v/RT) \tag{2.2}$$

(where Q_v is the activation energy for viscous flow and R is the gas constant) in the supercooled liquid state down to T_g. Such a deviation is generally called fragility or the dynamic fragility.

If we assume that the viscosity of a liquid should rise steeply when the liquid Fe volume approaches that of the crystal, take 1200 K as T_g (see Fig. 2.10) when the viscosity reaches about 10^{12} Pa·s and fit the viscosity of the equilibrium liquid Fe together with the point at T_g by the famous Vogel-Fulcher-Tammann-Hesse (VFTH) equation [37,111,112]:

$$\eta = \eta_0 \cdot \exp[D_f T_0/(T-T_0)] \tag{2.3}$$

where η_0, D_f and T_0 are the fitting parameters, we obtain the fragility parameter $D_f=2.1$. Thus, liquid Fe, as well as Ni [93, 94] should be more fragile than any metallic glass-former known so far, because the most fragile metallic glasses have $D_f\sim6$ [113]. It, again, indicates that pure metals, as very fragile liquids, should be very unstable against crystallization.

Above T_l the equilibrium viscosity of liquid pure metals generally obeys the Arrhenius law in a wide temperature range where the measurements were made [114]. Fragility (here and below dynamic) can also be defined using the parameter

$$m = d(\log(\eta))/d(T_g/T) \tag{2.4}$$

at T_g. The correlations between the thermodynamic (the difference in specific heat capacity C_p between a liquid and a glass) and dynamic fragilities were discussed [115]. Typical values of D and m are shown in Table 2.6 [116]. A liquid with a smaller m or larger D exhibits a smaller deviation from the Arrhenius temperature dependence of viscosity. Pd-, Pt-, and Ni-based liquid alloys are more fragile than the Mg-based and the Zr-based ones. A larger specific heat difference (liquid/crystal) is found for more fragile supercooled liquids, in general.

There is also a relation between the viscosity and the configurational entropy of a liquid, S_C [117],

$$\eta = \eta_0 \exp[C/(T \cdot S_C)]. \tag{2.5}$$

where, C is a constant and S_C is the configurational part of the entropy at a certain temperature.

Another representation of fragility is by assuming that the activation energy for viscous flow is not constant but changes as the structure of a liquid varies with temperature (hing Q_{vh} value for low temperature (supercooled) liquid and low Q_{vl} value for high temperature (equilibrium) liquid) [118]. Then the Doremus' fragility ratio (R_D) is given by:

$$R_D = Q_{vH}/Q_{vL} \tag{2.6}$$

Other methods are also used to describe and separate thermodynamic and kinetic fragility [119]. There are some reasons to believe that the fragility is also related to the global interatomic repulsion parameter (λ) [120]. Such correlation was also demonstrated in the subsequent work [121].

Table 2.6. Fragility parameters for different alloys from Ref. [116] rounded to integers.

Alloy	Df	m
$Au_{49}Cu_{26.9}Si_{16.3}Ag_{5.5}Pd_{2.3}$	21	46
$Cu_{47}Ti_{34}Zr_{11}Ni_8$	20	47
$Fe_{67}Mo_6Ni_{3.5}Cr_{3.5}P_{12}C_{5.5}B_{2.5}$	21	45
$Mg_{65}Cu_{25}Y_{10}$	22	45
$Ni_{69}Cr_{8.5}Nb_3P_{16.5}B_3$	15	57
$Pd_{40}Ni_{40}P_{20}$	15	56
$Pd_{40}Cu_{30}Ni_{10}P_{20}$	15	60
$Pt_{57.3}Cu_{14.6}Ni_{5.3}P_{22.8}$	14	62
$Pt_{60}Cu_{16}Co_2P_{22}$	12	69
$Zr_{46.75}Be_{27.5}Ti_{8.25}Cu_{7.5}Ni_{10}$	22	44
$Zr_{58.5}Cu_{15.6}Ni_{12.8}Al_{10.3}Nb_{2.8}$	20	48

The correlation observed between the critical cooling rate for glass formation R_c, liquid fragility index m or D_f, and T_{rg} [42] explains the low glass-forming ability (GFA) of a pure metal even if it has a high T_g. This is intuitively clear because according to the plot (Fig. 2.13) strong liquids, as a rule, have a higher viscosity above T_g, and thus, should be less prone to crystallization for kinetic reasons. It, however, does not include the thermodynamic aspects related to the driving force in terms of the Gibbs free energy difference between the liquid and crystalline phases [72]. A lower viscosity of the melt promotes both nucleation and growth processes of crystals [122] even at the same values of the reduced glass-transition temperature (T_{rg}). A more fragile liquid is going to have a lower viscosity in the entire range between T_l and T_g (see Fig. 2.13).

The origin of fragile behavior should be connected with structural changes in the supercooled liquid leading to the variation in the activation energy for viscous flow. For example, intensification of the covalent bonding between the metallic atoms and P was found in Pd-Cu-Ni-P melts in-situ cooled down to the glass-transition region [123] which was responsible for the changes in the structure of a liquid (clustering [2]), and thus, fragile behavior of this melt. Similar structural changes were observed in Zr-Cu [124] and Zr-Cu-Al [125] liquids on cooling. Although fragile liquids are generally predisposed to have a lower

GFA compared to strong liquids Pd-based alloys still have a good GFA owing to their exceptionally high T_{rg} values approaching 0.7 [52].

A structural parameter, γ, based on the shift of the first peak in the structure factor, $S(q_1)$ is used to characterize the liquid fragility through the rate of structural ordering with decreasing temperature [126]. The link between fragility and glass-forming ability arises from the coupling of fragility with structure and dynamics. Rather low fragility of the Zr-Pt metallic melts [127] is attributed to a combination of localized polar interatomic bonds between Zr and Pt atoms [128].

The reduced Arrhenius crossover temperature $\theta_A = T_A/T_g$, a crossover in the mean diffusion coefficient from the high-temperature Arrhenius to low-temperature non-Arrhenius behavior in metallic glass-forming liquids was found to take place approximately at $\theta_A = 2$ in the equilibrium liquid state [129]. In case of fragile molecular liquids, this crossover occurs at $\theta_A = 1.4$. The θ_A values for strong network liquids are higher than 2. It is suggested that the reduced crossover temperature depends strongly on the liquid fragility, and can be observed either in the supercooled state (molecular glass former) or in the equilibrium liquid state (metallic and network glass formers). The effective activation barrier of the high-temperature Arrhenius behavior was found to be a nearly universal value of $11 k_B T_g$ for nonpolar molecular and metallic liquids. Such correlations between the low and high-temperature parameters imply that T_g can be estimated from the high-temperature activation energy barrier and the fragility parameter m can be estimated from the reduced crossover temperature [127]. It should be noted again that a proper definition of Tg have to be used [130].

Also, a liquid, in order to be able to attain a high temperature and retain its liquid state boiling must be avoided. This is possible provided that the liquid is under an overcritical external pressure [131]. Below the critical point (the point above which one cannot distinguish between the liquid and gaseous phase) the high-temperature part of the viscosity plot cannot exist. Above this point the liquid/gaseous phase can reach high temperatures without undergoing boiling phase transformation before it transforms to plasma (Fig. 2.13).

The temperature dependence of the equilibrium and non-equilibrium viscosity below T_g has been studied extensively. Due to long relaxation times a non-equilibrium viscosity is often measured leading to a Non-Newtonian flow and at room temperature can be as low as 10^{17} Pa·s [132]. The extrapolated equilibrium viscosity at room temperature should be very high and it is not clear is there a change in the slope of the equilibrium viscosity at T_g or somewhere below it.

Following the variable activation energy of viscous flow Q_v in the equation of the equilibrium viscosity (eq. 2.2) [118] has two constant asymptotes – high Q_{vh} for low temperature (supercooled) liquid and low Q_{vl} for high temperature (equilibrium) liquid (Fig. 2.14). As it was mentioned, more generally the universal temperature relationship for the activation energy of viscous flow of liquids is:

$$Q_v(T) = Q_{vl} + RT \cdot \ln[1 + \exp(-S_d/R) \exp((Q_{vh} - Q_{vl})/RT)] \qquad (2.7)$$

which depends on the asymptotic energies and on the entropy of configurons S_d.

In situ measurements of the vibrational spectra of strong and fragile metallic glasses in the glass, liquid and crystalline phases were performed using inelastic neutron scattering. The similarity of the phonon density of states curves above and below T_g indicates that the excess vibrational entropy is unchanged on the glass transition. The excess entropy of the liquid and glass is dominated by configurational entropy in these metallic glasses while the excess vibrational entropy in the $Cu_{50}Zr_{50}$ and $Cu_{46}Zr_{46}Al_8$ alloys was found to be very small [133]. To the contrary molecular and network glasses indicate vibrational contribution to the excess liquid/glass entropy [134]. The configurational entropy increases with increasing temperature but at higher temperatures anharmonicity may cause thermodynamically significant changes in the vibrational dynamics of the liquid as well.

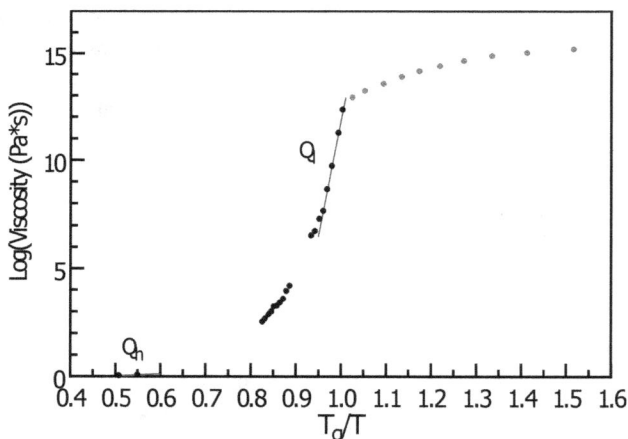

Fig. 2.14. Summarized temperature dependence of viscosity of the $Pd_{40}Ni_{40}P_{20}$ and $Pd_{40}Cu_{30}Ni_{10}P_{20}$ alloys with similar properties in the liquid, supercooled liquid and metallic glassy state. The data is taken from Refs. [135,136,137]. The values below T_g are for non-equilibrium viscosity.

Here one should mention that anomalous variation in a liquid viscosity in the vicinity of T_l and above it was observed for Fe- [138] and other alloys. The authors stated that liquid metallic alloys can have various structural states and significant overheating is required to dissolve clusters inherited from the solid state and to reach equilibrium liquid structure. The existence of liquid-liquid transitions in deeply supercooled state [139] and strong to fragile liquid transitions [140] should also be taken into account when describing fragility.

2.2 Crystal growth controlled mechanism

The $Fe_{48}Cr_{15}Mo_{14}C_{15}B_6Tm_2$ and $Fe_{48}Cr_{15}Mo_{14}C_{15}B_6Y_2$ alloys are the example of crystal growth mechanism limited BMGs because they contain pre-existed nuclei [141]. GFA of the Fe-Cr-Mo-C-B alloys is drastically improved by the addition of rare-earth (RE) metals (Fig. 2.15 (a)) [142]. The additions of Y [143] and Gd [144] were reported to shift the chemical composition closer to eutectic and to combine oxygen into the RE oxide. $(Fe,Co)_{50-x}Cr_{15}Mo_{14}C_{15}B_6Tm_x$ bulk metallic glasses showed exceptionally high glass-forming ability: the critical thickness of cast samples reached 1 cm [145,146]. The addition of Tm is more effective than Y in improving the glass-forming ability of Fe-based alloy as it leads to a lower T_l^* temperature, and thus to higher T_{rg} and γ parameters.

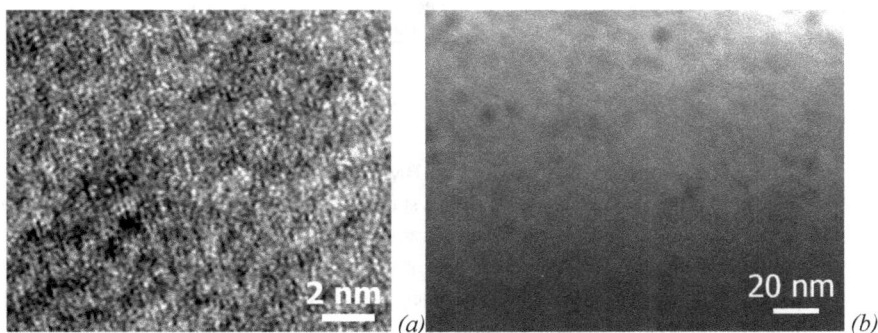

Fig. 2.15. (a) High-resolution TEM image of the $Fe_{48}Cr_{15}Mo_{14}C_{15}B_6Tm_2$ BMG alloy in the as-solidified state. (b) Bright -field TEM image of the $Fe_{48}Cr_{15}Mo_{14}C_{15}B_6Tm_2$ glassy alloy annealed at 893 K for 420 s.

The nanoscale size particles of the χ-$Fe_{36}Cr_{12}Mo_{10}$ phase were also found to precipitate on heating from the supercooled liquid state in the $Fe_{48}Cr_{15}Mo_{14}C_{15}B_6RE$ alloys. TEM of the

$Fe_{48}Cr_{15}Mo_{14}C_{15}B_6Tm_2$ alloy annealed at 893 K for 420 s shown in Fig. 2.15 (b) indicates the particles of about 10 nm size, while a wide (330) X-ray diffraction peak corresponds to a significantly smaller particles size of 3.1 nm which indicates a large amount of the internal strain in this phase.

The energy-dispersive X-ray (EDX) spectra obtained indicated that the distribution of Tm content is close to Gaussian. This indicates that Tm as well as Y [141] is nearly equally present in both the χ-$Fe_{36}Cr_{12}Mo_{10}$ phase and the residual glassy matrix. At the same time χ-$Fe_{36}Cr_{12}Mo_{10}$ phase has a higher Fe content than the glassy matrix. The diffusion coefficient (D) of RE metals at 900 K is about $4.2 \cdot 10^{-23}$ m^2/s [147]. Such a slow diffusion of RE metals drastically slows down the growth rate of χ-$Fe_{36}Cr_{12}Mo_{10}$ phase and explains good glass-forming ability of the $Fe_{48}Cr_{15}Mo_{14}C_{15}B_6Tm_2$ BMG.

The differential isothermal calorimetry investigation (Fig. 2.16) was carried out below T_x measured on continuous heating. Crystallization in the $Fe_{48}Cr_{15}Mo_{14}C_{15}B_6RE_2$ (RE=Tm and Y) alloys started without an incubation period. No clear incubation period was observed in Fig. 2.16 while the χ-$Fe_{36}Cr_{12}Mo_{10}$ phase precipitated. A clear incubation period is found for the $Fe_{50}Cr_{15}Mo_{14}C_{15}B_6$ BMG. Also, the existence of crystal-like clusters which can act as pre-existing nuclei in the $Fe_{48}Cr_{15}Mo_{14}C_{15}B_6Tm_2$ alloy was observed earlier [148]. On the contrary $Fe_{50}Cr_{15}Mo_{14}C_{15}B_6$, $Co_{48}Cr_{15}Mo_{14}C_{15}B_6Tm_2$, $Fe_{50}Cr_{15}Mo_{14}C_6B_{15}$ and $Fe_{48}Cr_{15}Mo_{14}C_6B_{15}Tm_2$ glassy alloys showed nucleation and growth behavior and crystallization started after the incubation period as usual nucleation rate controlled BMGs [149]. Destabilization of the competing crystalline phases, as it was found for marginal glass-formers [150] is considered to be the dominant reason for the improved GFA of these BMGs by the RE metals.

Another example of crystal growth-controlled BMG is the $Zr_{65}Al_{7.5}Ni_{10}Pd_{17.5}$ bulk glassy alloy [151] in which an icosahedral quasicrystalline phase is also formed without the incubation period. Some of Al-RE-TM (TM-transition metals) [152] system alloys (their critical thickness is about 0.5 mm) produced by rapid solidification techniques also showed growth controlled mechanism, especially those containing Cu [153]. Partial replacement of Ni for Cu in the $Al_{85}Y_8Ni_5Co_2$ metallic glass caused formation of the nanoscale Al particles, resulted in sharp drops in tensile strength, hardness, crystallization temperature and disappearance of the supercooled liquid prior to crystallization. Nevertheless, the absolute number of such growth-controlled BMGs is low and formation of the most of BMG is nucleation-controlled.

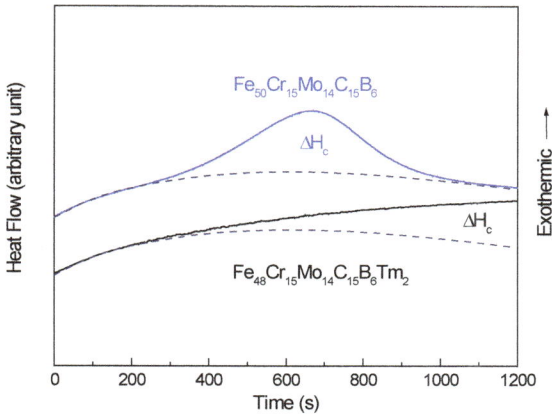

Fig. 2.16. Differential isothermal calorimetry traces of $Fe_{48}Cr_{15}Mo_{14}C_{15}B_6Tm_2$ at 868 K and $Fe_{50}Cr_{15}Mo_{14}C_{15}B_6$ alloy at 848 K. The dashed lines represent the baseline. The data is taken from Ref. [149] with permission of Elsevier.

2.3 Extrinsic factors influencing glass-forming ability

First group of purely extrinsic factors influencing the GFA are [91]: factors that modify the melt-mould heat transfer coefficient (h). They include mould surface quality and cleanliness, mould surface temperature, thermal conductivity of the mould, casting temperature, casting pressure, liquid metal turbulence during casting as well as the chamber pressure and atmosphere [154]. The presence or absence of turbulence and the degree of turbulence can have a strong influence on heat transfer and GFA. These factors are typically not reported in the literature. These factors may play a central role in the poor reproducibility of critical thickness for a given metallic glass composition that is often apparent in the literature. The influence of the extrinsic factors is often not known quantitatively, and only a qualitative discussion is given below.

The mold surface cleanliness and roughness affect the melt-mould heat transfer coefficient and the Biot number:

$$Bi=hL/\lambda_l, \hspace{5cm} (2.8)$$

(where L is a characteristic length scale equal to radius of the cast cylindrical rod, and λ_l is the thermal conductivity of the molten alloy being cast and h is the heat transfer coefficient at the interface). It indicates whether heat transfer through the melt or heat transfer through the melt/mould interface dominates in cooling (see Fig. 2.17 (a)). The value of the Biot number calculated for typical Cu- and Zr-based glass-forming alloys is about unity which implies that the temperature gradients inside the melt are non-negligible in accordance with Fig. 2.15 (b). Thus, both heat transfer through the melt and heat transfer through the melt/mould interface influence the cooling rate [155] of bulk metallic glasses.

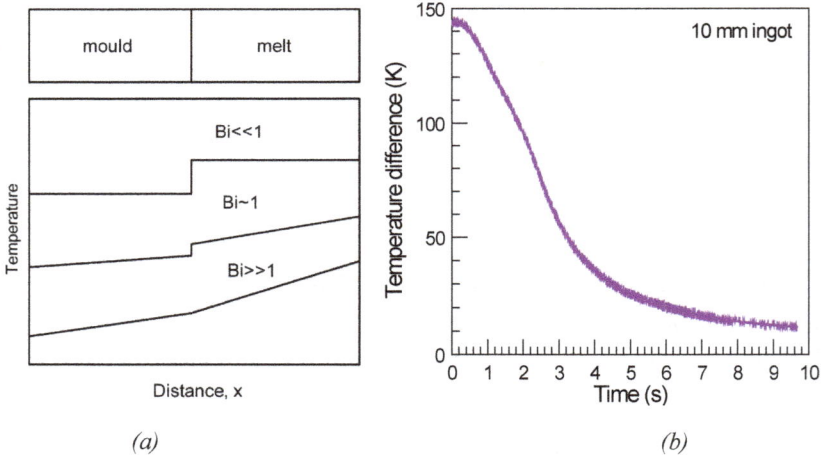

Fig. 2.17. (a) Schematic representation of temperature gradients close to the mould-melt interface depending upon the Biot number. Reproduced from [91] with permission of WILEY-VCH Verlag GmbH & Co. (b) The temperature difference between the center of the $Cu_{44}Ag_{15}Zr_{36}Ti_5$ ingot of 10 mm diameter and half distance to the mold upon casting of the melt.

The heat flow Q_f is proportional to the temperature gradient:

$$\frac{dQ_f}{dt} = -\lambda_l A \frac{dT}{dx}$$ (2.9)

where Q_f is the heat flow, t is time, λ_l is the thermal conductivity of a liquid, A is the transversal surface area, dT is the temperature gradient through which the heat is being transferred, dx is the thickness of the body of matter through which the heat is passing.

An ideal time-temperature relation for a cooling object is Newton's exponential decay law of cooling:

$$T(t) = T_s + (T_0 - T_s)e^{-t/\tau}$$ (2.10)

(where T is the body temperature, $T_0 = T$ at $t=0$, T_s is the temperature of the surrounding matter (mould), and $-1/\tau$ (or K) is a constant). Plotting $ln(T-T_s)$ versus t should therefore give a straight line with a slope $-1/\tau$ if cooling is influenced only by one kind of process. However, this law fulfils only if the Biot number is low enough. The cooling curve of the $Cu_{58.8}Zr_{26.2}Ti_{15}$ bulk glassy alloy (Fig. 2.18) can't be fit well with a single exponential decay function fitting (violet line). However, it can be fit very well (Fig. 2.18) if an exponential decay equation with two exponents (second order) is used:

$$T(t) = T_0 + A_1 e^{-t/\tau_1} + A_2 e^{-t/\tau_2}$$ (2.11)

where A_1, A_2, τ_1 and τ_2 are the constants at $T_0 = 423$ K, $A_1 = 356$ K, $\tau_1 = 0.074$ s, $A_2 = 571$ K, $\tau_2 = 0.478$ s. Thus, the cooling of the melt is influenced by both processes: the heat transfer through the melt as well as through the melt/mould interface.

Such a not-single exponential dependence of the cooling rate with time and the extrinsic factor connected with the heat transfer coefficient are likely responsible for the non-monotonic character of the dependence of the cooling rate on samples size even in logarithmic coordinates (Fig. 2.19). The estimated upper cooling rate (β) for the melt spun ribbon samples is 1 MK/s [156]. However, a better linear fit could be obtained if β at T_l and T_g was 20 and 3 MK/s, respectively [157].

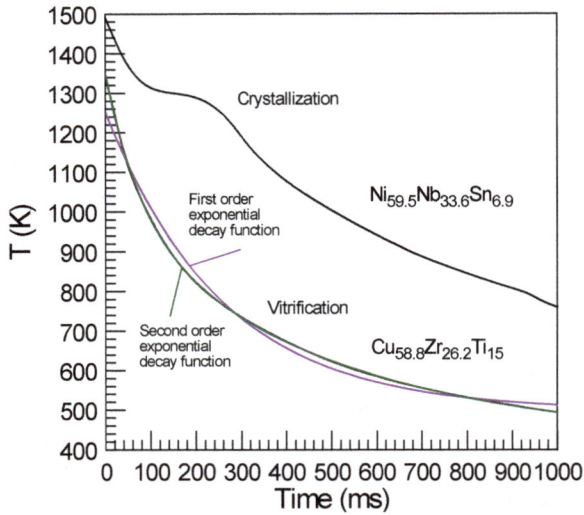

Fig. 2.18. The cooling curve of the $Cu_{58.8}Zr_{26.2}Ti_{15}$ bulk glassy alloy and $Ni_{59.5}Nb_{33.6}Sn_{6.9}$ alloy which crystallized. Violet line – the best fit with the first order exponential decay function. Green line – fit with the second order exponential decay function. Reproduced from [64] with permission of Cambridge University Press.

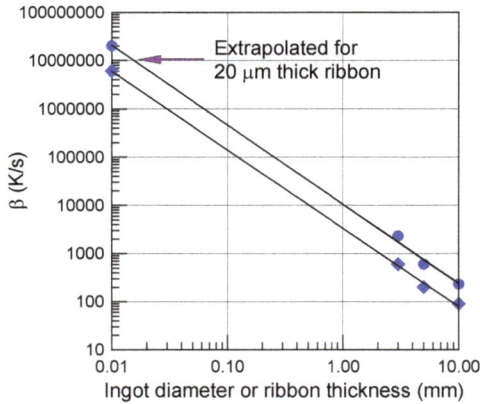

Fig. 2.19. Cooling rate (β) of the bulk samples having 3, 5 and 10 mm in diameter (•) at T_l and (♦) at T_g as a function of the ingot diameter in logarithmic coordinates. The data are taken from [158].

The second group of factors takes into account heterogeneous crystal nucleation as a potent source for decreasing GFA. Possible sources of heterogeneous nucleation include irregularities of the mould and impurities in the melt, for example, ceramic. An external alternating electromagnetic field [159] can also influence the nucleation process. Thus, for successful bulk glass formation one should take into account both intrinsic and extrinsic factors. For example, an alloy with high intrinsic GFA may not form a fully amorphous product due to limiting extrinsic GFA parameters related to purity of the alloying elements and the processing methodology.

2.4   Effect of flux treatment

Contamination by impurities or inclusions can be removed by flux treatment [160,161]. Fluxing by removing metallic oxide inclusions suppresses heterogeneous nucleation and thus inhibits crystallization. It favors homogeneous nucleation rather than heterogeneous nucleation [162] or reduces the nucleation rate in case of heterogeneous nucleation and drastically improves the critical diameter of Pd-based bulk glassy samples [163,164]. The $Pd_{40}Ni_{20}Cu_{20}P_{20}$ glassy alloy was prepared in a bulk form with a diameter up to 72 mm [165]. Similar to the Pd-Ni-Cu-P alloys some of Fe-based BMG were also successfully subjected to flux treatment used to improve their glass-forming ability [166]. A cyclic heating-cooling treatment with fluxing can significantly promote the purification effects, implying that impurities within the melting alloy can be driven to the surface by moving the solidification front [167]. Consequently, the critical cooling rate for glass formation can be drastically reduced.

Fig. 2.20 shows the DTA traces of the un-fluxed and fluxed $Pd_{40}Ni_{40}Si_4P_{16}$ bulk metallic glasses. The differently processed samples exhibit almost the same glass transition temperature T_g and approximate liquidus (nonequilibrium liquidus) temperature T_{lh}^* determined from the heating curves [168]. A similar behavior in glass transition and melting (in addition to the EDX compositional analysis) confirmed the compositional identity of the fluxed and not fluxed $Pd_{40}Ni_{40}Si_4P_{16}$ alloys. The supercooled liquid regions for both not fluxed and fluxed samples exceeds 100 K. From Fig. 2.20 and Table 2.7 one can see that crystallization of the fluxed $Pd_{40}Ni_{40}Si_4P_{16}$ alloy is much more sluggish than that of the not fluxed sample upon cooling. On continuous cooling crystallization of the fluxed sample started at 769 K while the crystallization of the not fluxed sample started at 872 K. The difference is 103 K. In addition, the supercooling ΔT (T_{lh}^*-T_{lc}^*, where T_{lc}^* is the temperature at which solidification on cooling starts) of the fluxed sample reaches 202 K. The supercooling ΔT (T_{lh}^*-T_{lc}^*) determines stability of a liquid, which is of importance to evaluate the glass-forming ability of alloys.

Table 2.7. Characteristic temperatures derived from the DTA traces shown in Fig. 2.19.

Alloys	T_g (±2 K)	T_x (±2 K)	T_{lh}^* (± 2K)	T_{lc}^* (± 2K)	ΔT (K)
fluxed	579	688	971	769	202
un-fluxed	579	683	971	872	99

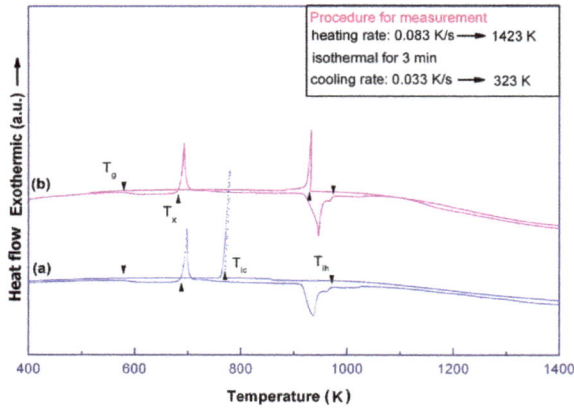

Fig. 2.20. DTA traces of (a) the fluxed and (b) the un-fluxed $Pd_{40}Ni_{40}Si_4P_{16}$ BMGs. Reproduced from [168] with permission of Elsevier.

Fig. 2.21 shows the crystalline eutectic colonies with a spherical morphology formed in not fluxed and fluxed samples. The density of precipitates related to the number of nucleation events is clearly different in two cases.

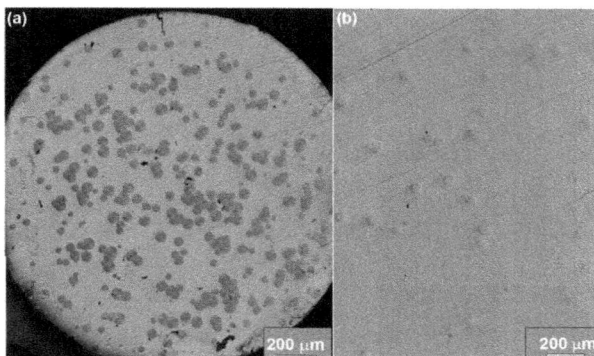

Fig. 2.21. SEM images of the polished cross section of not fluxed (a) and fluxed (b) samples annealed at 673 K for 9 minutes. Reproduced from [168] with permission of Elsevier.

The effect of small additions of Dy, Gd and Er to $Zr_{62.5}Cu_{22.5}Fe_5Al_{10}$ was also studied recently [169]. Ingots of the alloy without RE elements had a mixed glassy-crystalline structure and fully crystalline structure for the samples of 3 and 5 mm in diameter, respectively. The addition of 0.5 at.% Dy and Er did not promote formation of the amorphous structure, however the samples containing 1 at.% Dy and Er exhibited improved GFA leading to fully amorphous structure in both samples: 3 and 5 mm in diameter (Table 2.8). Gd containing samples with high γ parameter demonstrated amorphous structure at both diameters. The oxide particles are present in the structure of the alloys containing RE elements. RE elements formed oxides and inhibited oxygen-induced crystallization during cooling. Dy_2O_3, Gd_2O_3, Er_2O_3 with the average size of about 0.5 μm were uniformly distributed in the glassy phase. Their average volume fraction is 0.4 % in case of 0.5 at.% RE content and 0.7 % in case of 1.0 at. % RE content.

$Fe_{66}Nb_4B_{30}$ ferromagnetic bulk glassy alloy with high B content was also produced by fluxing and casting [170]. A record size (diameter 2.5 mm) $Al_{86}Ni_{6.75}Co_{2.25}Y_{3.25}La_{1.75}$ BMG alloy was produced by joint $MgCl_2$-$CaCl_2$ composite fluxing by the reduction of oxygen content [171].

Table 2.8. The values of T_g, T_x, T_s, T_l^* (approximate T_l) and GFA criteria for the studied alloys. Reproduced from [169] with permission of Elsevier.

Alloy	T_g (K)	T_x (K)	T_s (K)	T_l^* (K)	$T_{rg}=T_g/T_l$	$\gamma=T_x/(T_g+T_l)$
Base alloy	681	763	1143	1168	0.583	0.413
Dy 0.5 %	663	755	1066	1168	0.567	0.413
Dy 1 %	656	753	1070	1153	0.569	0.416
Gd 0.5 %	663	759	1067	1162	0.571	0.416
Gd 1 %	661	760	1085	1171	0.565	0.415
Er 0.5 %	662	758	1086	1161	0.570	0.416
Er 1 %	657	747	1084	1175	0.559	0.408

References

[1] T. Egami, Atomistic mechanism of bulk metallic glass formation, Journal of Non-Crystalline Solids, 317 (2003) 30-33.
https://doi.org/10.1016/S0022-3093(02)02003-3

[2] M. I. Ojovan, Ordering and structural changes at the glass–liquid transition, Journal of Non-Crystalline Solids, 382 (2013) 79-86.
https://doi.org/10.1016/j.jnoncrysol.2013.10.016

[3] H. S. Chen and D. Turnbull, Evidence of a glass–liquid transition in a gold–germanium–silicon alloy, J. Chem. Phys., 48 (1968) 2560.
https://doi.org/10.1063/1.1669483

[4] A. Inoue, High strength bulk amorphous alloys with low critical cooling rates, Mater. Trans. JIM, 36 (1995) 866-875. https://doi.org/10.2320/matertrans1989.36.866

[5] D. V. Louzguine, A. Inoue, Comparison of the long-term thermal stability of various metallic glasses under continuous heating, Scripta Mater, 47 (2002) 887-891. https://doi.org/10.1016/S1359-6462(02)00268-3

[6] J. Hafner, Theory of the formation of metallic glasses, Phys. Rev. B, 21 (1980) 406-426. https://doi.org/10.1103/PhysRevB.21.406

[7] D. V. Louzguine-Luzgin, D. B. Miracle, L. V. Louzguina-Luzgina and A. Inoue, Comparative analysis of glass-formation in binary, ternary and multicomponent alloys, Journal of Applied Physics, 108 (2010) 103511. https://doi.org/10.1063/1.3506687

[8] D. Xu, B. Lohwongwatana, G. Duan, W. L. Johnson and C. Garland, Bulk metallic glass formation in binary Cu-rich alloy series - $Cu_{100-x}Zr_x$ (x=34, 36, 38.2, 40 at.%) and mechanical properties of bulk $Cu_{64}Zr_{36}$ glass, Acta Mater., 52 (2004) 2621-2624. https://doi.org/10.1016/j.actamat.2004.02.009

[9] D. Wang, Y. Li, B. B. Sun, M. L. Sui, K. Lu and E. Ma, Bulk metallic glass formation in the binary Cu–Zr system, Appl. Phys. Lett., 84 (2004) 4029-4031. https://doi.org/10.1063/1.1751219

[10] Y.T. Sun, H.Y. Bai, M.Z. Li, W.H. Wang, Machine learning approach for prediction and understanding of glass-forming ability, J. Phys. Chem. Lett., 8 (2017) 3434-3439. https://doi.org/10.1021/acs.jpclett.7b01046

[11] L. Ward, S. C. O'Keeffe, J. Stevick, G. R. Jelbert, M. Aykol, C. Wolverton, A machine learning approach for engineering bulk metallic glass alloys, Acta Materialia, 159 (2018) 102-111. https://doi.org/10.1016/j.actamat.2018.08.002

[12] M. Gabski, M. Peterlechner, G. Wilde, Exploring the phase space of multi-principal-element alloys and predicting the formation of bulk metallic glasses, Entropy, 22 (2020) 292. https://doi.org/10.3390/e22030292

[13] T. Egami and Y. Waseda, Atomic size effect on the formability of metallic glasses, J. Non-Cryst Solids, 64 (1984) 113-134. https://doi.org/10.1016/0022-3093(84)90210-2

[14] D. B. Miracle and O. N. Senkov, Topological criterion for metallic glass formation, Mater. Sci. Eng. A, 347 (2003) 50-58. https://doi.org/10.1016/S0921-5093(02)00579-8

[15] R. Eiblert, J. Redinger and A. Neckel, Electronic structure, chemical bonding and spectral properties of the intermetallic compounds FeTi, CoTi and NiTi, J. Phys. F. Met. Phys., 17 (1987) 1533. https://doi.org/10.1088/0305-4608/17/7/011

[16] T. Egami, M. Ojha, D. M. Nicholson, D. V. Louzguine-Luzgin, N. Chen and A. Inoue, Glass formability and the Al-Au system, Philos. Mag., 92 (2012) 655-665. https://doi.org/10.1080/14786435.2011.630692

[17] D. B. Miracle, The efficient cluster packing model. An atomic structural model for metallic glasses, Acta Mater., 54 (2006) 4317-4336. https://doi.org/10.1016/j.actamat.2006.06.002

[18] H. W. Sheng, W. K. Luo, F. M. Alamgir, J. M. Bai and E. Ma, Atomic packing and short-to- medium-range order in metallic glasses, Nature, 439 (2006) 419-425. https://doi.org/10.1038/nature04421

[19] H. A. Davies, Metallic glass formation, in Amorphous Metallic Alloys (ed) F.E. Luborsky, London, Butterworths (1983) pp. 8–25. https://doi.org/10.1016/B978-0-408-11030-3.50007-8

[20] Y. W. Kim, H. M. Lin and T. F. Kelly, Amorphous solidification of pure metals in submicron spheres, Acta Metall. 37 (1989) 247-255. https://doi.org/10.1016/0001-6160(89)90283-6

[21] J. Brillo and I. Egry, Density determination of liquid copper, nickel and their alloys, International Journal of Thermophysics, 24 (2003) 1155-1170. https://doi.org/10.1023/A:1025021521945

[22] W. F. Gale, T.C. Totemeier, editors. Smithells Metals Reference Book 8-th Edition (Elsevier Butterworth-Heinemann Ltd., Oxford UK, 2004) p. 14-1.

[23] I.-R. Lu, G.P. Gorler, H.J. Fecht and R. Willnecker, Investigation of specific volume of glass-forming Pd-Cu-Ni-P alloy in the liquid, vitreous and crystalline state, Journal of Non-Crystalline Solids, 312–314 (2002) 547-551. https://doi.org/10.1016/S0022-3093(02)01767-2

[24] W. Kauzmann, The nature of the glassy state and the behavior of liquids at low temperatures, Chem. Rev., 43 (1948) 219–256. https://doi.org/10.1021/cr60135a002

[25] D. V. Louzguine-Luzgin, I. Seki, S. V. Ketov, L. V. Louzguina-Luzgina, V. I. Polkin, N. Chen, H. Fecht, A. N. Vasiliev and H. Kawaji, Glass-transition process in an Au-based metallic glass, Journal of Non-Crystalline Solids, 419 (2015) 12–15. https://doi.org/10.1016/j.jnoncrysol.2015.03.018

[26] L.-M. Martinez and C. A. Angell, A thermodynamic connection to the fragility of glass-forming liquids, Nature, 410 (2001) 663-667. https://doi.org/10.1038/35070517

[27] W. H. Wang, Bulk metallic glasses with functional physical properties, Adv. Mater., 21 (2009) 4524–4544. https://doi.org/10.1002/adma.200901053

[28] I. S. Klein and C. A. Angell, Excess thermodynamic properties of glassforming liquids. The rational scaling of heat capacities and the thermodynamic fragility dilemma resolved, Journal of Non-Crystalline Solids, 451 (2016) 116-123. https://doi.org/10.1016/j.jnoncrysol.2016.06.006

[29] C.T. Moynihan, A.J. Easteal, M.A. DeBolt, J. Tucker, Dependence of the fictive temperature of glass on cooling rate, J. Am. Ceram. Soc., 59 (1976) 12. https://doi.org/10.1111/j.1151-2916.1976.tb09376.x

[30] P. W. Anderson, Through the Glass Lightly, Science, 267 (1995) 1615. https://doi.org/10.1126/science.267.5204.1615-e

[31] C. A. Angell, The Glass Transition," Pergamon Encyclopedia of Materials: Science and Technology, Vol 4 (2001) 3365.

[32] J.H. Na, S.L. Corona, A. Hoff, W.L. Johnson, Observation of an apparent first-order glass transition in ultrafragile Pt–Cu–P bulk metallic glasses, Proceedings of the National Academy of Sciences of the United States of America, 117 (2020) 2779-2787. https://doi.org/10.1073/pnas.1916371117

[33] M. I. Ojovan and D. V. Louzguine-Luzgin, Revealing structural changes at glass transition via radial distribution functions, Journal of Physical Chemistry B, 124 (2020) 3186–3194. https://doi.org/10.1021/acs.jpcb.0c00214

[34] H. S. Chen, H. Kato and A. Inoue, A fictive stress model calculation of nonlinear viscoelastic behaviors in a Zr-based glassy alloy: stress growth and relaxation, Jpn. J. Appl. Phys., 39 (2000) 5184–5187. https://doi.org/10.1143/JJAP.39.5184

[35] R. F. Tournier, Fragile-to-fragile liquid transition at t_g and stable-glass phase nucleation rate maximum at the Kauzmann Temperature, Physica B, 454 (2014) 253-271. https://doi.org/10.1016/j.physb.2014.07.069

[36] D. Turnbull and M. H. Cohen, Free-volume model of the amorphous phase: glass transition, J. Chem. Phys., 34 (1961) 120-125. https://doi.org/10.1063/1.1731549

[37] P. G. Debenedetti and F. H. Stillinger, Supercooled liquids and the glass transition, Nature, 410 (2001) 259-267. https://doi.org/10.1038/35065704

[38] A. Inoue, Stabilization of metallic supercooled liquid and bulk amorphous alloys, Acta Mater., 48 (2000) 279-306. https://doi.org/10.1016/S1359-6454(99)00300-6

Materials Research Foundations **85** (2021)
https://doi.org/10.21741/9781644901014

[39] K. F. Kelton, Transient nucleation in glasses, Mater. Sci. Eng. B, 32 (1995) 145-151. https://doi.org/10.1016/0921-5107(95)80023-9

[40] Z. P. Lu and C. T. Liu, A new glass-forming ability criterion for bulk metallic glasses, Acta Materialia, 50 (2002) 3501-3512. https://doi.org/10.1016/S1359-6454(02)00166-0

[41] X. H. Du, J. C. Huang, C. T. Liu and Z. P. Lu, New criterion of glass forming ability for bulk metallic glasses, J. Appl. Phys., 101 (2007) 086108. https://doi.org/10.1063/1.2718286

[42] O. N. Senkov, Correlation between fragility and glass-forming ability of metallic alloys, Phys. Rev. B, 76 (2007) 104202. https://doi.org/10.1103/PhysRevB.76.104202

[43] W. L. Johnson, J. H. Na and M. D. Demetriou, Quantifying the origin of metallic glass formation, Nat. Commun., 7 (2016) 10313. https://doi.org/10.1038/ncomms10313

[44] D.V. Louzguine-Luzgin, G. Xie, Q. Zhang and A. Inoue, Devitrification behavior and crystal-glassy mixed-phase structures observed in partially crystallized Cu-based glassy alloys, Ceramic Transactions, 219 (2010) 3-8. https://doi.org/10.1002/9780470917145.ch1

[45] Q. S. Zhang, W. Zhang, D. V. Louzguine-Luzgin, A. Inoue, High glass-forming ability and unusual deformation behavior of new Zr-Cu-Fe-Al bulk metallic glasses, Materials Science Forum, 654-656 (2010) 1042-1045. https://doi.org/10.4028/www.scientific.net/MSF.654-656.1042

[46] F. Jia, W. Zhang, H.M. Kimura, A. Inoue, Effects of additional Ag on the thermal stability and glass-forming ability of La–Al–Cu bulk glassy alloys, Materials Science and Engineering B 148 (2008) 119–123. https://doi.org/10.1016/j.mseb.2007.09.002

[47] W. Zhang, F. Jia and A. Inoue, Formation and properties of new La-based bulk glassy alloys with diameters up to centimeter order, Materials Transactions, 48 (2007) 68 - 73. https://doi.org/10.2320/matertrans.48.68

[48] H. Tan, Y. Zhang, D. Ma, Y.P. Feng, Y. Li, Optimum glass formation at off-eutectic composition and its relation to skewed eutectic coupled zone in the La based La–Al–(Cu,Ni) pseudo ternary system, Acta Materialia 51 (2003) 4551–4561. https://doi.org/10.1016/S1359-6454(03)00291-X

[49] Y. Zhang, Y. Li, H. Tan, G.L. Chen, H.A. Davies, Glass forming ability criteria for La–Al–(Cu,Ni) alloys, Journal of Non-Crystalline Solids, 352 (2006) 5482–5486. https://doi.org/10.1016/j.jnoncrysol.2006.09.027

[50] E. R. Arata, F. H. Dalla Torre, J. F. Loffler, Identification of bulk metallic glass-forming compositions in La-based systems using ultrahigh gravitational acceleration, Acta Materialia 56 (2008) 651–658. https://doi.org/10.1016/j.actamat.2007.10.026

[51] Z.P. Lu, H. Tan, Y. Li and S.C. Ng, The correlation between reduced glass transition temperature and glass forming ability of bulk metallic glasses, Scripta mater. 42 (2000) 667–673. https://doi.org/10.1016/S1359-6462(99)00417-0

[52] N. Nishiyama and A. Inoue, Direct comparison between critical cooling rate and some quantitative parameters for evaluation of glass-forming ability in Pd-Cu-Ni-P alloys, Mater. Trans., 43 (2002) 1913-1917. https://doi.org/10.2320/matertrans.43.1913

[53] L. Zhang, Y. Wu, X. Bian, H. Li, W. Wang and S. Wu, Short-range and medium-range order in liquid and amorphous $Al_{90}Fe_5Ce_5$ alloys, J. Non-Cryst. Sol., 262 (2000) 169-176. https://doi.org/10.1016/S0022-3093(99)00699-7

[54] P.S. Popel and Y.E. Sidorov, Microheterogeneity of liquid metallic solutions and its influence on the structure and properties of rapidly quenched alloys, Materials Science and Engineering A, 226-228 (1997) 237-244. https://doi.org/10.1016/S0921-5093(96)10624-9

[55] I. W. Donald and H. A. Davies, Prediction of glass-forming ability for metallic systems, J. Non-Cryst. Solids, 30 (1978) 77. https://doi.org/10.1016/0022-3093(78)90058-3

[56] A. R. Yavari, Solving the hume-rothery eutectic puzzle using miracle glasses, Nature Mater., 4 (2005) 1-2. https://doi.org/10.1038/nmat1289

[57] D. V. Louzguine, L. V. Louzguina and A. Inoue, Factors influencing glass formation in rapidly solidified Si,Ge–Ni and Si,Ge–Ni–Nd alloys, Appl. Phys. Lett., 80 (2002) 1556-1558. https://doi.org/10.1063/1.1457522

[58] K.F. Yao, N. Chen, Pd-Si binary bulk metallic glass, Sci. China. Ser. G, 51 (2008) 414-420. https://doi.org/10.1007/s11433-008-0051-4

[59] N. Chen, H. A. Yang, A. Caron, P. C. Chen, Y. C. Lin, D. V. Louzguine-Luzgin, K. F. Yao, M. Esashi and A. Inoue, Glass-forming ability and thermoplastic formability of a $Pd_{40}Ni_{40}Si_4P_{16}$ glassy alloy, Journal of Materials Science, 46 (2011) 2091-2096. https://doi.org/10.1007/s10853-010-5043-x

[60] H. Tan, Y. Zhang, D. Ma, Y.P. Feng and Y. Li, Optimum glass formation at off-eutectic composition and its relation to skewed eutectic coupled zone in the La based

La–Al–(Cu,Ni) pseudo ternary system, Acta Materialia, 51 (2003) 4551-4561. https://doi.org/10.1016/S1359-6454(03)00291-X

[61] A. Zhu, G. Shiflet and D. B. Miracle, Glass forming ranges of Al–rare earth metal alloys: thermodynamic and kinetic analysis, Scripta Mater., 50 (2004) 987-991. https://doi.org/10.1016/j.scriptamat.2004.01.019

[62] S. Gorsse, G. Orveillon, O. N. Senkov and D. B. Miracle, Thermodynamic analysis of glass-forming ability in a Ca-Mg-Zn ternary alloy system, Phys. Rev. B, 73 (2006) 224202. https://doi.org/10.1016/j.scriptamat.2004.01.019

[63] D. V. Louzguine-Luzgin, A. D. Setyawan, H. Kato and A. Inoue, Thermal conductivity of an alloy in relation to the observed cooling rate and glass-forming ability, Phil. Mag., 87 (2007) 1845. https://doi.org/10.1080/14786430601096928

[64] D. V. Louzguine-Luzgin, T. Saito, J. Saida and A. Inoue, Thermal conductivity of metallic glassy alloys and its relationship to the glass forming ability and the observed cooling rates, J. Mater. Res., 23 (2008) 2283-2287. https://doi.org/10.1557/JMR.2008.0286

[65] D. V. Louzguine-Luzgin, T. Saito, J. Saida and A. Inoue, Influence of cooling rate on the structure and properties of a Cu–Zr–Ti–Ag glassy alloy, J. Mater. Res., 23 (2008) 515-522. https://doi.org/10.1557/JMR.2008.0066

[66] S. R. Nagel and J. Taue, Nearly-free-electron approach to the theory of metallic glass alloys, Phys. Rev. Lett., 35 (1975) 380. https://doi.org/10.1103/PhysRevLett.35.380

[67] D. V. Louzguine and A. Inoue, Investigation of structure and properties of the Al-Y-Ni-Co-Cu metallic glasses, J Mater Res., 17 (2002) 1014-1018. https://doi.org/10.1557/JMR.2002.0149

[68] J. H. Perepezko, S. D. Imhoff and R. J. Hebert, Nanostructure development during devitrification and deformation, J. Alloys Comp., 495 (2010) 360-364. https://doi.org/10.1016/j.jallcom.2009.10.051

[69] Z. Huang, J. Li, Q. Rao and Y. Zhou, Thermal stability and primary phase of Al–Ni(Cu)–La amorphous alloys, J. Alloys Compd., 463 (2008) 328-335. https://doi.org/10.1016/j.jallcom.2007.09.005

[70] V.I. Tkatch, S.G. Rassolov, V.V. Popov, V.V. Maksimov, V.V. Maslov, V.K. Nosenko, A.S. Aronin, G.E. Abrosimova and O.G. Rybchenko, Complex crystallization mode of amorphous/nanocrystalline composite $Al_{86}Ni_2Co_{5.8}Gd_{5.7}Si_{0.5}$, Journal of

Non-Crystalline Solids, 357 (2011) 1628–1632.
https://doi.org/10.1016/j.jnoncrysol.2011.02.029

[71] D. V. Louzguine-Luzgin and A. Inoue, A glance on the glass-transition phenomenon from the viewpoint of devitrification, J. Alloys Comp., 434-435 (2007) 121.
https://doi.org/10.1016/j.jallcom.2006.08.146

[72] D. V. Louzguine-Luzgin, N. Chen, A. Yu. Churymov, L. V. Louzguina-Luzgina, V. I. Polkin, L. Battezzati and A. R. Yavari, Role of different factors in the glass-forming ability of binary alloys, Journal of Materials Science, 50 (2015) 1783-1793.
https://doi.org/10.1007/s10853-014-8741-y

[73] A. A. Tsarkov, E. N. Zanaeva, A. Yu. Churyumov, S. V. Ketov and D. V. Louzguine-Luzgin, Crystallization kinetics of Mg–Cu–Yb–Ca–Ag metallic glasses, Materials Characterization, 111 (2016) 75–80.
https://doi.org/10.1016/j.matchar.2015.10.034

[74] J. D. Bernal, The structure of liquids, Proc. R. Soc. A., 280 (1964) 299.
https://doi.org/10.1098/rspa.1964.0147

[75] O. N. Senkov, D. B. Miracle and H. M. Mullens, Topological criteria for amorphization based on a thermodynamic approach, J. Appl. Phys., 97 (2005) 103502.
https://doi.org/10.1063/1.1896434

[76] D. B. Miracle, W. S. Sanders and O. N. Senkov, The influence of efficient atomic packing on the constitution of metallic glasses, Phil. Mag., 83 (2003) 2409.
https://doi.org/10.1080/1478643031000098828

[77] A. Inoue, T. Negishi, H. M. Kimura, T. Zhang and A. R. Yavari, High packing density of Zr- and Pd-based bulk amorphous alloys, Mater. Trans. JIM, 39 (1998) 318.
https://doi.org/10.2320/matertrans1989.39.318

[78] D. B. Miracle, D. V. Louzguine-Luzgin, L. V. Louzguina-Luzgina and A. Inoue, An assessment of binary metallic glasses correlations between structure, glass forming ability and stability, International Materials Reviews, 55 (2010) 218-256.
https://doi.org/10.1179/095066010X12646898728200

[79] T. Egami, Universal criterion for metallic glass formation, Mater. Sci. Eng. A, 226–228 (1997) 261-267. https://doi.org/10.1016/S0921-5093(97)80041-X

[80] D. Hobbs, J. Hafner and D. Spisak, Understanding the complex metallic element Mn. I. Crystalline and noncollinear magnetic structure of α-Mn, Phys. Rev. B, 68 (2003) 014407. https://doi.org/10.1103/PhysRevB.68.014407

[81] D. V. Louzguine-Luzgin, A. R. Yavari, G. Vaughan and A. Inoue, Clustered crystalline structures as glassy phase approximants, Intermetallics, 17 (2009) 477-480. https://doi.org/10.1016/j.intermet.2008.12.008

[82] D. B. Miracle, W. S. Sanders and O. N. Senkov, The influence of efficient atomic packing on the Constitution of metallic glasses, Philos Mag, 83 (2003) 2409-2428. https://doi.org/10.1080/1478643031000098828

[83] L. Xia, W. H. Li, S. S. Fang, B. C. Wei and Y. D. Dong, Binary Ni–Nb bulk metallic glasses, J Appl. Phys., 99 (2006) 026103. https://doi.org/10.1063/1.2158130

[84] D. V. Louzguine and A. Inoue, Electronegativity of the constituent rare-earth metals as a factor stabilizing the supercooled liquid region in Al-based metallic glasses, Appl. Phys. Lett., 79 (2001) 3410-3412. https://doi.org/10.1063/1.1420781

[85] D. V. Louzguine-Luzgin, A. Inoue and W. J. Botta, Reduced electronegativity difference as a factor leading to the formation of Al-based glassy alloys with a large supercooled liquid region of 50 K, Appl. Phys. Lett., 88 (2006) 011911-011913. https://doi.org/10.1063/1.2159420

[86] C.S. Ma, J. Zhang, X.C. Chang, W.L. Hou and J.Q. Wang, Electronegativity difference as a factor for evaluating the thermal stability of Al-rich metallic glasses, Philosophical Magazine Letters, 88 (2008) 917-924. https://doi.org/10.1080/09500830802526596

[87] C. Suryanarayana, I. Seki, A. Inoue, A critical analysis of the glass-forming ability of alloys, Journal of Non-Crystalline Solids, 355 (2009) 355-360. https://doi.org/10.1016/j.jnoncrysol.2008.12.009

[88] R. D. Sá Lisboa, C. Bolfarini, W. J. Botta F., and C. S. Kiminami, Topological instability as a criterion for design and selection of aluminum-based glass-former alloys, Appl. Phys. Lett. 86, 211904 (2005). https://doi.org/10.1063/1.1931047

[89] E. S. Park, D. H. Kim and W. T. Kim, Parameter for glass forming ability of ternary alloy systems, Appl. Phys. Lett., 86 (2005) 061907. https://doi.org/10.1063/1.1862790

[90] K. Song, X.F. Bian, X.Q. Lu, M.T. Xie and R. Jia, The correlations among the fragility of supercooled liquids, the fragility of superheated melts and the glass-forming ability for marginal metallic glasses, J. Appl. Phys., 105 (2009) 024304. https://doi.org/10.1063/1.3043632

[91] D. V. Louzguine-Luzgin, D. B. Miracle and A. Inoue, Intrinsic and extrinsic factors influencing the glass-forming ability of alloys, Adv. Eng. Mater., 10 (2008) 1008-1015. https://doi.org/10.1002/adem.200800134

[92] J. H. Perepezko, Nucleation in undercooled liquids, Mater. Sci. Eng., 65 (1984) 125-135. https://doi.org/10.1016/0025-5416(84)90206-4

[93] D. V. Louzguine-Luzgin and A. Inoue, An extended criterion for estimation of glass-forming ability of metals, J. Mater. Res., 22 (2007) 1378-1383. https://doi.org/10.1557/jmr.2007.0167

[94] D.V. Louzguine-Luzgin, Vitrification and devitrification processes in metallic glasses, Journal of Alloys and Compounds, 586 (2014) S2–S8. https://doi.org/10.1016/j.jallcom.2012.09.057

[95] A. R. Yavari, Small volume change on melting as a new criterion for easy formation of metallic glasses, Phys. Lett. A, 95 (1983) 165. https://doi.org/10.1016/0375-9601(83)90825-3

[96] Y. Li, Q. Guo, J. A. Kalb and C.V. Thompson, Matching glass-forming ability with the density of the amorphous phase, Science, 322 (2008) 1816. https://doi.org/10.1126/science.1163062

[97] N. Jakse and A. Pasturel, Local order and dynamic properties of liquid and undercooled Cu_xZr_{1-x} alloys by ab-initio molecular dynamics, Phys. Rev. B, 78 (2008) 214204.

[98] K. Russew, L. Stojanova, S. Yankova, E. Fazakas and L. K. Varga, Thermal behavior and melt fragility number of $Cu_{100-x}Zr_x$ glassy alloys in terms of crystallization and viscous flow, Journal of Physics Conference Series, 144 (2009) 012094. https://doi.org/10.1088/1742-6596/144/1/012094

[99] D. V. Louzguine-Luzgin, R. Belosludov, M. Saito, Y. Kawazoe and A. Inoue, Glass-transition behavior of Ni. Calculation, prediction and experiment, J. Appl. Phys., 104 (2008) 123529. https://doi.org/10.1063/1.3042240

[100] Y. Waseda, in Liquid Metals, edited by R. Evans and D.A. Greenwood, The Institute of Physics, Bristol, United Kingdom (1977) pp. 230-40.

[101] H. H. Wendt and F. F. Abraham, Empirical criterion for the glass transition region based on monte carlo simulations, Phys. Rev. Lett., 41 (1978) 1245-1247. https://doi.org/10.1103/PhysRevLett.41.1244

[102] S. Plimpton, Fast parallel algorithms for short-range molecular dynamics, J. Comp. Phys, 117 (1995) 1-19. https://doi.org/10.1006/jcph.1995.1039

[103] M.I. Mendelev, S. Han, D.J. Srolovitz, G.J. Ackland, D.Y. Sun and M. Asta, Development of new interatomic potentials appropriate for crystalline and liquid iron, Phil. Mag., 83 (2003) 3977-3994. https://doi.org/10.1080/14786430310001613264

[104] A. V. Evteev, A. T. Kosilov and E. V. Levchenko, Atomic mechanisms of pure iron vitrification, Journal of Experimental and Theoretical Physics, 99 (2004) 522–529. https://doi.org/10.1134/1.1809680

[105] W.T. Laughlin and D.R. Uhlmann, Viscous flow in simple organic liquids, J. Phys. Chem. 76 (1972) 2317-2325. https://doi.org/10.1021/j100660a023

[106] C.A. Angell, Spectroscopy simulation and scattering and the medium range order problem in glass, J. Non-Cryst. Solids, 73 (1985) 1-17. https://doi.org/10.1016/0022-3093(85)90334-5

[107] J. C. Mauro, Y.Z. Yue, A. J. Ellison, P. K. Gupta and D. C. Allan, Viscosity of glass-forming liquids, Proc. Natl. Acad. Sci. USA, 106 (2009) 19780-19784. https://doi.org/10.1073/pnas.0911705106

[108] W. L. Johnson, J. Lu and M. D. Demetriou, Deformation and flow in bulk metallic glasses and deeply undercooled glass forming liquids—a self consistent dynamic free volume model, Intermetallics, 10 (2002) 1039–1046. https://doi.org/10.1016/S0966-9795(02)00160-7

[109] C. A. Angell, Formation of glasses from liquids and biopolymers, Science, 267 (1995) 1924. https://doi.org/10.1126/science.267.5206.1924

[110] P. G. Debenedetti and F. H. Stillinger, Supercooled liquids and the glass transition, Nature, 410 (2001) 259-267. https://doi.org/10.1038/35065704

[111] M. H. Cohen and D. Turnbull, Molecular transport in liquids and glasses, J. Chem. Phys., 31 (1959) 1164. https://doi.org/10.1063/1.1730566

[112] M. H. Cohen and G. S. Grest, Liquid-glass transition, a free-volume approach, Phys. Rev., 20 (1979) 1077. https://doi.org/10.1103/PhysRevB.20.1077

[113] S. Mukherjee, J. Schroers, Z. Zhou, W.L. Johnson and W.-K. Rhim, Viscosity and specific volume of bulk metallic glass-forming alloys and their correlation with glass forming ability, Acta Materialia, 52 (2004) 3689–3695. https://doi.org/10.1016/j.actamat.2004.04.023

[114] L. Battezzati and A.L. Greer, The viscosity of liquid metals and alloys, Acta Metallurgica, 37 (1989) 1791-1802. https://doi.org/10.1016/0001-6160(89)90064-3

[115] D.-H. Huang and G.B. McKenna, New insights into the fragility dilemma in liquids, J. Chem. Phys., 114 (2001) 5621-5630. https://doi.org/10.1063/1.1348029

[116] I. Gallino, On the fragility of bulk metallic glass forming liquids, Entropy, 19 (2017) 483. https://doi.org/10.3390/e19090483

[117] G. Adam, J.H. Gibbs, On the temperature dependence of cooperative relaxation properties in glass-forming liquids, J. Chem. Phys. 43, (1965), 139–146. https://doi.org/10.1063/1.1696442

[118] M.I. Ojovan, About activation energy of viscous flow of glasses and melts, Materials Research Society Symposium Proceedings, 1757 (2015) 7-12. https://doi.org/10.1557/opl.2015.44

[119] K. F. Kelton, Kinetic and structural fragility - A correlation between structures and dynamics in metallic liquids and glasses, Journal of Physics Condensed Matter, 29 (2017) 023002. https://doi.org/10.1088/0953-8984/29/2/023002

[120] J. Krausser, K. Samwer and A. Zaccone, Interatomic repulsion softness controls the fragility of supercooled metallic melts, Proc. Natl Acad. Sci. USA, 112 (2015) 13762. https://doi.org/10.1073/pnas.1503741112

[121] C. E. Pueblo, M.H. Sun and K. F. Kelton, Strength of the repulsive part of the interatomic potential determines fragility in metallic liquids, Nature Materials, 16 (2017) 792–796. https://doi.org/10.1038/nmat4935

[122] D. R. Uhlmann, A kinetic treatment of glass formation, Journal of Non-crystalline Solids, 7 (1972) 337-348. https://doi.org/10.1016/0022-3093(72)90269-4

[123] D. V. Louzguine-Luzgin, R. Belosludov, A. R. Yavari, K. Georgarakis, G. Vaughan, Y. Kawazoe, T. Egami and A. Inoue, Structural basis for supercooled liquid fragility established by synchrotron-radiation method and computer simulation, Journal of Applied Physics, 110 (2011) 043519. https://doi.org/10.1063/1.3624745

[124] V. Wessels, A. K. Gangopadhyay, K. K. Sahu, R. W. Hyers, S. M. Canepari, J. R. Rogers, M. J. Kramer, A. I. Goldman, D. Robinson, J. W. Lee, J. R. Morris and K. F. Kelton, Rapid chemical and topological ordering in supercooled liquid $Cu_{46}Zr_{54}$, Phys. Rev. B, 83 (2011) 094116. https://doi.org/10.1103/PhysRevB.83.094116

[125] K. Georgarakis, L. Hennet, G.A. Evangelakis, J. Antonowicz, G.B. Bokas, V. Honkimaki, A. Bytchkov, M.W. Chen and A.R. Yavari, Probing the structure of a liquid metal during vitrification, Acta Materialia, 87 (2015) 174–186. https://doi.org/10.1016/j.actamat.2015.01.005

[126] K.F. Kelton, Kinetic and structural fragility - A correlation between structures and dynamics in metallic liquids and glasses, Journal of Physics Condensed Matter, 29 (2017) 023002. https://doi.org/10.1088/0953-8984/29/2/023002

[127] N. A. Mauro, M. Blodgett, M. L. Johnson, A. J. Vogt and K. F. Kelton, A structural signature of liquid fragility, Nature Communications, 5 (2014) 4616. https://doi.org/10.1038/ncomms5616

[128] H. Wang and E. A. Carter, Metal-metal bonding in Engel-Brewer intermetallics , anomalous, charge transfer in zirconium-platinum ($ZrPt_3$) J. Am. Chem. SOC, 115 (1993) 2357-2362. https://doi.org/10.1021/ja00059a034

[129] T. Iwashita, D. M. Nicholson and T. Egami, Elementary excitations and crossover phenomenon in liquids, Phys. Rev. Lett., 110 (2013) 205504. https://doi.org/10.1103/PhysRevLett.110.205504

[130] S. V. Nemilov Maxwell Equation and Classical Theories of Glass Transition as a Basis for Direct Calculation of Viscosity at Glass Transition Temperature, Glass Physics and Chemistry, 39 (2013) 609–623. https://doi.org/10.1134/S1087659613060084

[131] D. V. Louzguine-Luzgin, L. V. Louzguina-Luzgina and H. Fecht, On limitations of the viscosity versus temperature plot for glass-forming substances, Materials Letters, 182 (2016) 355–358. https://doi.org/10.1016/j.matlet.2016.07.006

[132] A. E. Berlev, O. P. Bobrov, V. A. Khonik, K. Csach, A. Jurıkova, J. Miskuf, H. Neuhauser and M. Yu. Yazvitsky, Viscosity of bulk and ribbon Zr-based glasses well below and in the vicinity of T_g. A comparative study, Physical Review B, 68 (2003) 132203. https://doi.org/10.1103/PhysRevB.68.132203

[133] H. L. Smith, C. W. Li, A. Hoff, G. R. Garrett, D. S. Kim, F. C. Yang, M. S. Lucas, T. Swan-Wood, J. Y. Y. Lin, M. B. Stone, D. L. Abernathy, M. D. Demetriou and B. Fultz, Separating the configurational and vibrational entropy contributions in metallic glasses, Nature Physics, 13 (2017) 4142-4146. https://doi.org/10.1038/nphys4142

[134] P. D. Gujrati and M. Goldstein, Viscous liquids and the glass transition. Nonconfigurational contributions to the excess entropy of disordered phases, J. Phys. Chem., 84 (1980) 869–873. https://doi.org/10.1021/j100445a013

[135] Y. Kawamura and A. Inoue, Newtonian viscosity of supercooled liquid in a $Pd_{40}Ni_{40}P_{20}$ metallic glass, Applied Physics Letters, 77 (2000) 1114-1116. https://doi.org/10.1063/1.1289502

[136] K. Csach, O. P. Bobrov, V. A. Khonik and S. A. Lyakhov and K. Kitagawa, Relationship between the shear viscosity and heating rate of metallic glasses below T_g, Physical Review B, 73 (2006) 092107. https://doi.org/10.1103/PhysRevB.73.092107

[137] G. Wilde, G. P. Görler, R. Willnecker and H. J. Fecht, Calorimetric, thermomechanical and rheological characterizations of bulk glass-forming $Pd_{40}Ni_{40}P_{20}$, J. Appl. Phys., 87 (2000) 1141-1152. https://doi.org/10.1063/1.371991

[138] U. Dahlborg, M. Calvo-Dahlborg, P.S. Popel, and V.E. Sidorov, Structure and properties of some glass-forming liquid alloys, Eur. Phys. J. B, 14 (2000) 639-648.

[139] M. Stolpe, I. Jonas, S. Wei, Z. Evenson, W. Hembree, F. Yang, A. Meyer, and R. Busch, Structural changes during a liquid-liquid transition in the deeply undercooled $Zr_{58.5}Cu_{15.6}Ni_{12.8}Al_{10.3}Nb_{2.8}$ bulk metallic glass forming melt, Phys. Rev. B, 93 (2016) 014201.

[140] I. Gallino, D. Cangialosi, Z. Evenson, L. Schmitt, S. Hechler, M. Stolpe, B. Ruta, Hierarchical aging pathways and reversible fragile-to-strong transition upon annealing of a metallic glass former, Acta Materialia, 144 (2018) 400-410. https://doi.org/10.1016/j.actamat.2017.10.060

[141] D. V. Louzguine-Luzgin, A. I. Bazlov, S. V. Ketov, A. L. Greer and A. Inoue, Crystal growth limitation as a critical factor for formation of Fe-based bulk metallic glasses, Acta Materialia, 82 (2015) 396–402. https://doi.org/10.1016/j.actamat.2014.09.025

[142] V. Ponnambalam, S. J. Poon and J. Shiflet, Synthesis of iron-based bulk metallic glasses as nonferromagnetic amorphous steel alloys, J. Mater. Res., 19 (2004) 3046-3053. https://doi.org/10.1557/JMR.2004.0374

[143] Z. P. Lu, C. T. Liu and W. D. Porter, Role of yttrium in glass formation of Fe-based bulk metallic glasses, Appl. Phys. Lett., 83 (2003) 2581-2583. https://doi.org/10.1063/1.1614833

[144] A. Bouchare, B. Bendjemil, R. Piccin and M. Baricco, Formation and thermal properties of Fe-based BMG's with Y or Gd addition, Int. J. Nanoelectronics and Materials, 3 (2010) 63-70.

[145] K. Amiya and A. Inoue, Fe-(Cr, Mo)-(C, B)-Tm bulk metallic glasses with high strength and high glass-forming ability, Rev. Adv. Mater. Sci., 18 (2008) 27-29.

[146] T. Zhang, Q. Yang, J. I. Fei, L. I. Ran, S. J. Pang, J. F. Wang and T. Xu, Centimeter-scale-diameter Co-based bulk metallic glasses with fracture strength exceeding 5000 MPa, Chinese Science Bulletin, 56 (2011) 3972-3977. https://doi.org/10.1007/s11434-011-4765-8

[147] C. Hin, B. D. Wirth and J. B. Neaton, Formation of Y_2O_3 nanoclusters in nanostructured ferritic alloys during isothermal and anisothermal heat treatment. A kinetic Monte Carlo study, Phys. Rev. B 80 (2009) 134118. https://doi.org/10.1103/PhysRevB.80.134118

[148] A. Hirata, Y. Hirotsu, K. Amiya and A. Inoue, Crystallization process and glass stability of an $Fe_{48}Cr_{15}Mo_{14}C_{15}B_6Tm_2$ bulk metallic glass, Physical Review B 78 (2008) 144205. https://doi.org/10.1103/PhysRevB.78.144205

[149] D.V. Louzguine-Luzgin, A.I. Bazlov, S.V. Ketov and A. Inoue, Crystallization behavior of Fe- and Co-based bulk metallic glasses and their glass-forming ability, Materials Chemistry and Physics, 162 (2015) 197–206. https://doi.org/10.1016/j.matchemphys.2015.05.058

[150] D. Turnbull, Under what conditions can a glass be formed?, Contemporary Physics 10 (1969) 473-488. https://doi.org/10.1080/00107516908204405

[151] D. V. Louzguine-Luzgin, Y. Zeng, A. D. H. Setyawan, N. Nishiyama, H. Kato, J. Saida and A. Inoue, Deformation behavior of Zr- and Ni-based bulk glassy alloys, Journal of Materials Research, 22 (2007) 1087-1092. https://doi.org/10.1557/jmr.2007.0126

[152] A. Inoue, K. Ohtera, A. P. Tsai and T. Masumoto, Glass transition behavior of Al-Y-Ni and Al-Ce-Ni amorphous alloys, Jpn. J. Appl. Phys., 27 (1988) 280-282. https://doi.org/10.1143/JJAP.27.L280

[153] D. V. Louzguine and A. Inoue, Investigation of structure and properties of the Al-Y-Ni-Co-Cu metallic glasses, Journal of Materials Research, 17 (2002) 1014-1018. https://doi.org/10.1557/JMR.2002.0149

[154] H. Kato, J. Saida and A. Inoue, Influence of hydrostatic pressure during casting on as cast structure and mechanical properties in $Zr_{65}Al_{7.5}Ni_{10}Cu_{17.5-x}Pd_x$ (x = 0, 17.5) alloys, Scripta Mater., 51 (2004) 1063-1068. https://doi.org/10.1016/j.scriptamat.2004.08.004

[155] K.J. Laws, B. Gun and M. Ferry, Influence of casting parameters on the critical casting size of bulk metallic glass, Metallurgical and Materials Transactions A 40 (2009) 2377-2387. https://doi.org/10.1007/s11661-009-9929-7

[156] R.C. Budhani, T.C. Goel and K.L. Chopra, Melt-spinning technique for preparation of metallic glasses. Bull. Mater. Sci., 4 (1982) 549–561. https://doi.org/10.1007/BF02824962

[157] D. V. Louzguine-Luzgin, G. Xie, Q. Zhang, C. Suryanarayana and A. Inoue, Formation, structure, and crystallization behavior of Cu-based bulk glass-forming alloys, Metallurgical and Materials Transactions A, 41 (2010) 1664-1669. https://doi.org/10.1007/s11661-009-0087-8

[158] D. V. Louzguine-Luzgin, G. Xie, Q. Zhang, C. Suryanarayana and A. Inoue, Formation, structure, and crystallization behavior of Cu-based bulk glass-forming alloys, Metallurgical and Materials Transactions A, 41 (2010) 1664-1669. https://doi.org/10.1007/s11661-009-0087-8

[159] T. Tamura, K. Amiya, R. S. Rachmat, Y. Mizutani and K. Miwa, Electromagnetic vibration process for producing bulk metallic glasses, Nature Mater., 4 (2005) 289-292. https://doi.org/10.1038/nmat1341

[160] H. W. Kui, A. L. Greer and D. Turnbull, Formation of bulk metallic glass by fluxing, Appl. Phys. Lett., 45 (1984) 615-619. https://doi.org/10.1063/1.95330

[161] Y. He, R. B. Schwarz and J. I. Archuleta, Bulk glass formation in the Pd–Ni–P system, Appl. Phys. Lett., 69 (1996) 1861-1863. https://doi.org/10.1063/1.117458

[162] G. Wilde, J. L. Sebright and J. H. Perepezko, Bulk liquid undercooling and nucleation in gold, Acta Mater., 54 (2006) 4759-4769. https://doi.org/10.1016/j.actamat.2006.06.007

[163] N. Nishiyama and A. Inoue, Glass-forming ability of $Pd_{42.5}Cu_{30}Ni_{7.5}P_{20}Pd_{42.5}Cu_{30}Ni_{7.5}P_{20}$ alloy with a low critical cooling rate of 0.067 K/s, Appl. Phys. Lett., 80 (2002) 568-570. https://doi.org/10.1063/1.1445475

[164] I.-R. Lu, G. Wilde, G.P. Görler and R. Willnecker, Thermodynamic properties of Pd-based glass-forming alloys, J. Non-Cryst. Solids, 250–252 (1999) 577-581. https://doi.org/10.1016/S0022-3093(99)00135-0

[165] N. Nishiyama and A. Inoue, Flux treated Pd-Cu-Ni-P amorphous alloy having low critical cooling rate, Mater Trans JIM, 38 (1997) 464-472. https://doi.org/10.2320/matertrans1989.38.464

[166] K. F. Yao and C. Q. Zhang, Fe-based bulk metallic glass with high plasticity, Appl. Phys. Lett., 90 (2007) 061901. https://doi.org/10.1063/1.2437722

[167] T.D. Shen and R.B. Schwarz, Lowering critical cooling rate for forming bulk metallic glass, Appl. Phys. Lett., 88 (2006) 091903. https://doi.org/10.1063/1.2172160

[168] N. Chen, L. Gu, G. Q. Xie, D. V. Louzguine-Luzgin, A. R. Yavari, G. Vaughan, S. D. Imhoff, J. H. Perepezko, T. Abe and A. Inoue, Flux-induced structural modification and phase transformations in a $Pd_{40}Ni_{40}Si_4P_{16}$ bulk-glassy alloy, Acta Materialia, 58 (2010) 5886-5897. https://doi.org/10.1016/j.actamat.2010.07.003

[169] A.Yu. Churyumov, A.I. Bazlov, A.A. Tsarkov, A.N. Solonin and D.V. Louzguine-Luzgin, Microstructure, mechanical properties and crystallization behavior of Zr-based bulk metallic glasses prepared under a low vacuum, Journal of Alloys and Compounds, 654 (2016) 87–94. https://doi.org/10.1016/j.jallcom.2015.09.003

[170] M. Stoica, S. Kumar, S. Roth, S. Ram, J. Eckert, G. Vaughan and A. R. Yavari, Crystallization kinetics and magnetic properties of $Fe_{66}Nb_4B_{30}$ bulk metallic glass, Journal of Alloys and Compounds, 483 (2009) 632-637. https://doi.org/10.1016/j.jallcom.2007.11.150

[171] B. J. Yang, W. Y. Lu, J. L. Zhang, J. Q. Wang & E. Ma, Melt fluxing to elevate the forming ability of Al-based bulk metallic glasses, Scientific Reports, 7 (2017) 11053. https://doi.org/10.1038/s41598-017-11504-6

CHAPTER 3

Structure of Liquids and Glasses and Structural Changes in Metallic Liquids on Cooling

Crystals are characterized by a unit cell and existence of a translational symmetry. Quasicrystals do not have such a cell in three-dimensional space but exhibit a rotational symmetry. The structure of liquids and glasses is less well understood. As the atomic structure of bulk metallic glasses defines their properties, in particular mechanical, research on the structure of glassy alloys is an important topic of modern physical metallurgy.

Contents

3.1   Structure of liquids and glasses

Structure of liquids and glasses was studied by X-ray diffraction (small angle and wide angle methods) [1,2], including synchrotron radiation X-ray experiments [3,4], neutron diffraction [5] (especially useful when alloys contain light elements like B, Si, C, P etc…), transmission electron microscopy [6] and fluctuation electron microscopy (FEM) [7] experiments. It was also successfully studied by modeling: the reverse Monte Carlo modeling (RMC) [8], classical [9] and the quantum mechanics molecular dynamics simulation (MD) [10]. The

Lennard-Jones (LJ) potential utilized for simulating the systems with van der Waals interactions [9], in general, is not suitable for modeling metallic glasses. The embedded-atom method (EAM) many-body potential [11] is more efficient and usually applicable for metals and alloys.

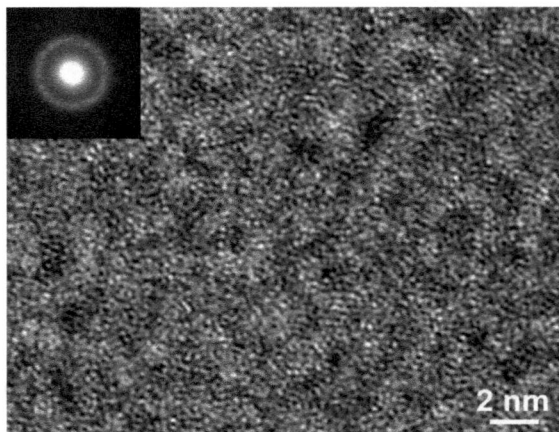

Fig. 3.1. High-resolution TEM image of a Zr-based BMG. The insert exhibits a nanobeam diffraction pattern.

The structure of metallic glasses looks completely disordered at first glance as can be seen from the high-resolution TEM image in Fig. 3.1. However, for a glassy structure it is possible to define the degree of topological (TSRO) and chemical short range order (CSRO). CSRO depends on chemical environment of the nearest-neighbor atoms as deviation from the expectation of a random distribution [12]. In general, a certain atom in a metallic glass attracts atoms of different kind as the nearest neighbors. A typical example is related to the metalloid centered clusters observed in metal-metalloid glasses, for example in Ni-P metallic glasses [13].

Conventional X-ray diffraction pattern produces a series of broad maxima (Fig. 3.2 (a)) while selected-area and nanobeam electron diffraction patterns produce a series of diffraction rings (Fig. 3.1 insert).

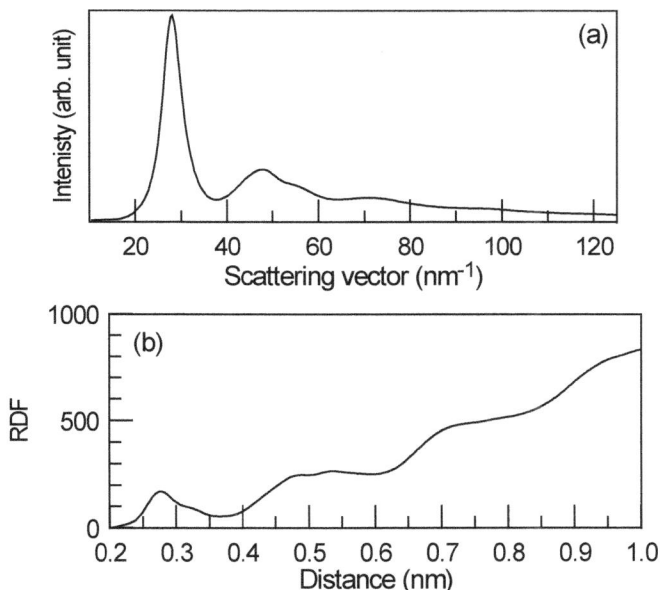

Fig. 3.2. Scattering intensity (a) and radial distribution function (b) of the $Cu_{60}Zr_{30}Ti_{10}$ alloy.

In order to produce the radial distribution function (RDF) shown in Fig. 3.2 (b) correction for the scattering from the sample container, air scattering, polarization, absorption [14], and Compton scattering [15] must be done. The measured intensity can be converted to electron units per atom with the generalized Krogh-Moe-Norman method [16], using the X-ray atomic scattering factors and anomalous dispersion corrections [17]. The total structure factor $S(Q)$ and the interference function $Q_i(Q) = Q[S(Q) – 1]$ (scattering vector $Q = 4\pi \cdot sin\theta/\lambda$, θ is the diffraction angle) can be obtained from the coherent scattering intensity by using atomic scattering factors. The radial distribution $RDF(R)$ and pair distribution functions $PDF(R)$ can be subsequently obtained by the Fourier transform of $Q_i(Q)$:

$$RDF(R) = 4\pi r^2 \rho(R) = 4\pi r^2 \rho_0 + 2R/\pi \int_0^{Qmax} Q(S(Q) - 1)\sin(QR)dQ \qquad (3.1)$$

where $\rho(R)$ is the total radial number density function and ρ_0 is the average number density of the sample. RDF also shows a series of the maxima as a function of the interatomic distance (Fig. 3.2b) indicative of the short and medium range order.

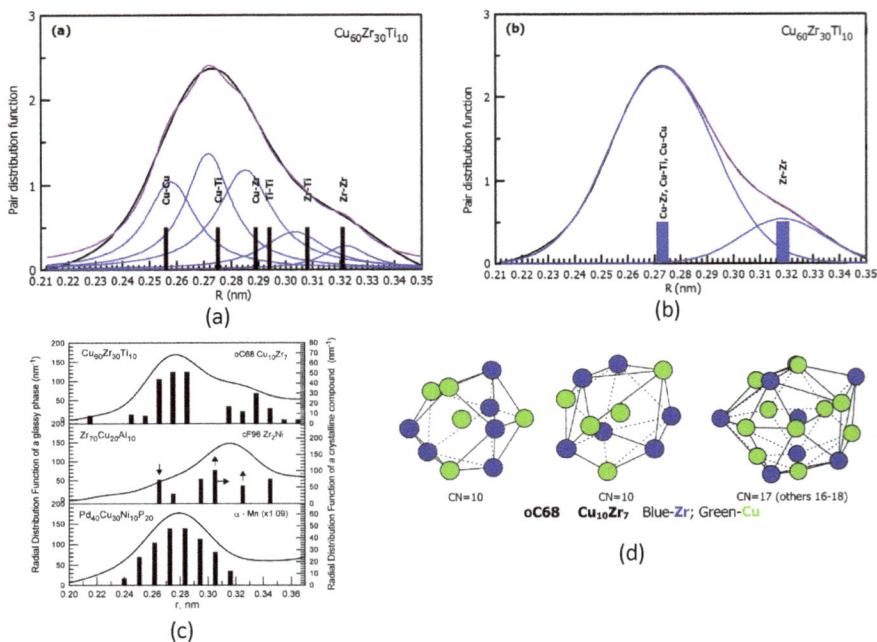

Fig. 3.3. Fitting of the pair distribution function of the $Cu_{60}Zr_{30}Ti_{10}$ alloy with 5 (a) and 2 (b) Gaussian peaks. Reproduced from [18] with permission of Elsevier. (c) Radial distribution functions of some glassy alloys representing first coordination shell. Bars show the corresponding radial distribution functions of the intermetallic compounds integrated at 0.01 nm step. (d) Atomic clusters found in the Cu10Zr7 phase. Reproduced from [19] with permission of Elsevier.

Lorentzian function fitting of the first maximum of *PDF(R)* from 0.21 to 0.35 nm of $Cu_{60}Zr_{30}Ti_{10}$ showed good correspondence to the Cu-Cu, Cu-Ti, Cu-Zr, Zr-Ti and Zr-Zr interatomic distances produced a reasonable correspondence with the original *PDF(R)* data of the $Cu_{60}Zr_{30}Ti_{10}$ alloy (Fig. 3.3(a-c)) [18]. As the maximum of the left shoulder (0.273 nm) corresponds to lower value than the predicted Cu-Zr interatomic distance (0.289 nm) one can assume that this distance in the studied glassy alloy is lower than that predicted by the Goldschmidt atomic radii due to negative mixing enthalpy (attractive interaction) in this atomic pair which is quite common in the intermetallic compounds [19]. The integration of the main RDF(R)^c maximum (both shoulders from 0.20 nm to 0.37 nm) (Fig. 3.3) gave the area under the peak in $Cu_{60}Zr_{30}Ti_{10}$ alloy of 13.6.

An orthorhombic oC68 $Cu_{10}Zr_7$ phase forming in the $Cu_{60}Zr_{30}Ti_{10}$ alloy on heating is the main structural constituent in this alloy after complete crystallization. This phase also dissolves a large amount of Ti. Some clusters in this phase (Fig. 3.3(d)) contain very short Cu-Cu distances as small as about 0.22 nm and Cu-Zr interatomic distances are about 0.243-0.244 nm which are significantly smaller than 0.256 nm and 0.289 nm predicted by the Goldschmidt atomic radii [19]. A similar decrease in the interatomic distances was observed in Zr-based alloys [20]. The $Cu_{10}(Zr,Ti)_7$ crystalline phase can be used as an approximant to explain the structural features of this glassy alloy in the first coordination shell. oC68 $Cu_{10}Zr_7$ phase has a number of different polyhedron units (Fig. 3.3 (d)). The average coordination number is close to that of the glassy state and these clusters can be realized in the glassy alloy. However, in the glassy state there is a wider variation of the interatomic distances than in a crystal. Integration of the number interatomic distances at 0.01 nm step produces peaks corresponding quite well to the *RDF(R)* in Fig. 3.3 (c). As one can see the structure of the $Cu_{60}Zr_{30}Ti_{10}$ glassy alloy in the first coordination shell can be perfectly described by that of the oC68 $Cu_{10}Zr_7$ compound provided that there is a shift of about ±0.005 nm in the atomic positions. Similar considerations are applicable to Zr- and Pd-based alloys. The structure of Zr-based alloys having composition close to $Zr_2(TM,Al)$ might be described as an aggregation of the atomic clusters of a metastable cF96 Zr_2Ni compound (a=1.227 nm; Ni has nearly the same atomic radius as Cu) which is quite common metastable phase formed in Zr-based glassy alloys (Fig. 3.3(c)). The *RDF(R)* of the $Pd_{40}Cu_{30}Ni_{10}P_{20}$ glassy alloy can be described with a complex cI58 α-Mn structure provided that the interatomic distances are integrated at 0.01 nm step and shifted towards larger R values by a factor of 1.09 in accordance with a larger atomic size of Pd (Fig. 3.3(c)) [19].

An icosahedral-type medium range order which is expected to be a dominant local configuration in liquids and metallic glasses in terms of the distorted icosahedra [21] may be a good explanation why no short-range order crystal-like atomic arrays are observed in the HRTEM images (Fig. 3.1) of Cu-Zr- and Zr-Cu-based metallic glasses. The structure of Cu-Zr glassy alloys is one of the most widely studied by various methods including MD simulation [22,23,24]. The structure of Zr–Cu binary metallic glasses except for the $Zr_{67}Cu_{33}$ one can be modeled by statistics of Cu–Zr, Cu–Cu, and Zr–Zr nearest neighbor populations in an "ideal solid solution" [25]. The existence of icosahedral structure was found in Cu-Zr metallic glasses [26,27]. On the other hand, the atomic structure of Cu–Zr glasses analyzed by means of the RMC, XRD, neutron diffraction and extended X-ray absorption fine structure (EXAFS) methods did not show a dominant special atomic structure [28].

More detailed structural information can be obtained by using anomalous X-ray scattering (AXS) [29], the X-ray absorption fine structure (XAFS) [30], including EXAFS [31,32] and X-ray absorption near edge structure (XANES) [33] when environmental RDFs for certain atomic pairs can be obtained. Upon anomalous X-ray scattering experiments energy dependencies of the anomalous dispersion terms should be used for each constituent element. The difference between the scattering intensities for the element A, $\Delta i_A(Q, E_1, E_2)$, measured at two energies of E_1 and E_2 that are usually 300 and 25 eV lower than the energies of the corresponding absorption edges, respectively is a function of the scattering vector Q are:

$$\Delta i_A(Q, E_1, E_2) \cong \frac{\{I_{obs}(Q, E_1) - \langle f^2(Q, E_1) \rangle\} - \{I_{obs}(Q, E_2) - \langle f^2(Q, E_2) \rangle\}}{c_A \{f'_A(E_1) - f'_A(E_2)\} W(Q, E_1, E_2)} =$$

$$= \sum_{m=1}^{n} \frac{c_m \Re\{j_m(Q, E_1) + j_m(Q, E_2)\}}{W(Q, E_1, E_2)} \{a_m(Q) - 1\},$$ (3.2)

where

$$W(Q, E_1, E_2) = \sum_{m=1}^{n} c_m \Re\{f_m(Q, E_1) + f_m(Q, E_2)\}$$

where n is the number of elements, the values c_m and f_m are the atomic fraction and the X-ray atomic scattering factor of the m-th element, respectively. $\Re$ denotes the real part of the values in brackets.

The method is illustrated using the research data for the $Ge_{50}Al_{40}Cr_{10}$ alloy [29]. The partial pair distribution functions are shown in Fig. 3.4.

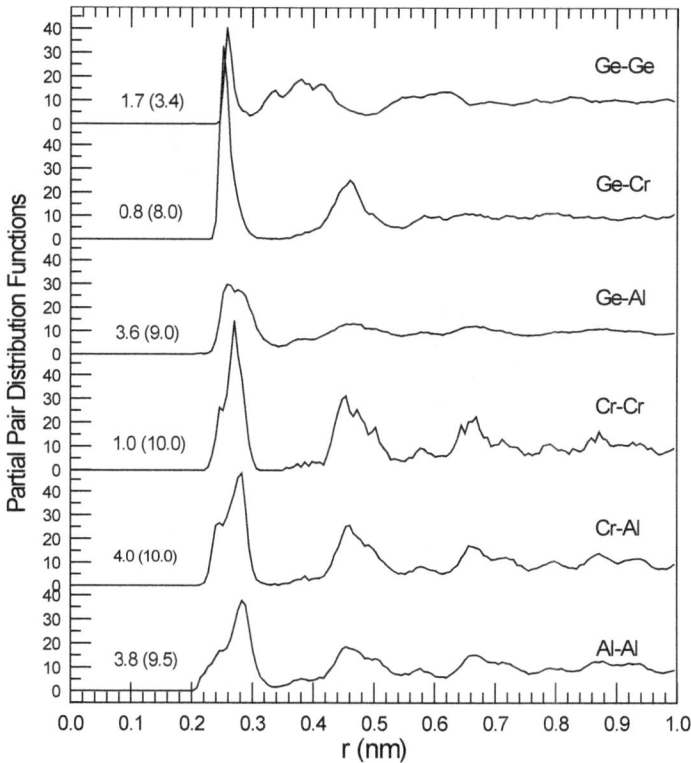

Fig. 3.4. Partial pair distribution functions of the $Ge_{50}Al_{40}Cr_{10}$ alloy. Reproduced from [29] with permission of the Physical Society of Japan.

The results obtained in the present study indicate that the structure of the $Ge_{50}Al_{40}Cr_{10}$ alloy is inhomogeneous at nanoscale. Some medium range order (MRO) zones of about 1 nm were observed in the high-resolution image. A similar structure is formed in the $Ge_{55}Al_{30}Cr_{10}La_{5}$ metallic glass [34]. Nanoscale zones produce distinct spots in the dark-field TEM images (Figure 3.5), even though their size is estimated at about 1-2 nm. This may indicate separation to Ge-rich and Ge-poor zones producing a double type halo pattern with clearly visible two haloes one corresponding to $Q=23.5$ nm^{-1} another to 30.2 nm^{-1}. One of these regions may be a precursor for further nanocrystallization observed on heating.

Fig. 3.5. (a) bright-field image of the Ge$_{55}$Al$_{30}$Cr$_{10}$La$_5$ metallic glass in as-solidified state with the selected-area electron diffraction pattern as an inset. (b and c) dark-field images, respectively, taken from the first and the second ring of the separated halo pattern(the inset). The first and the second rings are marked as (1) and (2).

The partial pair distribution functions (see Fig. 3.4) indicate that corresponding pairs of the alloying elements have their own preferential interatomic distances. Only the Ge-Ge and Ge-Cr pairs have one sharp peak in the first coordination shell, whereas corresponding functions of the Ge-Al, Cr-Cr, Cr-Al and Al-Al atomic pairs show the two sub-peaks. The interatomic distance of about 0.255 nm is usual around Ge atoms, whereas partial pair distribution functions around Al atoms show two distances of about 0.246 and 0.280 nm. The partial coordination number in the Ge-Ge pair of 3.4 is not far from that value of 4 typical for amorphous Ge. On the other hand the values for other atomic pairs are closer to that of usual metallic glasses. Thus, Ge forms a framework like a-Ge which is different from the distribution of the other elements. The results obtained are in agreement with the idea of separation over a short range into Ge-rich and Al-rich amorphous phases. The interatomic distances are shortened owing to strong Ge-Cr interactions. Also, in the Al-Fe-Ce system alloys the Al-Fe distances are significantly shortened [35] and a strong interaction between Al and Fe leads to chemical and topological short-range ordering. It indicates that the Al–Fe pairs (and in general Al-TM pairs) [36] involve covalent-like interaction originating from the hybridization of the $3d$-states of Fe with the sp-sates of Al [37]. Local chemical and topological order was studied in Al–Tb alloys and found to be inhomogeneous on nanoscale [38]. MD simulation also showed pronounced chemical SRO in the Al-Mn and Al-Ni alloys [39].

Pair distribution functions can be also obtained by using the selected-area electron diffraction patters (SAED) representing two-dimensional intensity distribution in the reciprocal space. For example, *PDF(R)* of the $Fe_{48}Cr_{15}Mo_{14}C_{15}B_6Y_2$ glassy alloy in as-cast state is shown in Fig. 3.6. One-dimensional intensity distribution function was obtained by circular integration of the scattered intensity over 180 degrees around the center of the spot produced by the transmitted beam (000).

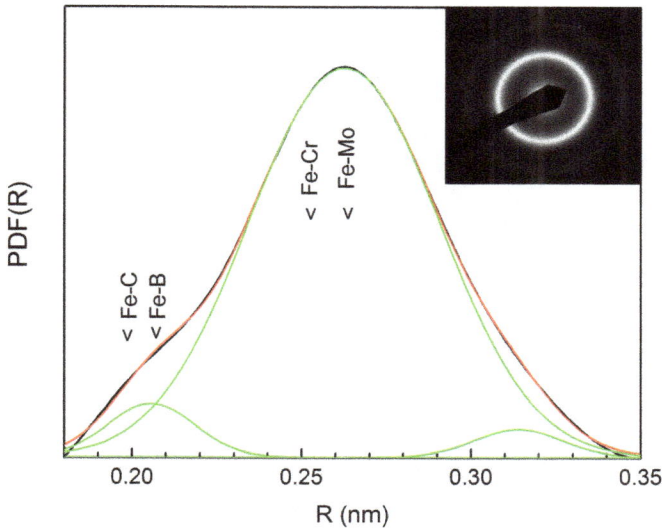

Fig. 3.6. PDF(R) of the Fe₄₈Cr₁₅Mo₁₄C₁₅B₆Y₂ glassy alloy (black curve fitted with 3 Gaussian functions shown in green). The resulted fitting curve is marked in red. The corresponding SAED pattern is shown in the inset.

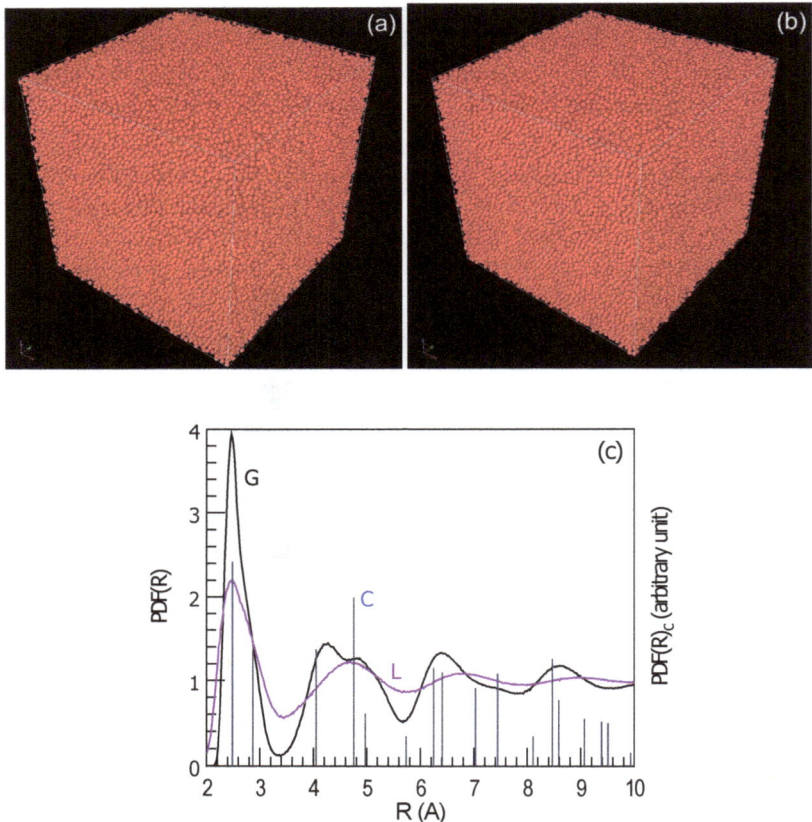

Fig. 3.7. MD simulation of liquid Fe preformed by LAMMPS package at nearly constant pressure with an embedded atom potential. (a) An atomic cell at 2500 K and (b) the same atomic cell at 300 K after cooling. (c) PDFs of crystalline (C), liquid (L) and glassy (G) phase. Note that although (a) and (b) look very similar their PDFs (c) are significantly different.

By using the interatomic forces calculations of the atomic structure and of the thermodynamic properties of the crystalline, liquid, and amorphous phases have been performed since long ago [40]. The interrelation between the formation of topologically close-packed intermetallic compounds, eutectics, and metallic glasses was found to depend on optimal embedding of the neighboring atoms. MD simulation of Fe was preformed by LAMMPS package [41,42] at constant pressure with an embedded atom method potential [43] and used to model the structure of liquid (Fig. 3.7 (a)) and glassy (Fig. 3.7 (b)) iron. The atomic structure exhibits no crystals though the positions of the maxima in the *PDF(R)* correspond quite well to those of crystalline α-Fe (Fig. 3.7 (c)). Visually the structures Fig. 3.7 (a) and (b) look very similar but the *PDFs(R)* show that the structure of a glass is more ordered than that of a liquid indicating a stronger PDF peak. Another clear difference is splitting of the second PDF maximum observed in the case of glass. This is a common feature observed in many glasses. The first PDF maximum is also asymmetrical consisting of two peaks.

The coordination number determined from the experimental PDF of the first peak is used as a measure of local order and compared with the results of simulation. The bond-orientational order [44], the common neighbor [45], the Voronoi polyhedral analysis displayed in Fig. 3.8 (a) [46] and other methods can be used to study the atomic structures of glasses produced by simulation.

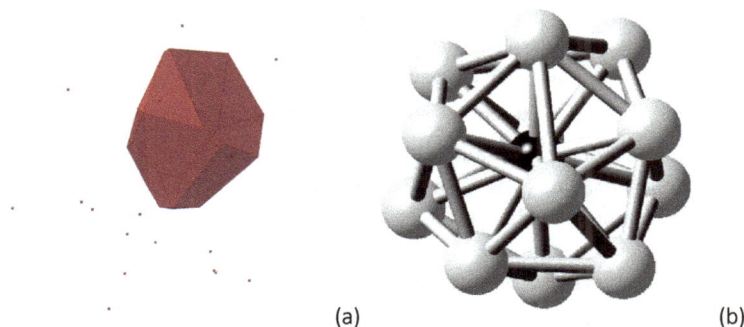

Fig. 3.8. A Voronoi cell produced by sections of planes located at the centers between the points in space (a) (courtesy of A. Tsarkov), distorted icosahedron in a metallic glass (b).

The structure of metallic liquids as a package of hard spheres (dense random packing) was proposed by Bernal [47] using a tetrahedron, octahedron, tetragonal dodecahedron, capped trigonal prism and Archimedean capped antiprism. However, dense random packing method cannot fill out the space efficiently. The maximum packing fraction is significantly lower than the packing fraction of the closed packed crystalline lattices. Moreover, metallic atoms are not "hard" but rather "soft" spheres. Owing to a small density difference between bulk metallic glassy and the corresponding crystalline structure [48] it become clear that the metallic glasses are quite well packed in expense of their short and medium-range order. According to modern viewpoint the structure of metallic glasses consists of the efficiently packed atomic clusters rather than dense random packing of atoms. Cluster-based models of glassy structure have been issued [49,50] and the importance of an efficient atomic packing for the formation of metallic glasses is emphasized [51,52]. It explains high relative densities of metallic glasses close to those of the crystalline counterparts [53].

It was shown that Pd- and Ni-based bulk glassy alloys of Pd-Si [54], Pd-Ni-P [55,56], Pd-Cu-Si [57] and Ni-Pd-P [58] systems contain clear medium-range order zones. The structure models created for the Fe-B glassy alloys showed existence of the prismatical clusters typical for Fe-B compounds formed around B atoms [2,59]. Frank-Kasper polyhedra and defective icosahedral clusters (Fig. 3.8 (b)) were found in metallic glasses [60]. Quasicrystalline-type short-range order found in Zr-based metallic glasses consisting of distorted icosahedra was demonstrated by sub-nanometer beam electron diffraction [61]. The local atomic structures around Pt as well as Zr in the amorphous and quasicrystalline Zr-Al-Ni-Pt alloys determined by the anomalous X-ray scattering method [62] showed possible existence of the chemical short range order clusters in the glassy state similar to those of a quasicrystal. This idea is in a good consistency with the formation of the nanoscale icosahedral phase observed in a Cu-Zr-Ti-Pd alloy on devitrification [63]. However, as has been already mentioned there are various metallic glasses, for example, Fe-based and Al-based where icosahedral clusters are not the dominant structure units.

The reduced supercooling (undercooling) before crystallization from the melt was found to be the largest for crystalline phases, smaller for crystal approximants and the smallest for quasicrystals indicating possible structural similarity to metallic glassy phase [64]. It is also reasonable to assume that complex structural units of quasicrystals beyond the first shell [65] also exist in metallic glasses. Such a research topic deserves further investigation.

A parameter called flexibility volume

$$v_{flex} = <r^2>/a_i^2 \cdot \Omega_a,$$ \hfill (3.3)

where $<r^2>$ is the mean root squared displacement, a_i is the interatomic distance and Ω_a is the average atomic volume was introduced as a universal indicator, to bridge the structural state the of metallic glasses with their properties, on both atomic and macroscopic levels [66]. The flexibility volume combines static atomic volume with dynamics information via atomic vibrations that probe local configurational space and interaction between the neighboring atoms. The flexibility volume can quantitatively predict the shear modulus and it correlates strongly with atomic packing topology, and with the activation energy for thermally activated relaxation.

A simple empirical model for the estimation of the atomic cluster size was suggested, and the value obtained (about 0.8 nm) was in good agreement with the corresponding scanning tunneling microscopy (STM) observation. At a low resolution the surface structure of the $Ni_{62}Nb_{38}$ metallic glass was also found to consist of atomic clusters (Fig. 3.9) [67]. The results were compared with those obtained by transmission electron microscopy and nano-beam diffraction measurements.

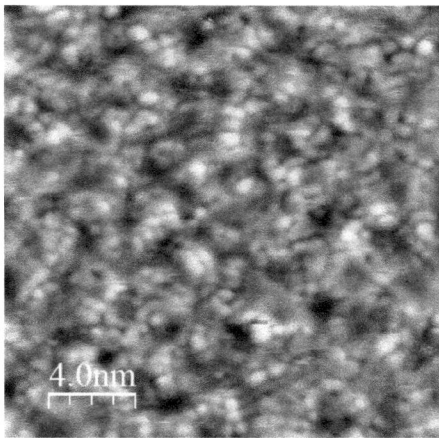

Fig. 3.9. STM an image of Ni-Nb metallic glass indicating a clustered structure. Courtesy of A. Oreshkin.

A high-resolution surface image of the as-cast structure of metallic glass obtained later (Fig. 3.10 (a)) reveals the atomic structure of the sample [68]. Singular atoms are resolved. A typical atomic landscape profile shown in Fig. 3.10 (b) clearly represents the atomic-scale distances while 3D topological representation is given in Fig 3.10 (c). On the other hand the interatomic distances measured using the surface image corresponded to the 1st and 2nd peaks in the first coordination shell and to the distances between these two peaks in the 3D landscape.

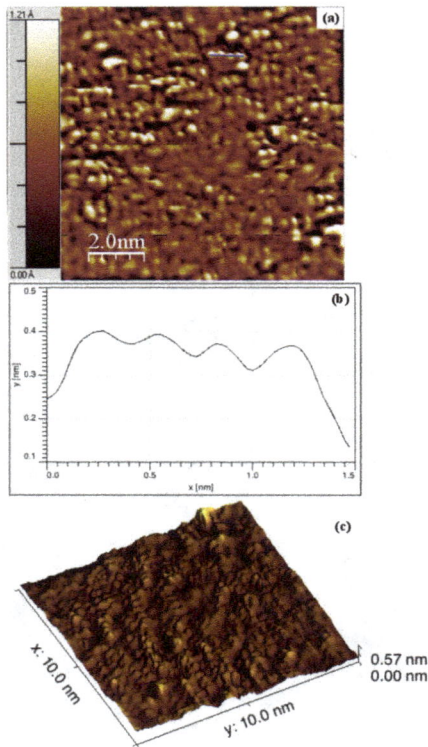

Fig. 3.10. UHV STM image of the studied BMG sample (10x10 nm size) (a), the atomic profile in a densely packed direction (b) as marked in (a) and 3D topological image (c). Reproduced from [68] with permission of Elsevier.

The electronic structure of the $Ni_{63.5}Nb_{36.5}$ glassy alloy was studied by computer simulation (Fig. 3.11 (a)) and STM images of different surface configurations were generated (Fig. 3.11 (b,c)) from the atomic cells (Fig. 3.11 (d,e)). Figure 3.11a shows the results for the total density of states (DOS) and partial density of states (PDOS), the calculated STM images as well as the atomic configuration of the metallic glassy surface. The present theoretical results show that Nb atoms appear as bright spots at positive bias voltage (Fig. 3.11 (c)) while at negative bias voltage the Ni atoms on the STM images appear as bright spots (Fig. 3.11 (b)). The two selected theoretical images correlate well with PDOS (Fig. 3.11 (a)) which show the larger contribution of *3d* states of Ni atoms below the Fermi energy level while the region of electron density above the Fermi level is mostly formed by states of Nb atoms. Application of negative bias voltage is preferable in STM experiments due to greater contribution of DOS below the Fermi level. Thus the atomic structure visible in the high resolution STM image (Fig. 1 (a)) corresponds to Ni atoms on the surface, that is brighter spots correspond to Ni atoms. Nb atoms remain invisible at negative bias voltage. These facts lead to larger interatomic distances observed.

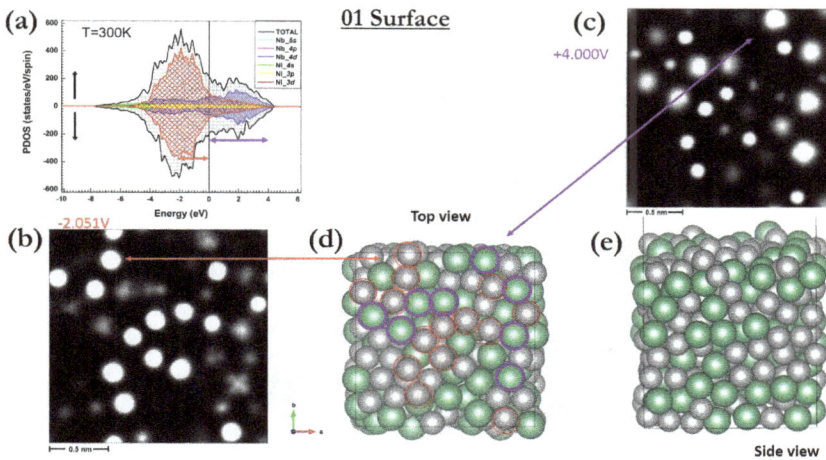

Fig. 3.11. Total and partial density of states (a), as well as the STM images calculated at the bias of (b) -2.051 V and (c) +4.00 V. (d) and (e) are top and side views of atomic configuration of the Ni-Nb glassy alloy surface. Reproduced from [68] with permission of Elsevier.

The as-cast $Au_{49}Cu_{26.9}Ag_{5.5}Pd_{2.3}Si_{16.3}$ [69] BMG rods contained a micrometer-scale crystalline surface layer containing the Au-based solid solution grains embedded in the amorphous phase (with a small fraction of an intermetallic phase) [70], which was responsible for nice yellow color of the samples surface. Internal part of the ingot is grey metallic owing to relatively high Si content. The origin of such an effect could be surface nucleation and growth of Au crystal layer. It was found that in the liquid state somewhat above the eutectic temperature this Au-based alloy also contains a very thin layer of a crystalline phase [71]. The thickness of this crystalline layer depends on the casting temperature. In the samples cast from 720 K the thickness was 80 μm while when casted from 1570 K the thickness was only about 10 μm. The ribbon samples had no such a layer owing to significantly higher cooling rate (about 10^6 K/s).

Recently-discovered materials called nanoglasses or nano-structured metallic glasses consisting of nanoscale spheroid glassy particles separated by glassy interfaces represent a new class of metallic glasses which demonstrate good mechanical and chemical properties. The former material is produced by the compaction of nanometer-sized amorphous spheres formed by the inert gas condensation [72] while the latter one is formed by magnetron sputtering (Fig. 3.12) [73,74].

Fig. 3.12. SEM image of the $Zr_{62.5}Pd_{37.5}$ nanostructured metallic glass obtained in secondary electrons.

3.2   Thermal expansion detected by X-ray diffraction

The thermal expansion of the glassy alloys and melts upon heating or cooling was studied by Synchrotron radiation X-ray diffraction for the $Fe_{40}Ni_{40}P_{14}B_6$ [75], $Pd_{40}Cu_{30}Ni_{10}P_{20}$ [76], $Zr_{55}Cu_{30}Ni_5Al_{10}$ [77,78], various La-based [79,80], $Cu_{55}Hf_{25}Ti_{15}Pd_5$ and $Cu_{55}Zr_{30}Ti_{10}Ni_5$ [81] compositions. As an example, shift in the first $PDF(R)$ maximum of the $Cu_{55}Hf_{25}Ti_{15}Pd_5$ glassy alloy with temperature on heating is shown in Fig. 3.13 [82]. The least squares fitting gives the thermal expansion coefficient (β) of $1.6 \cdot 10^{-5}$ K^{-1}. The linear coefficient of thermal expansion measured by thermo-mechanical analysis in the range of 300-530 K was 1.5 ± 0.1 10^{-5} K^{-1} [81].

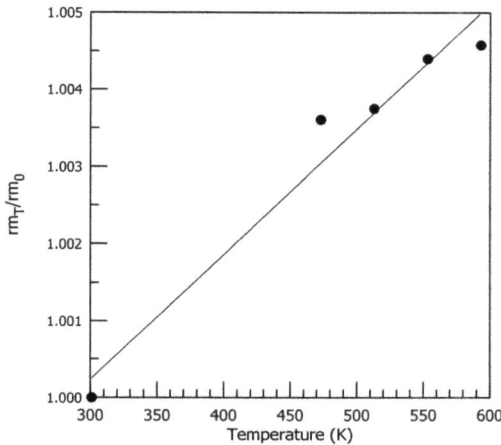

Fig. 3.13. Location of the main peak of PDF(r) rm_T divided by that at room temperature rm_0 as a function of temperature. Reproduced from [82] with permission of AIP Publishing.

3.3   Structural changes on glass formation from the liquid state

The evolution of atomic structure of liquids [83,84] including metals [85,86] on undercooling/supercooling was studied. For example, a liquid structural change in liquid Ga is discovered at the critical temperature of about 1000 K [87]. Below this temperature, the liquid Ga has a large increasing slope of peak position shift of the first peak of the structure factor. Above this temperature, a small reducing slope of peak-position shift of the first peak of the structural factor and pair-distribution functions. Liquid polymorphism was also reported to be observed in the case of Te-Ge alloys [88] and In [89] as a function of temperature.

Materials Research Forum LLC

https://doi.org/10.21741/9781644901014

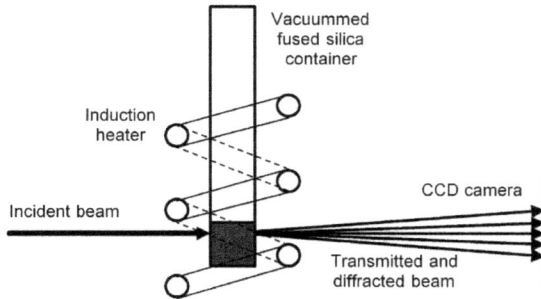

Fig. 3.14. Schematic representation of the experimental setup. Reproduced from [90] with permission of Elsevier.

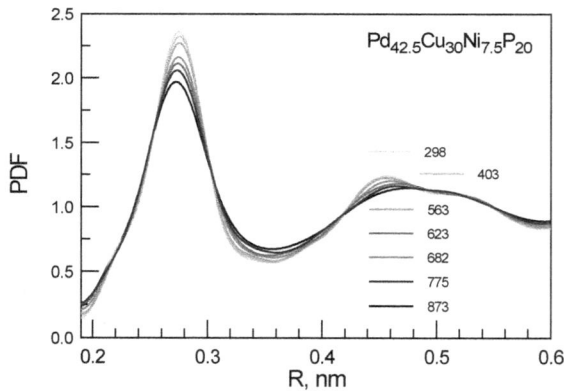

Fig. 3.15. PDF(R) functions of the studied alloy obtained at different temperatures given in Kelvin. Reproduced from [90] with permission of Elsevier.

The structural changes and vitrification of the $Pd_{42.5}Cu_{30}Ni_{7.5}P_{20}$ liquid alloy on cooling from above the equilibrium liquidus temperature were also studied in-situ by synchrotron radiation X-ray diffraction and compared with the results of *ab-initio* computer simulation [90,91]. $Pd_{42.5}Cu_{30}Ni_{7.5}P_{20}$ forms a relatively fragile melt (in terms of its temperature dependence of the viscosity). Its fragility index m is close to 60 [92]. The experimental setup is schematically shown in Fig. 3.14. Several sets of heating and cooling experiments were performed. As a result a series of *PDF(R)* spectra were recorded in-situ at different temperatures (Fig. 3.15).

Intensification of the shoulder in the second coordination peak found in the $Pd_{42.5}Cu_{30}Ni_{7.5}P_{20}$ liquid alloy at about 0.45 nm (Fig. 3.15) on cooling was also observed in the $Fe_{80}B_{20}$ liquid alloy at 0.415 nm (Fig.3.16 (a)) [93].

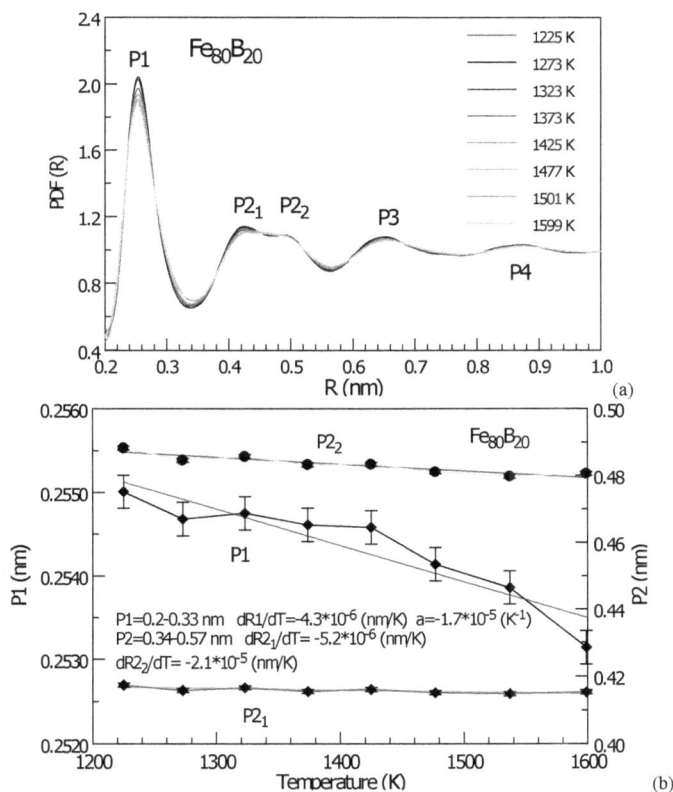

Fig. 3.16. (a) PDF and the corresponding maxima of the $Fe_{80}B_{20}$ liquid alloy in the 1st, 2nd, 3rd and 4th coordination shells. (b) Displacement of P1 and P2 (two sub-peaks/shoulders of P2 are marked as $P2_1$ and $P2_2$) with temperature. The curves are drawn through the experimental points, while linear fits were used to calculate the average thermal expansion coefficient only. Reproduced from [93] with permission of Elsevier.

The thermal expansion coefficient is negative in the case of the two $PDF(R)$ peaks the $Fe_{80}B_{20}$ alloy (Fig. 3.16 (b) while for pure Fe it is negative only for the first peak. The positions of peak 3 and 4 of $PDF(R)$ exhibit a positive shift with temperature. This behavior argues for continuous structural changes in the supercooled liquids. Intensive changes in the short and medium range order in the supercooled liquid are also reflected in the different thermal expansion coefficients obtained for the subsequent atomic coordination shell [94,95]. The integration of the first $RDF(R)$ maximum of the $Pd_{42.5}Cu_{30}Ni_{7.5}P_{20}$ liquid alloy (both sub-peaks from 0.20 nm to 0.35 nm) (not shown here) produced the area under the peak (A) corresponding to the coordination number (CN) in the first coordination shell at room temperature of 12.6 (Fig. 3.17). It indicates formation of rather a densely packed structure.

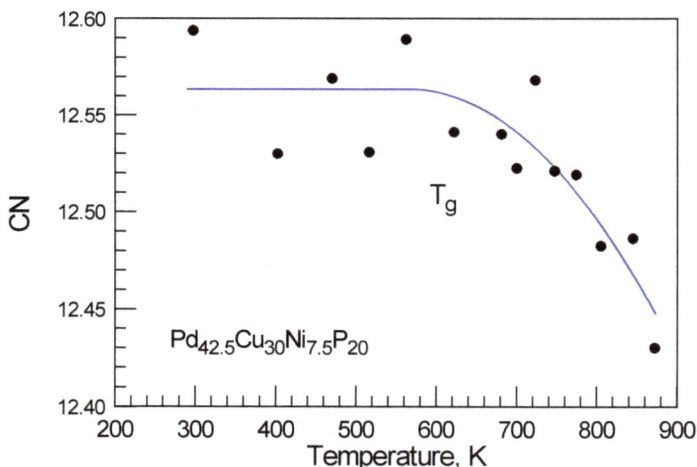

Fig. 3.17. Coordination number (CN) from the RDF(R) in the first coordination shell (0.2-0.35 nm) as a function of temperature on cooling as well as two fits – linear below T_g and exponential above it. Reproduced from [90] with permission of Elsevier.

Fitting of the first $PDF(R)$ maximum from 0.2 to 0.35 nm (baseline was corrected to make both outermost points be equal to zero) using two Gaussian peak functions produced a reasonable correspondence with the original $PDF(R)$ plot (Fig. 3.18). According to the interatomic distances, the first sub-peak (P1) corresponds to nearest Cu-P or Ni-P atomic pairs while the second sub-peak (P2) mostly corresponds to the mixture of Pd-Pd, Pd-Cu and

Pd-Ni pairs. Radial distribution functions for three different Pd-based alloys were compared in Ref. [96] and their comparative analysis was performed.

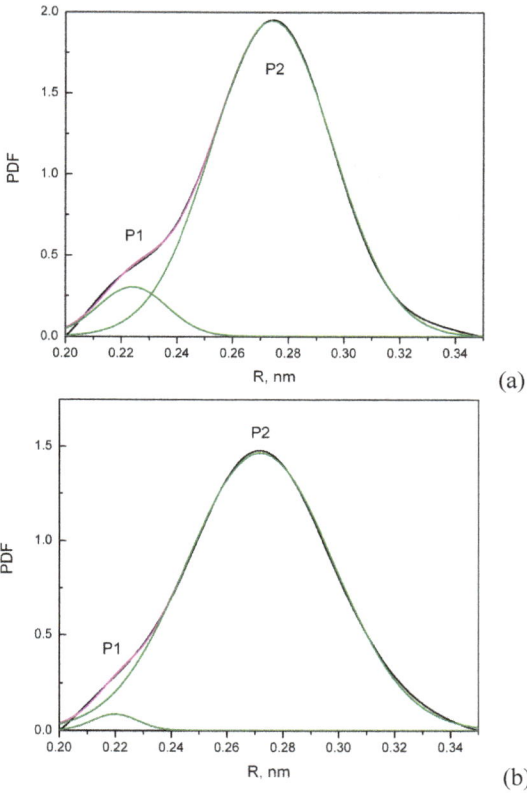

Fig. 3.18. Fitting of the first PDF(R) maximum (black curve) from 0.2 to 0.35 nm using two Gaussian functions related to P1 (left) and P2 (right). The curves in (a) correspond to 298 K while the curves in (b) to 873 K. The baseline was corrected prior to fitting. Reproduced from Ref. [91] with permission from the American Institute of Physics.

Fig. 3.19 shows the area under the sub-peak P1 divided by the area under the subpeak P2 obtained on cooling. A drastic change in this ratio above T_g is observed.

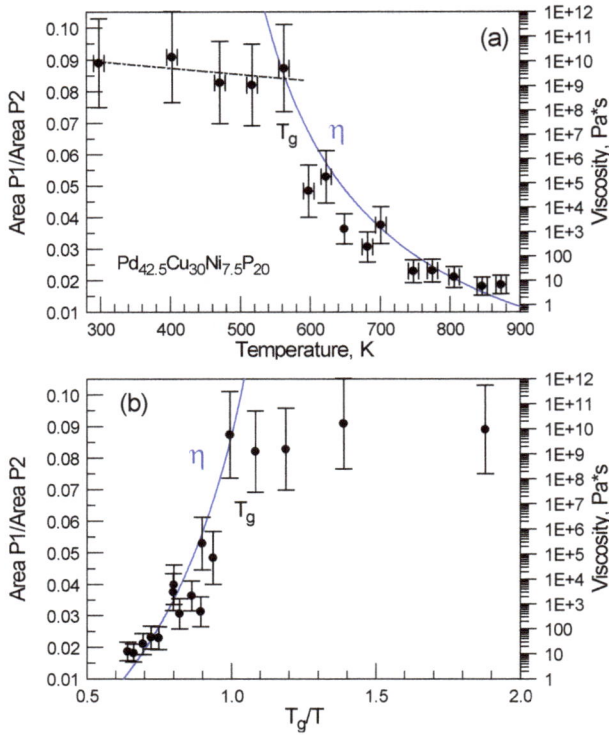

Fig. 3.19. (a) The area under the sub-peak P1 (A(P1)) divided by the area under P2 (A(P2)) in the first coordination shell obtained on cooling (solid circles) as well as the viscosity of the studied alloy (solid line) as a function of temperature. (b) The same plot using the reciprocal temperature scale normalized by T_g. Reproduced from Ref. [91] with permission American Institute of Physics.

A high degree of the short-range order was observed consistent with models which predict that metallic glasses are not made of random packing of atoms but consists of dense packing of clusters [49] which is also consistent with the small volume difference between glassy and the corresponding crystalline phases [97,98].

Partial *PDF(R)* for the Ni-P and Cu-P atomic pairs obtained by MD simulation also indicated an increase in the number of such atomic bonds on cooling in agreement with the experimental results. Analysis of the atomic and electronic structure of the alloy in the liquid and glassy states reveals rapid formation of chemical bonds with *p-d* hybridization between P and metal atoms due to charge transfer from a metal to phosphorous in the supercooled liquid close to the glass-transition region. The calculated partial densities of states projected on the *3p* state of P atoms and the *3d* and *4d* states of Ni, Cu and Pd atoms show noticeable changes in the electronic structures of the $Pd_{42.5}Cu_{30}Ni_{7.5}P_{20}$ alloy towards a chemical short-range ordering in the glassy state (Fig. 3.20). Thus covalently bonded Ni,Cu-P clusters are formed on cooling towards T_g.

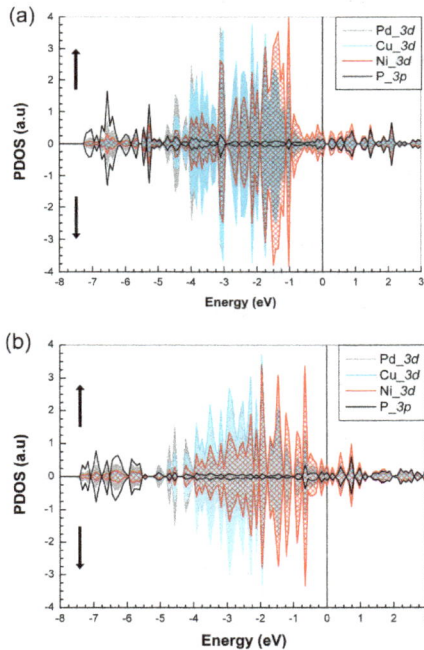

Fig. 3.20. The calculated partial densities of states for the spin-up (↑) and spin-down (↓) 3d electrons of the Pd (gray), Cu (blue), Ni (red) and 3p electron P (black) atoms, at T= 950K (a) and at T= 550K (b) respectively. The Fermi level (vertical line) has been chosen as zero energy. Reproduced from Ref. [91] with permission American Institute of Physics.

Metallic Glasses and Their Composites – 2nd Edition Materials Research Forum LLC
Materials Research Foundations 85 (2021) https://doi.org/10.21741/9781644901014

As shown in Fig. 3.19 with a decrease in temperature the integrated area under P2 (AP2) changes slightly while that under P1 (AP1) increases significantly and the ratio AP1/AP2 increases about 5 times on cooling from the melt to T_g. It indicates that the number of Ni-P and Cu-P atomic pairs, and thus chemical short range order (CSRO) around P, increases on cooling the melt and then remains nearly constant in the glassy state.

The temperature dependence of viscosity of a liquid can be written as:

$$\eta = (Nh/Vm) \cdot \exp(Q_a/RT) \tag{3.4}$$

where Q_a is the activation energy for viscous flow, V_m is molar volume, N is the Avogadro's number, h is the Planck's constant and R is the universal gas constant. The activation energy Q_a depends on the structure of the liquid and supports the idea that the structural changes affect the viscosity. This fact is responsible for a relatively high fragility the of the Pd-Ni-Cu-P liquid in the supercooled state compared to other metallic glasses as the temperature dependence of the viscosity of this bulk metallic glass-former strongly deviates from Arrhenius behavior (Fig. 3.19). This is also reflected in its high fragility index m [92,99], additional to D which is also used [92] (see Chapter 2). The link between the activation energy for viscous flow and structure of the melt is discussed in [100]. Such an argument was also used for interpretation of the glass-formation in Ni [101].

As has been also mentioned in Chapter 2 Ni-Nb alloys behave in a similar way [102,103]. Larger values of γ parameter related to the shift in the peak in the structure factor $S(Q)$, introduced in Ref. [104], indicate a more fragile liquid. For the fragile $Ni_{62}Nb_{38}$ liquid γ is equal to 8.63, while for the stronger Zr-based Vit 106a liquid it is 3.85. Significant structural changes in metallic glasses in the supercooled liquid state have been also observed in the $Ti_{39.5}Zr_{39.5}Ni_{21}$ liquid [105] and predicted in Cu-Zr-Al glasses by MD computer simulation [106]. A link between the structure and viscosity of the $Cu_{64}Zr_{36}$ alloy was also suggested in Ref [107].

The $Zr_{55}Cu_{30}Ni_5Al_{10}$ liquid alloy showed a similar behavior of two peaks in the first PDF(R) maximum (Fig. 3.21) [108] to those found in the $Pd_{42.5}Cu_{30}Ni_{7.5}P_{20}$ alloy (see Figs. 3.18 and 3.19): expansion on cooling and the changes in the peak intensity ratio $AP1/AP2$. Two peaks (P1 and P2) here correspond to the Zr-Cu and Zr-Zr interactions. These findings imply the possible structural changes and atomic redistributions within and between the atomic coordination shells as schematically illustrated in Fig. 3.21b.

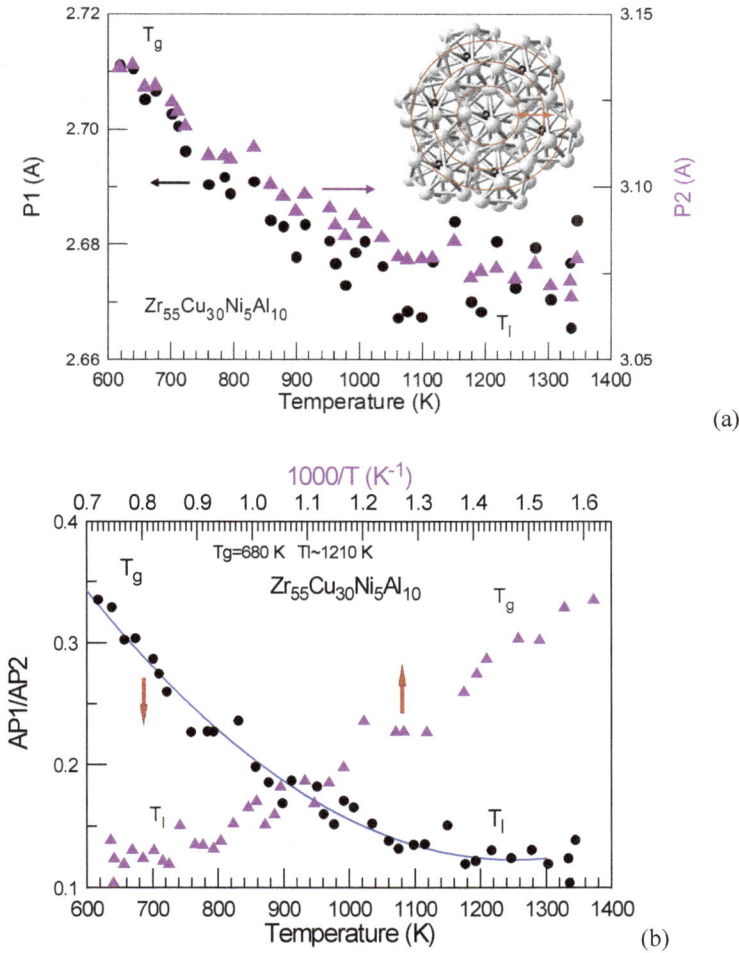

Fig. 3.21. Positions of P1 and P2 in the first coordination shell of the $Zr_{55}Cu_{30}Ni_5Al_{10}$ liquid alloy (a). The ratio of the first to second peak areas (AP1/AP2) (b). The insert in (a) is a schematic representation of atomic redistribution within and between the first and higher order coordination shells leading to continuous structure changes in metallic liquids on heating and cooling leading to redistribution of the atomic number density as indicated by the red double side arrow. Reproduced from [108] with permission of Elsevier.

However, the changes in $AP1/AP2$ ratio (APR) for the $Zr_{55}Cu_{30}Ni_5Al_{10}$ liquid alloy from T_l to T_g (the ratio in Fig. 3.21 is 0.30/0.12=2.5) are smaller in the absolute values compared to those found in the $Pd_{40}Cu_{30}Ni_{10}P_{20}$ liquid alloy (0.09/0.02=4.5). The $AP1/AP2$ ratio values normalized per Kelvin (APR^n) according to the supercooled liquid region on cooling (T_l-T_g) are 0.005 (APR^n=2.5/530 K^{-1}) and 0.017 (APR^n=4.5/270 K^{-1}), respectively. This fact is in line with a lower fragility (m~45) of the $Zr_{55}Cu_{30}Ni_5Al_{10}$ melt compared to that of the $Pd_{40}Cu_{30}Ni_{10}P_{20}$ one (m~60). Continuous structural changes in the supercooled liquid leading to the formation of atomic clusters likely change the activation energy for viscous flow and determine fragility of this glass-forming liquid through the equation 3.4. MD simulation was found to support the observed structure variation as a function of temperature and indicated intensification of Zr-Cu, Zr-Al and Zr-Zr interactions on cooling.

Fragility of liquids is also found to correlate with the difference in the specific heat capacity ($\Delta C_p^{l\text{-}g}$) at T_g between the liquid and glassy phases [109]. The $\Delta C_p^{l\text{-}g}$ values of 10 and 17 J/mol·K for $Zr_{55}Cu_{30}Ni_5Al_{10}$ and $Pd_{42.5}Cu_{30}Ni_{7.5}P_{20}$ alloys, respectively, also correlate with a larger fragility of the latter alloy [108]. The calculated thermodynamic fragility m_t parameter [109] is 58 for the $Zr_{55}Cu_{30}Ni_5Al_{10}$ alloy. The m_t parameter for the $Pd_{42.5}Cu_{30}Ni_{7.5}P_{20}$ alloy is 104. The structural changes on heating leading to structural relaxation and crystallization will be discussed in detail in Chapter 4.

3.4   Structural changes on elastic mechanical loading

Anelastic and viscoplastic strains can have distinct, even opposite, effects on properties. For example, relaxation enthalpy was found to decrease after indentation [110] and in an as-cast Pd-based BMG subjected to shot peening [111]. On the other hand, shot peening - induced rejuvenation of the annealed glass. In a Zr-based BMG cold-rolled to thickness reductions up to 50%, relaxation enthalpy first increased and then decreased [112]. A similar effect was observed in the case of a Zr-based BMG deformed in compression [113,114]. An ultrasonic vibrational treatment was applied to change the energy state of a Zr-based BMG and induced structural rejuvenation [115]. The rejuvenated BMG has a more heterogeneous structure and a higher room temperature plasticity. Excess volume is also increased leading to much larger plasticity of the BMGs.

The first demonstration that elastic-cycling effects extend to macroscopic loading, and are not limited to nanoindentation, was by Caron et al. [116]. They loaded a Zr-based bulk metallic glass (BMG) in uniaxial compression at $(37\pm15)\%$ of the yield stress σ_y. As a result of this cycling, the MG had a lower relaxation enthalpy as measured in DSC, indicating a more relaxed state, consistent with the observed decreases in internal friction. Load-free samples showed a residual contraction along the loading axis, taken to indicate a densification, though there was no measurable change in the Young's modulus (measured by the ultrasonic pulse-echo technique). The structural changes underlying these effects were attributed to an accumulation of anelastic strains associated with the cycle periods being much shorter than the anelastic recovery time. The cyclic loading was found to induce partial nanocrystallization of the metallic glass (Fig. 3.22) forming a crystalline phase. Further work on the same BMG at significantly higher stress values confirmed that cyclic loading can induce crystallization, but of a metastable phase [117]. There was a detectable shift of the first peak in X-ray diffraction patterns of the MG, and it was noted that this could be associated with anisotropic structural change. Cyclic loading was found to lead to a 0.5% decrease in density of the MG and increase in the crystallization enthalpy, opposite to the change expected for relaxation. This could be the result of initial nanoscale heterogeneity in the BMG sample [118].

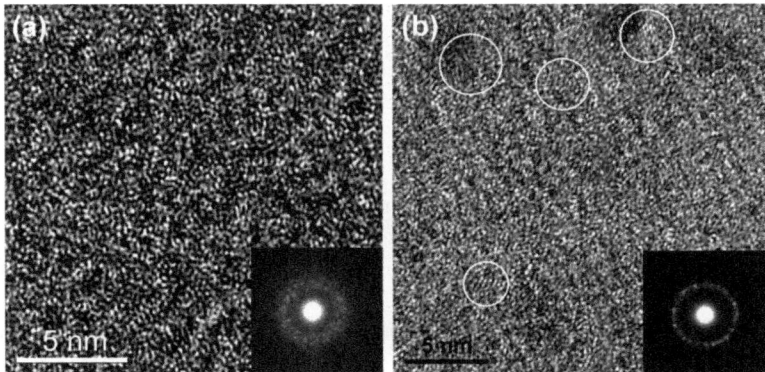

Fig. 3.22. High-resolution TEM images of a Zr-based cube sample in (a) as-cast and (b) after mechanical cycling at 586 ±242 MPa stress. The inserts represent corresponding NBD patterns. Reproduced from Ref. [116] with permission from American Institute of Physics.

As noted above, elastic static loading in nanoindentation has no detectable effect on MGs though the loading times are short. Static loading can be easily maintained for long time. BMG samples elastically loaded for a long time period in uniaxial compression show significant changes in properties [119,120], even though there is typically only a very small residual contraction along the loading axis. The relaxation enthalpy, measured in DSC, is increased, indicating that the loading has induced rejuvenation (i.e. the opposite of relaxation and ageing). This should be associated with a decrease in density, and such a decrease was later confirmed by direct measurement [121]. The rejuvenated samples show increases in plasticity and decreases in the elastic bulk and shear moduli [119,120,121].

It is clear from the works reviewed above that cyclic elastic loading has many effects on BMGs [122]. Although the stress states in nanoindentation and in macroscopic uniaxial compression are very different, there are some analogous effects of cycling in the two cases. MD simulations which, as noted above, can reproduce the observed effects of cycling in nanoindentation, have failed to detect any clear changes in the glassy structure [122] though such changes are clearly observed within the shear bands on plastic deformation [123]. The property changes indicate that the cycling induces both relaxation (consistent with observed hardening) or rejuvenation (consistent with lowered density and the increased mobility). It is possible, that mechanical treatments such as elastic cycling induce structural changes that are qualitatively different from those normally associated with relaxation and rejuvenation (such as would be achieved by annealing or by forming the glass from the liquid by quenching at different rates). For example, the relative changes in topological and in chemical short-range order [124] may be different in the mechanical and thermal treatments. These effects will also be discussed in Chapter 5.

3.5 Pressure and irradiation effects, polyamorphism

The structure of a metallic liquid was studied using molecular dynamics simulation. It found that isothermal compression slows down dynamics of the metallic liquid leading to glass transition [125]. The increase of the kinetic fragility with increasing pressure is accompanied by structural changes. With increasing pressure the liquid dynamics was found to become more heterogeneous. Local ordering at higher pressure and lower temperature leads to the increased five-fold symmetry. The irradiation effects were studied in Ref. [126].

The phenomenon of liquid-liquid transition between two thermodynamically distinct phases – a low-density liquid and high-density liquid below the critical temperature [127] and polyamorphism which is well known in the case of the amorphous ice [128,129] is also observed in inorganic solids [130], the $Ce_{55}Al_{45}$ metallic glass [131] and the $La_{32}Ce_{32}Al_{16}Ni_5Cu_{15}$ one bulk metallic glass [132] as a function of hydrostatic pressure. Significant increase of density was found on compression of the $Ce_{55}Al_{45}$ metallic glass while the $La_{32}Ce_{32}Al_{16}Ni_5Cu_{15}$ exhibited a sudden change in compressibility at about 14 GPa. According to MD simulations and X-ray absorption experiments in a $Ce_{75}Al_{25}$ [133] the origin of this transition is the delocalization of $4f$ electrons in Ce. After delocalization the bonds become shorter than those between the Ce atoms with localized $4f$ electrons.

References

[1] K. Suzuki, Glasses, in Methods in Experimental Physics, 23 (1987) 243-302. https://doi.org/10.1016/S0076-695X(08)60572-4

[2] Y. Waseda, H. S. Chen, A structural study of metallic glasses containing boron (Fe-B, Co-B, and Ni-B), Phys. Stat. Solidi., 49 (1978) 387-396. https://doi.org/10.1002/pssa.2210490149

[3] N. Mattern, H. Hermann, S. Roth, J. Sakowski, M.P. Macht, P. Jovari, J. Jiang, Structural behavior of $Pd_{40}Cu_{30}Ni_{10}P_{20}$ bulk metallic glass below and above the glass transition, Appl. Phys. Lett., 82 (2003) 2589-2591. https://doi.org/10.1063/1.1567457

[4] A.R. Yavari, N. Nikolov, N. Nishiyama, T. Zhang, A. Inoue, J.L. Uriarte and G. Heunen, The glass transition of bulk metallic glasses studied by real-time diffraction in transmission using high-energy synchrotron radiation, Mater. Sci. Eng. A, 375–377 (2004) 709-712. https://doi.org/10.1016/j.msea.2003.10.224

[5] P. Lamparter, Structure of metallic glasses, Physica Scripta, T57 (1995) 45-63. https://doi.org/10.1088/0031-8949/1995/T57/008

[6] A. Hirata, Y. Hirotsu, T. Ohkubo, N. Tanaka, T.G. Nieh, Local atomic structure of Pd–Ni–P bulk metallic glass examined by high-resolution electron microscopy and electron diffraction, Intermetallics, 14 (2006) 903-917. https://doi.org/10.1016/j.intermet.2006.01.007

[7] P.M. Voyles, J.M. Gibson, M.M.J. Treacy, Fluctuation microscopy a probe of atomic correlations in disordered materials, J. Electron Microsc., 49 (2000) 259–266. https://doi.org/10.1093/oxfordjournals.jmicro.a023805

[8] J. Hwang, A.M. Clausen, H. Cao, P.M. Voyles, Reverse Monte Carlo structural model for a zirconium-based metallic glass incorporating fluctuation microscopy medium-range order, data J. Mater. Res., 24 (2009) 3121–3129. https://doi.org/10.1557/jmr.2009.0386

[9] D. C. Rapaport, The art of molecular dynamics simulation, Cambridge University Press, The Edinburgh Building, Cambridge, UK, 2004.

[10] G. Kresse and J. Hafner, Ab initio molecular dynamics for liquid metals, Phys. Rev. B, 47 (1993) 558–561. https://doi.org/10.1103/PhysRevB.47.558

[11] M.S. Daw, M.I. Baskes, Embedded-atom method – derivation and application to impurities, surfaces, and other defects in metals, Phys. Rev. B, 29 (1984) 6443–6453. https://doi.org/10.1103/PhysRevB.29.6443

[12] G.S. Cargill and F. Spaepen, Description of chemical ordering in amorphous-alloys, J. Non-Cryst. Solids, 43 (1981) 91–97. https://doi.org/10.1016/0022-3093(81)90174-5

[13] H.W. Sheng, W.K. Luo, F.M. Alamgir, J.M. Bai, E. Ma, Atomic packing and short-to-medium-range order in metallic glasses, Nature, 439 (2006) 419–425. https://doi.org/10.1038/nature04421

[14] C. N. J. Wagner, Direct methods for the determination of atomic-scale structure of amorphous solids (X-ray, electron, and neutron scattering), J. Non-Cryst. Solids, 31 (1978) 1-27. https://doi.org/10.1016/0022-3093(78)90097-2

[15] D. T. Cromer, Compton scattering factors for aspherical free atoms, J. Chem. Phys., 47 (1969) 4857. https://doi.org/10.1063/1.1670980

[16] Y. Waseda, The Structure of Non-Crystalline Materials, McGraw-Hill, New York, 1980, p 670.

[17] International tables for X-ray Crystallography, ed. J. A. Ibers and W. C. Hamilton, Kynoch, Birmingem, 1974, Vol. 4, p.1.

[18] D. V. Louzguine-Luzgin, J. Antonowicz, K. Georgarakis, G. Vaughan, A. R. Yavari and A. Inoue, Real-space structural studies of Cu-Zr-Ti glassy alloy, Journal of Alloys and Compounds, 466 (2008) 106-110. https://doi.org/10.1016/j.jallcom.2007.11.039

[19] D. V. Louzguine-Luzgin, A. R. Yavari, G. Vaughan, A. Inoue, Clustered crystalline structures as glassy phase approximants, Intermetallics, 17 (2009) 477–480. https://doi.org/10.1016/j.intermet.2008.12.008

[20] S. Sato, T. Sanada, J. Saida, M. Imafuku, E. Matsubara and A. Inoue, Effect of Al on local structures of Zr-Ni and Zr-Cu metallic glasses, Materials Transactions, 46 (2005) 2893-2897. https://doi.org/10.2320/matertrans.46.2893

[21] K. F. Kelton, G.W. Lee, A. K. Gangopadhyay, R. W. Hyers, T. J. Rathz, J. R. Rogers, M. B. Robinson, and D. S. Robinson, First X-ray scattering studies on electrostatically levitated metallic liquids: demonstrated influence of local icosahedral order on the nucleation barrier, Phys. Rev. Lett., 90 (2003) 195504. https://doi.org/10.1103/PhysRevLett.90.195504

[22] Y. Q. Cheng, H. W. Sheng and E. Ma, Relationship between structure, dynamics, and mechanical properties in metallic glass-forming alloys, Phys. Rev. B, 78 (2008) 014207. https://doi.org/10.1103/PhysRevB.78.014207

[23] V. Wessels, A. K. Gangopadhyay, K. Sahu, R. W. Hyers, S. M. Canepari, J. R. S. Rogers, M. J. Kramer, A. I. Goldman, D. Robinson, J. W. Lee, J. Morris, K. F. Kelton, Rapid chemical and topological ordering in supercooled liquid $Cu_{46}Zr_{54}$, Phys. Rev. B, 83 (2011) 094116. https://doi.org/10.1103/PhysRevB.83.094116

[24] N. Mattern, J. Bednarcik, M. Stoica, J. Eckert, Temperature dependence of the short-range order of $Cu_{65}Zr_{35}$ metallic glass, Intermetallics, 32 (2013) 51-56. https://doi.org/10.1016/j.intermet.2012.08.024

[25] K. Georgarakis, A. R. Yavari, D. V. Louzguine-Luzgin, J. Antonowicz, M. Stoica, Y. Li, M. Satta, A. LeMoulec, G. Vaughan and A. Inoue, Atomic structure of Zr-Cu glassy alloys and detection of deviations from ideal solution behavior with Al addition by x-ray diffraction using synchrotron light in transmission, Applied Physics Letters, 94 (2009) 191912. https://doi.org/10.1063/1.3136428

[26] Ch.E. Lekka, A. Ibenskas, A.R. Yavari, G.A. Evangelakis, Tensile deformation accommodation in microscopic metallic glasses via subnanocluster reconstructions, Appl. Phys. Lett., 91 (2007) 214103. https://doi.org/10.1063/1.2816912

[27] K.W. Park, J.J. Jang, M. Wakeda, Y. Shibutani and J.C. Lee, Atomic packing density and its influence on the properties of Cu–Zr amorphous alloys, Scripta Mater., 57 (2007) 805–808. https://doi.org/10.1016/j.scriptamat.2007.07.019

[28] N. Mattern, P. Jóvári, I. Kaban, S. Gruner, A. Elsner, V. Kokotin, H. Franz, B. Beuneu, and J. Eckert, Short-range order of Cu–Zr metallic glasses, J. Alloys Compd., 485 (2009) 163–169. https://doi.org/10.1016/j.jallcom.2009.05.111

[29] D. V. Louzguine, M. Saito, Y. Waseda and A. Inoue, Structural study of amorphous $Ge_{50}Al_{40}Cr_{10}$ alloy, Journal of the Physical Society of Japan, 68 (1999) 2298-2303. https://doi.org/10.1143/JPSJ.68.2298

[30] J. Antonowicz, A. Pietnoczka, K. Pękała, J. Latuch, G.A. Evangelakis, Local atomic order, electronic structure and electron transport properties of Cu-Zr metallic glasses, J. Appl. Phys., 115 (2014) 203714. https://doi.org/10.1063/1.4879903

[31] W.K. Luo, E. Ma, EXAFS measurements and reverse Monte Carlo modeling of atomic structure in amorphous $Ni_{80}P_{20}$ alloys, J. Non-Cryst Solids, 354 (2008) 945–955. https://doi.org/10.1016/j.jnoncrysol.2007.08.028

[32] J. Antonowicz, A. Pietnoczka, W. Zalewski, R. Bacewicz, M. Stoica, K. Georgarakis, A.R. Yavari, Local atomic structure of Zr–Cu and Zr–Cu–Al amorphous alloys investigated by EXAFS method, J. Alloys Compd., 509S (2011) S34. https://doi.org/10.1016/j.jallcom.2010.10.105

[33] A. L. Ankudinov, B. Ravel, J. J. Rehr, and S. D. Conradson, Real-space multiple-scattering calculation and interpretation of X-ray-absorption near-edge structure, Phys. Rev. B, 58 (1998) 7565–7576. https://doi.org/10.1103/PhysRevB.58.7565

[34] D. V. Louzguine and A. Inoue, Ge-Al-Cr-La amorphous alloys containing crystalline-like zones, NanoStructured Materials, 11 (1999) 115-123. https://doi.org/10.1016/S0965-9773(99)00024-0

[35] H.Y. Hsieh, B. H. Toby, T. Egami, Y. He, S. J. Poon and G. J. Shiflet, Atomic-structure of amorphous $Al_{90}Fe_xCe_{10-x}$, J. Mater. Res., 5 (1990) 2807–2812. https://doi.org/10.1557/JMR.1990.2807

[36] A.N. Mansour, C.P. Wong, R.A. Brizzolara, Atomic-structure of amorphous $Al_{100-2x}Co_xCe_x$ (x = 8, 9, and 10) and $Al_{80}Fe_{10}Ce_{10}$ alloys – an XAFS study, Phys. Rev. B, 50 (1994) 12401–12412. https://doi.org/10.1103/PhysRevB.50.12401

[37] T. Egami, The atomic structure of aluminum based metallic glasses and universal criterion for glass formation, J Non-Cryst Solids, 207 (1996) 575–582. https://doi.org/10.1016/S0022-3093(96)00277-3

[38] Y.E. Kalay, I. Kalay, J.W. Hwang, P.M. Voyles, M.J. Kramer, Local chemical and topological order in Al–Tb and its role in controlling nanocrystal formation, Acta Materialia, 60 (2012) 994–1003. https://doi.org/10.1016/j.actamat.2011.11.008

[39] N. Jakse, O. Lebacq, A. Pasturel, Ab initio molecular-dynamics simulations of short-range order in liquid $Al_{80}Mn_{20}$ and $Al_{80}Ni_{20}$ alloys, Phys. Rev. Lett., 93 (2004) 207801. https://doi.org/10.1103/PhysRevLett.93.207801

[40] J. Hafner, Theory of the formation of metallic glasses, Phys. Rev. B, 21 (1980) 406-418. https://doi.org/10.1103/PhysRevB.21.406

[41] http //lammps.sandia.gov/

Metallic Glasses and Their Composites – 2ⁿᵈ Edition   Materials Research Forum LLC
Materials Research Foundations **85** (2021)   https://doi.org/10.21741/9781644901014

[42] S. Plimpton, Fast parallel algorithms for short-range molecular dynamics, J. Comp. Phys., 117 (1995) 1-19. https://doi.org/10.1006/jcph.1995.1039

[43] M.I. Mendelev, S. Han, D.J. Srolovitz, G.J. Ackland, D.Y. Sun, and M. Asta, Development of new interatomic potentials appropriate for crystalline and liquid iron, Phil. Mag., 83 (2003) 3977-3994. https://doi.org/10.1080/14786430310001613264

[44] W. Mickel, S. C. Kapfer, G. E. Schröder-Turk, and K. Mecke, Shortcomings of the bond orientational order parameters for the analysis of disordered particulate matter, Journal of Chemical Physics, 138 (2013) 044501. https://doi.org/10.1063/1.4774084

[45] J. D. Honeycutt and H. C. Andersen, Molecular dynamics study of mel, ting and freezing of small Lennard-Jones clusters, J. Phys. Chem. 91 (1987) 4950. https://doi.org/10.1021/j100303a014

[46] F. Aurenhammer, Voronoi diagrams – A survey of a fundamental geometric data structure, ACM Computing Surveys, 23 (1991) 345–405. https://doi.org/10.1145/116873.116880

[47] J.D. Bernal, Geometry of the structure of monatomic liquids, Nature, 185 (1960) 68–70. https://doi.org/10.1038/185068a0

[48] A. Inoue, T. Negishi, H.M. Kimura, T. Zhang, A.R. Yavari, High packing density of Zr- and Pd-based bulk amorphous alloys, Mater. Trans. JIM, 39 (1998) 318–321. https://doi.org/10.2320/matertrans1989.39.318

[49] D. B. Miracle, The efficient cluster packing model – An atomic structural model for metallic glasses, Acta Mater., 54 (2006) 4317-4437. https://doi.org/10.1016/j.actamat.2006.06.002

[50] H. W. Sheng, W. K. Luo, F. M. Alamgir, J. M. Bai and E. Ma, Atomic packing and short-to-medium-range order in metallic glasses, Nature, 439 (2006) 419-425. https://doi.org/10.1038/nature04421

[51] D. B. Miracle, W. S. Sanders, and O. N. Senkov, The influence of efficient atomic packing on the constitution of metallic glasses, Phil. Mag., 83 (2003) 2409-2428. https://doi.org/10.1080/1478643031000098828

[52] Y. Q. Cheng, E. Ma, Atomic-level structure and structure–property relationship in metallic glasses, Prog. Mater. Sci., 56 (2011) 379–473. https://doi.org/10.1016/j.pmatsci.2010.12.002

[53] D. V. Louzguine-Luzgin, A. R. Yavari, M. Fukuhara, K. Ota, G. Xie, G. Vaughan and A. Inoue, Free volume and elastic properties changes in Cu-Zr-Ti-Pd bulk glassy alloy on heating, Journal of Alloys and Compounds, 431 (2007) 136-140. https://doi.org/10.1016/j.jallcom.2006.05.069

[54] T. Ohkubo and Y. Hirotsu, Electron diffraction and high-resolution electron microscopy study of an amorphous $Pd_{82}Si_{18}$ alloy with nanoscale phase separation, Phys. Rev. B, 67 (2003) 094201. https://doi.org/10.1103/PhysRevB.67.094201

[55] O. Haruyama, K. Sugiyama, M. Sakurai, Y. Waseda, A local structure change of bulk $Pd_{40}Ni_{40}P_{20}$ glass during full relaxation, J. Non-Crystalline Solids, 353 (2007) 3053-3065. https://doi.org/10.1016/j.jnoncrysol.2007.05.071

[56] A. Hirata, Y. Hirotsu, T. Ohkubo, N. Tanaka, T. G. Nieh, Local atomic structure of Pd–Ni–P bulk metallic glass examined by high-resolution electron microscopy and electron diffraction, Intermetallics, 14 (2006) 903-920. https://doi.org/10.1016/j.intermet.2006.01.007

[57] Y. Hirotsu, M. Uehara and, M. Ueno, Microcrystalline domains in amorphous palladium-copper-silicon ($Pd_{77.5}Cu_6Si_{16.5}$) alloys studied by high-resolution electron microscopy, J. Appl. Phys., 59 (1986) 3081. https://doi.org/10.1063/1.336932

[58] D. V. Louzguine-Luzgin, Y. H. Zeng, A. D. Setyawan, N. Nishiyama, H. Kato, J. Saida and, A. Inoue, Deformation behavior of Zr- and Ni-based bulk glassy alloys, Journal of Materials Research, 22 (2007) 1087. https://doi.org/10.1557/jmr.2007.0126

[59] A. Hirata, Y. Hirotsu, T. Ohkubo, E. Matsubara, T. Ohkubo and K. Hono, Mechanism of nanocrystalline microstructure formation in amorphous Fe–Nb–B alloys, Phys. Rev. B, 74 (2006) 214206. https://doi.org/10.1103/PhysRevB.74.184204

[60] D. W. Qi and S. Wang, Icosahedral order and defects in metallic liquids and glasses, Phys. Rev. B, 44 (1991) 884–887. https://doi.org/10.1103/PhysRevB.44.884

[61] A. Hirata, L. J. Kang, T. Fujita, B. Klumov, K. Matsue, M. Kotani, A. R. Yavari, M. W. Chen, Geometric frustration of icosahedron in metallic glasses, Science, 341 (2013) 376–379. https://doi.org/10.1126/science.1232450

[62] E. Matsubara, M. Sakurai, T. Nakamura, M. Imafuku, S. Sato, J. Saida and A. Inoue, environmental structural studies in amorphous and quasicrystalline $Zr_{70}Al_6Ni_{10}Pt_{14}$ alloys, Scripta Mater., 44 (2001) 2297. https://doi.org/10.1016/S1359-6462(01)00895-8

[63] D. V. Louzguine and A. Inoue, Nanoparticles with icosahedral symmetry in Cu-based bulk glass former induced by Pd addition, Scripta Mater., 48 (2003) 1325-1329. https://doi.org/10.1016/S1359-6462(03)00018-6

[64] K. F. Kelton, Crystallization of liquids and glasses to quasicrystals, J. Non-Cryst. Sol., 334-335 (2004) 253. https://doi.org/10.1016/j.jnoncrysol.2003.11.052

[65] S. Ranganathan and A. Inoue, An application of Pettifor structure maps for the identification of pseudo-binary quasicrystalline intermetallics, Acta Materialia, 54 (2006) 3647–3656. https://doi.org/10.1016/j.actamat.2006.01.041

[66] J. Ding, Y.Q. Cheng, H. Sheng, M. Asta, R. O. Ritchie & E. Ma, Universal structural parameter to quantitatively predict metallic glass properties, Nature Communications, 7 (2016) 13733. https://doi.org/10.1038/ncomms13733

[67] A.I. Oreshkin, V.N. Mantsevich, S.V. Savinov, S.I. Oreshkin, V.I. Panov, A.R. Yavari, D.B. Miracle, and D.V. Louzguine-Luzgin, In situ visualization of Ni–Nb bulk metallic glasses phase transition, Acta Materialia, 61 (2013) 5216–5222. https://doi.org/10.1016/j.actamat.2013.05.014

[68] R.V. Belosludov, A.I. Oreshkin, S.I. Oreshkin, D.A. Muzychenko, H. Kato, D.V. Louzguine-Luzgin, The atomic structure of a bulk metallic glass resolved by scanning tunneling microscopy and ab-initio molecular dynamics simulation, Journal of Alloys and Compounds, 816 (2020) 152680. https://doi.org/10.1016/j.jallcom.2019.152680

[69] J. Schroers, B. Lohwongwatana, W. L. Johnson, and A. Peker, Gold based bulk metallic glass, Appl. Phys. Lett., 87 (2005) 061912. https://doi.org/10.1063/1.2008374

[70] S. V. Ketov, N. Chen, A. Caron, A. Inoue, and D. V. Louzguine-Luzgin, Structural features and high quasi-static strain rate sensitivity of $Au_{49}Cu_{26.9}Ag_{5.5}Pd_{2.3}Si_{16.3}$ bulk metallic glass, Applied Physics Letters, 101 (2012) 241905. https://doi.org/10.1063/1.4770072

[71] S. Mechler, E. Yahel, P. S. Pershan, M. Meron, and B. Lin, Crystalline monolayer surface of liquid Au–Cu–Si–Ag–Pd: Metallic glass former, Applied Physics Letters, 98 (2011) 251915. https://doi.org/10.1063/1.3599515

[72] J. X. Fang, U. Vainio, W. Puff, R. Würschum, X. L. Wang, D. Wang, M. Ghafari, F. Jiang, J. Sun, H. Hahn, and H. Gleiter, Atomic structure and structural stability of $Sc_{75}Fe_{25}$ nanoglasses, Nano Lett., 12 (2012) 458−463. https://doi.org/10.1021/nl2038216

[73] N. Chen, R. Frank, N. Asao, D. V. Louzguine, J. Q. Wang, X. Q. Xie, Y. Ishikawa, N. Hatakeyama, P. Sharma, Y. C. Lin, M. Esashi, Y. Yamamoto and A. Inoue, Formation and properties of Au-based nanograined metallic glasses, Acta Mater., 59 (2011) 6433. https://doi.org/10.1016/j.actamat.2011.07.007

[74] S. V. Ketov, X.T. Shi, G.Q. Xie, R. Kumashiro, A. Yu. Churyumov, A. I. Bazlov, N. Chen, Y. Ishikawa, N. Asao, H.K. Wu and D. V. Louzguine-Luzgin, Nanostructured Zr-Pd metallic glass thin film for biochemical applications, Scientific Reports, 5 (2015) 7799. https://doi.org/10.1038/srep07799

[75] T. Egami, Spectroscopic diffraction measurement with white x-rays, J. Mater. Sci., 13 (1978) 2587. https://doi.org/10.1007/BF02402745

[76] N. Mattern, H. Hermann, S. Roth, J. Sakowski, P. Macht, P. Jovari and J. Jiang, Structural behavior of $Pd_{40}Cu_{30}Ni_{10}P_{20}$ bulk metallic glass below and above the glass transition, Appl. Phys. Lett., 82 (2003) 2589. https://doi.org/10.1063/1.1567457

[77] A. R. Yavari, N. Nikolov, N. Nishiyama, T. Zhang, A. Inoue, J.L. Uriarte, G. Heunen, The glass transition of bulk metallic glasses studied by real time diffraction in transmission usinghigh energy synchrotron radiation, Mater. Sci. Eng. A, 375–377 (2004) 709. https://doi.org/10.1016/j.msea.2003.10.224

[78] A. R. Yavari, R Le Moulec, A. Inoue, N. Nishiyama, N. Lupu, E. Matsubara, W. J. Botta, G. Vaughan, M. Michiel, A. Kvick, Excess free volume in metallic glasses measured by X-ray diffraction, Acta Materialia, 53 (2005) 1611. https://doi.org/10.1016/j.actamat.2004.12.011

[79] Q. K. Jiang, Z. Y. Chang, X. D. Wang, J. Z. Jiang, Structures at glassy, supercooled liquid, and liquid states in La-based bulk metallic glasses, Metallurgical and Materials Transactions A, 41A (2010) 1634. https://doi.org/10.1007/s11661-009-0041-9

[80] M. Stoica, J. Das, G. Bednarčik, J. Wang, G. Vaughan, W. H. Wang, J. Eckert, Mechanical response of metallic glasses Insights from in-situ high energy X-ray diffraction, J. of Metals, 62 (2010) 76-82. https://doi.org/10.1007/s11837-010-0037-3

[81] D. V. Louzguine, A. R. Yavari, K. Ota, G. Vaughan, A. Inoue, Synchrotron X-ray radiation diffraction studies of thermal expansion, free volume change and glass transition phenomenon in Cu-based glassy and nanocomposite alloys on heating, J. Non-Cryst. Sol., 351 (2005) 1639. https://doi.org/10.1016/j.jnoncrysol.2005.04.054

[82] D. V. Louzguine-Luzgin, A. Inoue, A. R. Yavari, and G. Vaughan, Thermal expansion of a glassy alloy studied using a real-space pair distribution function, Applied Physics Letters, 88 (2006) 121926. https://doi.org/10.1063/1.2187955

[83] O. Mishima, H.E. Stanley, The relationship between liquid, supercooled and glassy water, Nature, 396 (1998) 329–335. https://doi.org/10.1038/24540

[84] J. Ding and E. Ma, Computational modeling sheds light on structural evolution in metallic glasses and supercooled liquids, NPJ Computational Materials, 3 (2017) 9. https://doi.org/10.1038/s41524-017-0007-1

[85] N. Jakse, A. Pasturel, Local order of liquid and supercooled zirconium by ab initio molecular dynamics, Phys. Rev. Lett., 91 (2003) 195501. https://doi.org/10.1103/PhysRevLett.91.195501

[86] H.B. Lou, X.D. Wang, Q.P. Cao, D.X. Zhang, J. Zhang, T.D. Hu, H.K. Mao, J.Z. Jiang, Negative expansions of interatomic distances in metallic melts, Proc. Natl. Acad. Sci., 110 (2013) 10068–10072. https://doi.org/10.1073/pnas.1307967110

[87] L.H. Xiong, X.D. Wang, Q. Yu, H. Zhang, F. Zhang, Y. Sun, Q.P. Cao, H.L. Xie, T.Q. Xiao, D.X. Zhang, C.Z. Wang, K.M. Ho, Y. Ren, J.Z. Jiang, Temperature-dependent structure evolution in liquid gallium, Acta Materialia, 128 (2017) 304-312. https://doi.org/10.1016/j.actamat.2017.02.038

[88] L. Battezzati and A. L. Greer, Thermodynamics of $Te_{80}Ge_{20-x}Pb_x$ glass-forming alloys, J. Mater. Res., 3 (1988) 570. https://doi.org/10.1557/JMR.1988.0570

[89] S.J. Cheng, X.F. Bian, J.-X. Zhang, X.B. Qin, Z.H. Wang, Correlation of viscosity and structural changes of indium melt, Materials Letters, 57 (2003) 4191–4195. https://doi.org/10.1016/S0167-577X(03)00288-X

[90] K. Georgarakis, D.V. Louzguine-Luzgin, J. Antonowicz, G. Vaughan, A.R. Yavari, T. Egami, A. Inoue, Variations in atomic structural features of a supercooled Pd-Ni-Cu-P glass forming liquid during in situ vitrification, Acta Mater., 59 (2011) 708–716. https://doi.org/10.1016/j.actamat.2010.10.009

[91] D. V. Louzguine-Luzgin, R. Belosludov, A. R. Yavari, K. Georgarakis, G. Vaughan, Y. Kawazoe, T. Egami and A. Inoue, Structural basis for supercooled liquid fragility established by synchrotron-radiation method and computer simulation, J. Appl. Phys., 110 (2011) 043519. https://doi.org/10.1063/1.3624745

[92] G. J. Fan, J. F. Löffler, R. K. Wunderlich and H.-J. Fecht, Thermodynamics, enthalpy relaxation and fragility of the bulk metallic glass-forming liquid $Pd_{43}Ni_{10}Cu_{27}P_{20}$, Acta Materialia, 52 (2004) 667. https://doi.org/10.1016/j.actamat.2003.10.003

[93] D. V. Louzguine-Luzgin, K. Georgarakis, A. Tsarkov, A. Solonin, V. Honkimaki, L. Hennet and A. R. Yavari, Structural changes in liquid Fe and Fe–B alloy on cooling, Journal of Molecular Liquids, 209 (2015) 233–238. https://doi.org/10.1016/j.molliq.2015.05.062

[94] N. Mattern, H. Hermann, S. Roth, J. Sakowski, M.P. Macht, P. Jovari, J. Jiang, Structural behavior of $Pd_{40}Cu_{30}Ni_{10}P_{20}$ bulk metallic glass below and above the glass transition, Appl. Phys. Lett., 82 (2003) 2589. https://doi.org/10.1063/1.1567457

[95] K. Georgarakis, D. V. Louzguine-Luzgin, J. Antonowicz, G. Vaughan, A. R. Yavari, T. Egami and A. Inoue, Variations in atomic structural features of a supercooled Pd–Ni–Cu–P glass forming liquid during in situ vitrification, Acta Materialia, 59 (2011) 708–716. https://doi.org/10.1016/j.actamat.2010.10.009

[96] D. V. Louzguine-Luzgin, A. I. Bazlov, A. Yu. Churyumov, K. Georgarakis, and A. R. Yavari, Comparative analysis of the structure of palladium-based bulk metallic glasses prepared by treatment of melts with flux, Physics of the Solid State, 55 (2013) 1985–1990. https://doi.org/10.1134/S1063783413100211

[97] A. Inoue, T. Negishi, H. M. Kimura, T. Zhang and A. R. Yavari, High packing density of Zr- and Pd-based bulk amorphous alloys, Mater. Trans. JIM 39 (1998) 318. https://doi.org/10.2320/matertrans1989.39.318

[98] D. V. Louzguine-Luzgin, A. R. Yavari, M. Fukuhara, K. Ota, G. Xie, G. Vaughan and A. Inoue, Free volume and elastic properties changes in Cu-Zr-Ti-Pd bulk glassy alloy on heating, J. Alloys and Comp. 431 (2007) 136. https://doi.org/10.1016/j.jallcom.2006.05.069

[99] R. Bohmer, K. L. Ngai, C. A. Angell, D. J. Plazek, Nonexponential relaxations in strong and fragile glass formers. J. Chem. Phys., 99 (1993) 4201. https://doi.org/10.1063/1.466117

[100] M. I. Ojovan, About activation energy of viscous flow of glasses and melts, Mater. Res. Soc. Symp. Proc., 15 (2015) 10.1557/opl.2015.44. https://doi.org/10.1557/opl.2015.44

[101] D. V. Louzguine-Luzgin, Vitrification and devitrification processes in metallic glasses, Journal of Alloys and Compounds, 586 (2014) 2. https://doi.org/10.1016/j.jallcom.2012.09.057

[102] N. A. Mauro, M. L. Johnson, J. C. Bendert, K. F. Kelton, Structural evolution in Ni–Nb and Ni–Nb–Ta liquids and glasses — A measure of liquid fragility, Journal of Non-Crystalline Solids 362 (2013) 237. https://doi.org/10.1016/j.jnoncrysol.2012.11.022

[103] K. F. Kelton, Kinetic and structural fragility—a correlation between structures and dynamics in metallic liquids and glasses, Journal of Physics Condensed Matter, 29 (2017) 023002. https://doi.org/10.1088/0953-8984/29/2/023002

[104] N. A. Mauro, M. Blodgett, M. L. Johnson, A. J. Vogt and K. F. Kelton, A structural signature of liquid fragility, Nat. Commun., 5 (2014) 4616. https://doi.org/10.1038/ncomms5616

[105] K. F. Kelton, G. W. Lee, A. K. Gangopadhyay, R. W. Hyers, T. J. Rathz, J. R. Rogers, M. B. Robinson and D. S. Robinson, A structural signature of liquid fragility, Phys. Rev. Lett., 90 (2003) 195504. https://doi.org/10.1103/PhysRevLett.90.195504

[106] Y. Q. Cheng, E. Ma and H. W. Sheng, Alloying strongly influences the structure, dynamics, and glass forming ability of metallic supercooled liquids, Appl. Phys. Lett., 93 (2008) 111913. https://doi.org/10.1063/1.2987727

[107] N. Jakse and A. Pasturel, Glass forming ability and short-range order in a binary bulk metallic glass by ab initio molecular dynamics, Applied Physics Letters, 93 (2008) 113104. https://doi.org/10.1063/1.2976428

[108] D.V. Louzguine-Luzgin, K. Georgarakis, J. Andrieux, L. Hennet, T. Morishita, K. Nishio, R.V. Belosludov, An atomistic study of the structural changes in a Zr–Cu–Ni–Al glass-forming liquid on vitrification monitored in-situ by X-ray diffraction and molecular dynamics simulation, Intermetallics, 122 (2020) 106795. https://doi.org/10.1016/j.intermet.2020.106795

[109] I.S. Klein and C.A. Angell, Excess thermodynamic properties of glassforming liquids: The rational scaling of heat capacities, and the thermodynamic fragility dilemma resolved, Journal of Non-Crystalline Solids, 451 (2016) 116–123. https://doi.org/10.1016/j.jnoncrysol.2016.06.006

[110] R. Bhowmick, R. Raghavan, K. Chattopadhyay and U. Ramamurty, Plastic flow softening in a bulk metallic glass, Acta Materialia, 54 (2006) 4221–4228. https://doi.org/10.1016/j.actamat.2006.05.011

[111] F.O. Méar, B. Lenk, Y. Zhang, and A.L. Greer, Structural relaxation in a heavily cold-worked metallic glass, Scripta Materialia, 59 (2008) 1243–1246. https://doi.org/10.1016/j.scriptamat.2008.08.023

[112] K.M. Flores, E. Sherer, A. Bharathula, H. Chen and Y.C. Jean, Sub-nanometer open volume regions in a bulk metallic glass investigated by positron annihilation, Acta Materialia, 55 (2007) 3403–3411. https://doi.org/10.1016/j.actamat.2007.01.040

[113] O. Haruyama, K. Kisara, A. Yamashita, K. Kogure, Y. Yokoyama and K. Sugiyama, Characterization of free volume in cold-rolled $Zr_{55}Cu_{30}Ni_5Al_{10}$ bulk metallic glasses, Acta Materialia, 61 (2013) 3224–3232. https://doi.org/10.1016/j.actamat.2013.02.010

[114] D.V. Louzguine-Luzgin, S.V. Ketov, Z. Wang, M.J. Miyama, A.A. Tsarkov and A.Yu. Churyumov, Plastic deformation studies of Zr-based bulk metallic glassy samples with a low aspect ratio, Mater. Sci. Eng. A, 616 (2014) 288–296. https://doi.org/10.1016/j.msea.2014.08.006

[115] Y. Lou, X. Liu, X. Yang, L.-C. Zhang, Z. Liu, Fast rejuvenation in bulk metallic glass induced by ultrasonic vibration precompression, Intermetallics, 118 (2020) 106687. https://doi.org/10.1016/j.intermet.2019.106687

[116] A. Caron, A. Kawashima, H.-J. Fecht, D.V. Louzguine-Luzgin and A. Inoue, On the anelasticity and strain induced structural changes in a Zr-based bulk metallic glass, Appl. Phys. Lett., 99 (2011) 171907. https://doi.org/10.1063/1.3655999

[117] A.Yu. Churyumov, A.I. Bazlov, V.Yu. Zadorozhnyy, A.N. Solonin, A. Caron and D.V. Louzguine-Luzgin, Phase transformations in Zr-based bulk metallic glass cyclically loaded before plastic yielding, Mater. Sci. Eng. A, 550 (2012) 358–362. https://doi.org/10.1016/j.msea.2012.04.087

[118] J. Ding, S. Patinet, M.L. Falk, Y.Q. Cheng and E. Ma, Soft spots and their structural signature in a metallic glass, Proc. Natl. Acad. Sci., 111 (2014) 14052–14056. https://doi.org/10.1073/pnas.1412095111

[119] S.-J. Lee, B.-G. Yoo, J.-I. Jang and J.-C. Lee, Irreversible structural change induced by elastostatic stress imposed on an amorphous alloy and its influence on the mechanical

properties, Metal. Mater. Int., 14 (2008) 9–13.
https://doi.org/10.3365/met.mat.2008.02.009

[120] K.-W. Park, C.-M. Lee, M. Wakeda, Y. Shibutani, M.L. Falk and J.-C. Lee, Elastostatically induced structural disordering in amorphous alloys, Acta Mater., 56 (2008) 5440–5454. https://doi.org/10.1016/j.actamat.2008.07.033

[121] H.B. Ke, P. Wen, H.L. Peng, W.H. Wang and A.L. Greer, Homogeneous deformation of metallic glass at room temperature reveals large dilatation, Scripta Mater., 64 (2011) 966–969. https://doi.org/10.1016/j.scriptamat.2011.01.047

[122] A.L. Greer and Y.H. Sun, Stored energy in metallic glasses due to strains within the elastic limit, Philos. Mag., 96 (2016) 1643–1663. https://doi.org/10.1080/14786435.2016.1177231

[123] A.J. Cao, Y.Q. Cheng and E. Ma, Structural processes that initiate shear localization in metallic glass, Acta Mater., 57 (2009) 5146–5155. https://doi.org/10.1016/j.actamat.2009.07.016

[124] Y. Ritter and K. Albe, Chemical and topological order in shear bands of $Cu_{64}Zr_{36}$ and $Cu_{36}Zr_{64}$ glasses, J. Appl. Phys., 111 (2012) 103527. https://doi.org/10.1063/1.4717748

[125] Y.C. Hu, P.F. Guan, Q. Wang, Y. Yang, H.Y. Bai and W.H. Wang, Pressure effects on structure and dynamics of metallic glass-forming liquid, The Journal of Chemical Physics, 146, (2017) 024507. https://doi.org/10.1063/1.4973919

[126] T. Nagase, A. Nino, T. Hosokawa and Y. Umakoshi, Materials Transactions, 48 (2007) 1651 - 1658. https://doi.org/10.2320/matertrans.MJ200701

[127] N. Giovambattista, T. Loerting, B. R. Lukanov and F. W. Starr, Interplay of the glass transition and the liquid-liquid phase transition in water, Scientific Reports, 2 (2012) 390. https://doi.org/10.1038/srep00390

[128] O. Mishima, L.D. Calvert, E. Whalley, 'Melting ice' I at 77 K and 10 kbar: a new method of making amorphous solids, Nature, 310 (1984) 393-395. https://doi.org/10.1038/310393a0

[129] R. V. Belosludov, K. V. Gets, O. S. Subbotin, R. K. Zhdanov, Y. Y. Bozhko, V. R. Belosludov, J.I. Kudoh, Modeling the polymorphic transformations in amorphous solid ice, Journal of Alloys and Compounds, 707 (2017) 108-113. https://doi.org/10.1016/j.jallcom.2016.12.197

Materials Research Forum LLC
https://doi.org/10.21741/9781644901014

[130] P. H. Poole, T. Grandea, F. Sciortinod, H.E. Stanley and C. A. Angel, Amorphous polymorphism, Computational Materials Science, 4 (1995) 373-382. https://doi.org/10.1016/0927-0256(95)00044-9

[131] H. W. Sheng, H. Z. Liu, Y. Q. Cheng, J. Wen, P. L. Lee, W. K. Luo, S. D. Shastri and E. Ma. Polyamorphism in a metallic glass, Nature Materials, 6 (2007) 192 – 197. https://doi.org/10.1038/nmat1839

[132] Q.S. Zeng, Y. C. Li, C. M. Feng, P. Liermann, M. Somayazulu, G. Y. Shen, H. K. Mao, R. Yang, J. Liu, T.D. Hu and J. Z. Jian, Anomalous compression behavior in lanthanum/cerium-based metallic glass under high pressure, Proc. Nat'l. Acad. Sci. USA, 104 (2007) 13565–13568. https://doi.org/10.1073/pnas.0705999104

[133] Q.-S. Zeng, Y. Ding, W. L. Mao, W. Yang, S. V. Sinogeikin, J. Shu, H.-K. Mao, and J. Z. Jiang, Origin of pressure-induced polyamorphism in $Ce_{75}Al_{25}$ metallic glass, Phys. Rev. Lett., 104 (2010) 105702. https://doi.org/10.1103/PhysRevLett.104.105702

CHAPTER 4

Structural Changes in the Glassy Phase on Heating

Glassy alloys are metastable at room temperature and devitrify/crystallize on heating above the temperature called T_x which, as well as T_g, varies upon the heating rate used. Devitrification is a wider term which includes crystallization as well as the formation of the supercooled liquid on heating; however, both these terms are used in the literature. Glassy alloys may also crystallize even at room temperature but for the majority of glassy alloys the process is very slow. In general, amorphous alloys devitrify directly on heating while glassy alloys upon fast enough heating initially transform to a supercooled liquid which later crystallizes. This process is a subject of discussion in the present chapter.

Contents

4.1   Relaxation processes in metallic glasses.

First of all one should separate two types of relaxations observed in metallic glasses on heating. One is a response of a metallic glass to an external action, for example mechanical loading, called "relaxation". Another one is "structural relaxation" which is connected with gradual structure changes on heating towards a more stable glass starting at the beginning of relaxation temperature (T_{br}) accompanied with exothermic heat release (Fig. 4.1 (a)). This heat release is stopped at the beginning of glass transition (T_{bg}) when the heat capacity of the glassy phase starts to rise towards that of a liquid as it was discussed in Chapter 2 (see Fig. 2.2 for example). The structural state of a glass is connected with the fictive temperature connected with the cooling rate at which the glass was formed. The glass cooled at a high rate has an excess volume and tends to become denser when thermal activation enhances atomic mobility.

Glass relaxation can be separated to a) slow high-temperature α-relaxation (involving the entire glass) responsible for the glass transition process (glass→liquid transition on heating) and b) fast low-temperature β-relaxation [1] taking place in the localized regions on a shorter timescale responsible for plastic flow and mechanical properties at room temperature. There are some data indicating that β-relaxation could be connected to α-relaxation [2].

Dynamic relaxation processes are studied by dynamic mechanical analysis (DMA). A strain of 0.01 % is a typical value to measure the storage modulus (E'), loss modulus (E'') and the internal friction (Q^{-1}) of materials [3] according to the formula:

$$Q^{-1} = \tan\delta = E''/E', \qquad (4.1)$$

where δ is the phase lag between the applied cyclic stress (σ) and the resulting strain (ε).

$$\sigma = \sigma_0 \cos(\omega t) \tag{4.2}$$

$$\varepsilon = \varepsilon_0 \cos(\omega t - \delta) \tag{4.3}$$

where ω is the angular frequency $\omega = 2\pi f$ and f is the oscillation frequency usually ranging from 0.01 to 10 Hz.

Fast β-relaxation peak becomes clearly visible either as a peak separated from α relaxation, mostly in La-based BMGs, or as a low-temperature shoulder of α relaxation (see a schematic view in Fig. 4.1(b)). Structural heterogeneity in metallic glasses was tested by atomic force microscopy (AFM). The microstructural heterogeneity lengthscale of different BMGs is found to be well correlated with their β-relaxation behavior [4]. It is maximized in La-based BMGs with a lengthscale of 200 nm. High degree of structural heterogeneity was also observed in the $Pd_{40}Ni_{10}Cu_{30}P_{20}$ BMG.

4.1.1 Relaxation effects on heating.

DSC (Fig. 4.1(c)) and DMA (Fig. 4.1(d)) of the $Zr_{65}Cu_5Ni_{10}Al_{7.5}Pd_{12.5}$ bulk metallic glassy sample indicate that the loss factor (tanδ) related to the viscoelastic component of the deformation exhibits a maximum at a characteristic temperature T_α of about 710 K [5]. This temperature is above the glass transition temperature but close to the onset temperature of the first exothermic peak called T_x. At higher temperature the loss factor shows a decrease (Fig. 4.1c) associated with crystallization (here formation of a quasicrystalline phase) [5]. However, the internal friction Q^{-1} or tanδ increases again at the higher temperature region owing to the decreasing viscosity of the residual supercooled liquid phase until it starts to crystallize.

Mechanical loading at room temperature can induce structural changes within the glassy phase which in turn alternates the relaxation behavior. The thermal and mechanical relaxation spectra of a $Zr_{64}Cu_{21}Fe_5Al_{10}$ bulk metallic glass were studied before and after room-temperature cyclic loading up to 0.4 % of elastic strain used to induce the structural changes within the glassy phase. It caused partial nanocrystallization after 10000 elastic cycles, which in turn influenced the relaxation behavior. A double-stage structure relaxation behavior was monitored using DMA at a lower strain (Fig. 4.2) [6].

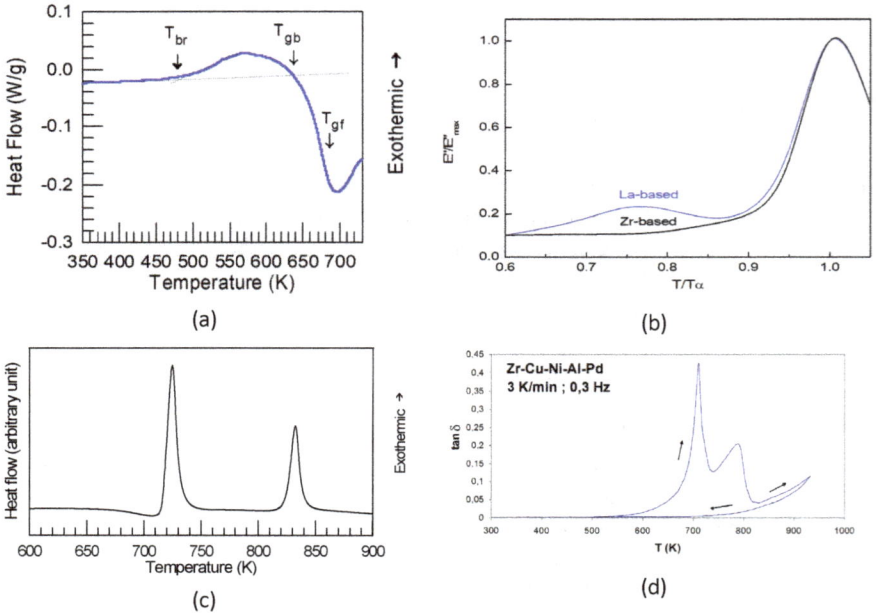

Fig. 4.1. (a) DSC trace of a Zr-Cu-Fe-Al bulk metallic glass indicating the heat effect of structural relaxation and glass transition processes. (b) A schematic dependence of E''/E''_{max} as a function of T/T_α for two types of BMGs. (c) DSC and (d) DMA traces of the $Zr_{65}Cu_5Ni_{10}Al_{7.5}Pd_{12.5}$ metallic glass. (c,d) Reproduced from [5] with permission of Elsevier.

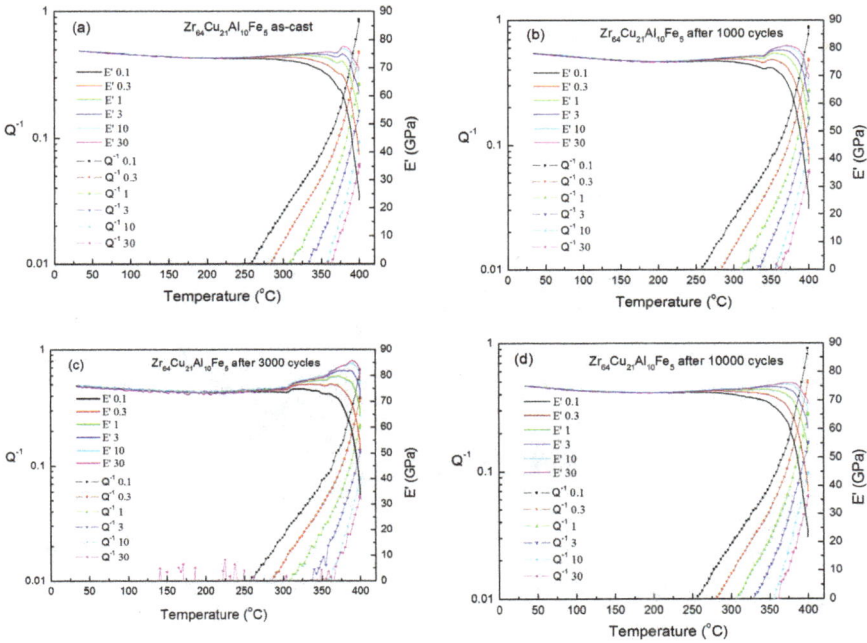

Fig. 4.2. DMA traces upon heating up to 400 °C at the heating rate of 1 °C/min. E' (Storage Modulus) and Q^{-1} (internal friction) as a function of temperature for the as-cast sample (a) and the samples preliminary cycled mechanically at room temperature at 3 Hz for 0.4 % of bending deformation for 1000 (b), 3000 (c) and 10000 (d) cycles. The DMA spectra were obtained at the frequencies of 0.1-30 Hz as indicated. Reproduced from [6] with permission of Elsevier.

Thermally induced structural relaxation starts at about 200 °C. At this temperature the storage modulus E' which initially decreased with temperature owing to thermal softening began to rise (Fig. 4.2(a)). However, there is another point of sudden increase in E' at a higher temperature. Its onset temperature decreases to a lower temperature for the samples subjected to 1000 and 3000 loading cycles at room temperature and disappears for the sample subjected to 10000 cycles (Figs. 4.2(b-d)). Also, this transition temperature is independent of the frequency used. The preliminary mechanical cycling at room temperature leads to some pronounced changes in the atomic structure of the BMG samples in turn leading to the changes in their relaxation behavior. Although the exothermic peak related to the structural relaxation of the metallic glass studied by DSC has a smooth character, mechanical relaxation measured by DMA in terms of the E' modulus has a two-stage behavior (first increment starting at 200 °C and another one close to the glass-transition temperature) in case of the as-cast and preliminary cycled samples, except for the sample preliminary cycled for 10000 cycles at room temperature. Room-temperature mechanical cycling lead to the formation of nuclei and nanoparticles of a metastable phase which disappear on subsequent heating before thermal crystallization (Fig. 4.3).

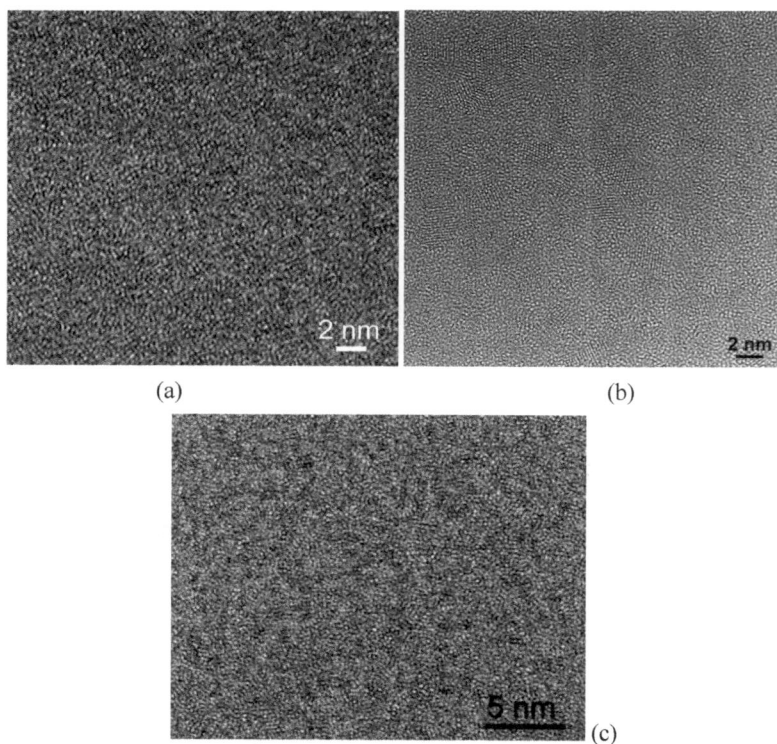

(a)   (b)

(c)

Fig. 4.3. HRTEM images of the samples after 1000 (a) and after 10000 bending cycles (ε_0 = 0.4 %) at room temperature (b) as well as after 10000 cycles and subsequent DMA test on heating up to 400 °C (c). Reproduced from [6] with permission of Elsevier.

4.1.2 Structural relaxation and rejuvenation.

Glasses vitrified at different cooling rates have different T_g and properties. Upon heating, as the heating rate is usually lower than the cooling one from the melt, the structure of metallic glasses relaxes leading to their densification and decrease in the excess volume [7]. This process leads to the changes in various properties including TCR as it was shown in Table 1.3 (Chapter 1). This process also affects many other properties including atomic diffusivity, mechanical properties, especially plasticity (see Chapter 5), etc...

Heat of structural relaxation can be readily measured below T_g (Fig. 4.1 (a)) starting at T_{br}. The glass transition on heating takes place in the temperature range between the conditional beginning (T_{gb}) and finish (T_{gf}) of glass transition temperatures (the area of change of C_p) as monitored in Fig. 4.1(a) by the deflection of the DSC curve of a Zr-Cu-Fe-Al alloy on heating. Usually the structure relaxation produces a single broad exothermic peak. However, often the relaxation process is separated into two processes: a high and a low-temperature one [8]. Structural relaxation process requires a higher cooling rate that a heating one. When the cooling rate of a BMG forming alloy is equal to the heating one (dashed red line) no relaxation enthalpy is observed (Fig. 4.4(a)) [9]. As it was shown in Chapter 2 here again a clear double-stage glass transition process is observed on heating when the process of structural relaxation is suppressed.

The thermal expansion and volume changes due to structural relaxation during heat treatment in a glassy $Cu_{55}Hf_{25}Ti_{15}Pd_5$ alloy were studied in situ using synchrotron X-ray radiation diffractometry. Good correspondence of the thermal expansion coefficients was obtained independently by the thermo-mechanical length change measurements and synchrotron X-ray radiation diffractometry. Good correlation was also found for the glass transition temperature obtained by DSC and Synchrotron X-ray radiation diffractometry [10].

The slope of the shift in the main peak position in the reciprocal space gives the linear coefficient of thermal expansion $\beta_g = 1.2 \cdot 10^{-5}$ K^{-1} before the structural relaxation and $\beta_g = 1.4 \cdot 10^{-5}$ K^{-1} after structural relaxation as well as volume coefficient of thermal expansion $\alpha_g = 3.6 \cdot 10^{-5}$ K^{-1} before the structural relaxation and $\alpha_g = 4.2 \cdot 10^{-5}$ K^{-1} after the structural relaxation. At temperatures higher than 600 K, structural relaxation of the glass towards equilibrium causes deviation from linear expansion as the excess free volume anneals out in the temperature range of 600-750 K. In the supercooled liquid state $\beta_l = 4.2 \cdot 10^{-5}$ K^{-1}, while $\alpha_l = 1.26 \cdot 10^{-4}$ K^{-1}, the values common for liquid metals. The linear coefficient of thermal expansion was also measured by the thermal mechanical analysis to be $1.5 \cdot 10^{-5}$ K^{-1} corresponding to those obtained from the Synchrotron X-ray radiation diffraction data. Thermal expansion coefficients below 10^{-5} K^{-1} were observed in Fe-based BMGs [11].

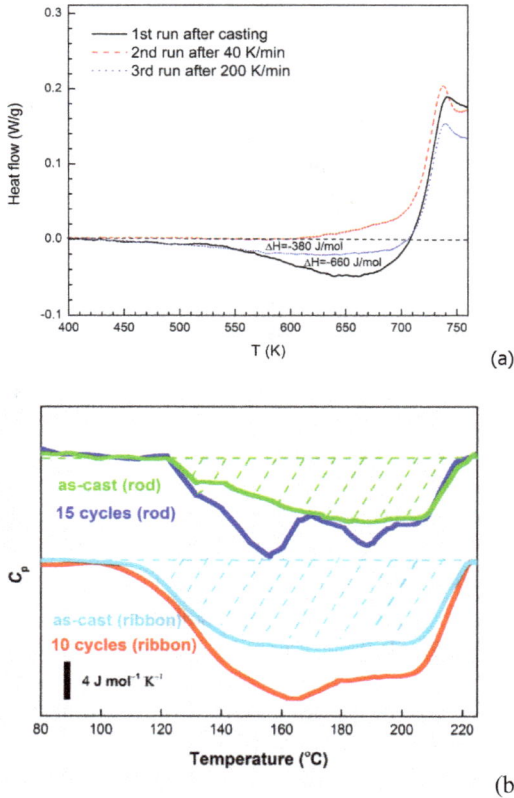

Fig. 4.4. (a) DSC traces of the $Cu_{47}Zr_{45}Al_8$ glassy alloy obtained at 40 K/min (0.67 K/s). The initial as-cast sample cooled at about 10^3 K/s was heated up to about 760 K (in the supercooled liquid state) 3 times: as prepared, after cooling at 40 K/min in a DSC cell and after cooling at 200 K/min in the same cell. The relaxation enthalpy was 660 and 380 J/mol in the as-prepared state and after cooling at 200 K/min, respectively. Reproduced from [9] with permission of Elsevier. Here exothermic reactions are plotted as negative direction peaks. (b) Differential scanning calorimetry traces of melt-spun ribbons of the $La_{55}Ni_{20}Al_{25}$ and bulk rods of the $La_{55}Ni_{10}Al_{35}$ metallic glasses indicating the increased heat of structural relaxation. Reproduced from [19] with permission of Nature Publishing Group.

The thermal expansion of a glassy $Cu_{55}Hf_{25}Ti_{15}Pd_5$ alloy studied by using the reciprocal space functions was also verified using a real-space pair distribution function [12]. The plot the atomic distance value for peak maximum of the *PDF(R)* at a given temperature divided by room temperature value for the $Cu_{55}Hf_{25}Ti_{15}Pd_5$ glassy alloy indicated the thermal expansion of the sample due to an increase in the interatomic distances. The linear coefficient of thermal expansion, $\beta_g = 1.6 \cdot 10^{-5}$ K^{-1}, is close to that obtained from change in the position of the center of mass of the main diffraction maximum in the reciprocal space and it is very close to that measured by the thermo-mechanical analysis.

The $Cu_{55}Hf_{25}Ti_{15}Pd_5$ and $Cu_{55}Zr_{30}Ti_{10}Pd_5$ glassy alloys were found to exhibit a large amount of the excess volume of about 0.8 % [13] significantly larger than the volume changes upon the structure relaxation of conventional bulk glassy alloys of about 0.1-0.3 % [14,15]. The density values of the $Cu_{55}Zr_{30}Ti_{10}Pd_5$ bulk glassy sample were 7518 kg/m^3 in as-prepared state, 7580 kg/m^3 for the glass relaxed at 740 K, 120 s and 7592 kg/m^3 after precipitation of the nanoscale icosahedral phase (after annealing at785 K, for 300 s).

Structural rejuvenation is an opposite process. It can be achieved by heating the glass above T_g and fast enough cooling thus restoring the initial glassy structure. However, some part of the incubation period for crystallization has already passed in the sample and its thermal stability will be damaged. Structural rejuvenation can also be done by ion irradiation [16], mechanical [17], thermal [18] or cryogenic cycling [19]. Cryogenic thermal cycling of metallic glasses induces rejuvenation, reaching the states of higher energy (Fig. 4.4b). It originates from the heterogeneous dynamics in liquids leading to the heterogeneities in the glasses. Thermal cycling is a non-destructive, isotropic and controllable method which induces no macroscopic residual stresses.

4.2 Phase separation

Some glassy alloys exhibit phase separation by spinodal or binodal mechanism [20] owing to solid- or liquid-state immiscibility among the constituent elements. Spinodal decomposition is characterized by nearly regular distribution of two phases and continuous variation of composition with time until equilibrium compositions are reached. Each phase parts are interconnected. The interface between phases being initially diffuse eventually sharpens with time. Opposite to the spinodal decomposition mechanism binodal separation takes place by nucleation and growth. Binodal separation is characterized by invariance of precipitating phase composition with time at constant temperature. Interface between the phases has same quality during growth. The growing particles are randomly distributed in the parent phase. The daughter phase has nearly spherical morphology.

4.2.1 Phase separation on cooling liquids

Some glassy alloys exhibit phase separation [20] either by spinodal or binodal mechanism owing to solid- or liquid-state immiscibility among the constituent elements. The solid-state immiscibility between Zr and Y caused spinodal phase separation in the Zr-Y-Al-Ni system during heating upon the glass transition [21]. Zr-rich/La-rich glassy phases were obtained in the melt-spun Zr-La-Al-Ni-Cu alloy [22]. The structure of multi-phase metallic glass strongly depends on the alloy composition as well as the processing conditions. The $Zr_{30}RE_{30}Al_{15}Ni_{25}$ and $Zr_{15}RE_{45}Al_{15}Ni_{15}$ (RE-rare earth metal) alloys also showed similar behavior [23]. A globular phase in the $Zr_{30}Nd_{30}Al_{15}Ni_{25}$ alloy is Nd-rich (Fig. 4.5(a)) while the surrounding matrix phase is a Zr-rich one. The elongated Zr-rich phase with a nominal atomic composition of $Zr_{53.0}Nd_{9.2}Al_{13.9}Ni_{23.9}$ produces larger halo ring of SAED pattern, Nd-rich second glassy phase has the $Zr_{17.8}Nd_{43.4}Al_{16.1}Ni_{22.3}$ composition.

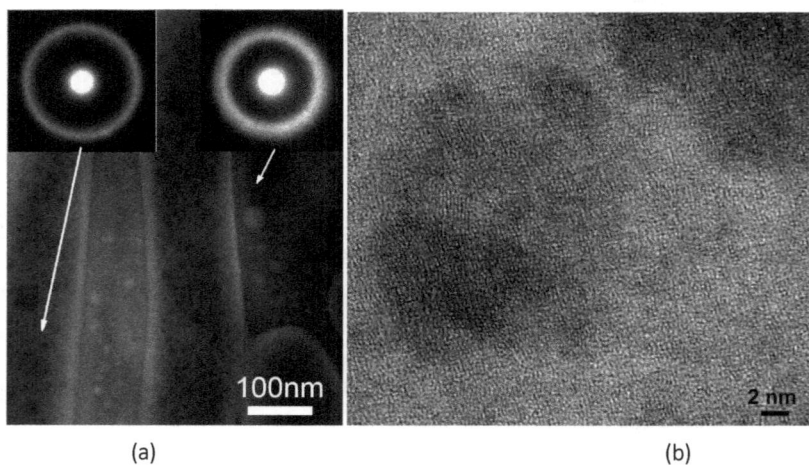

(a)   (b)

Fig. 4.5. (a) Bright-field transmission electron micrograph showing the two-phase structure of the $Zr_{30}Nd_{30}Al_{15}Ni_{25}$ sample and SAED patterns taken from the corresponding areas. Reproduced from [23] with permission of Elsevier. (b) HRTEM image of the $Cu_{36}Zr_{48}Al_8Ag_8$ glassy alloy sample treated for 25 ks at 675 K.

Surprisingly phase separation was also observed in the $Fe_{90}Zr_{10}$ metallic glass [24]. The observed evolution of the structure from metallic glass to the two-phase crystal structure could be the solute partitioning preceding eutectic crystallization. A nonuniform structure with two amorphous phases one depleted another enriched in yttrium was found in the Al-Ni-Y metallic glass [25].

4.2.2 Phase separation on heating metallic glasses

Phase separation competes with crystallization. However, the driving force is usually smaller and it takes place at a lower temperature. Binodal phase separation was observed in the $Cu_{36}Zr_{48}Al_8Ag_8$ glassy alloy [26] in the glassy/supercooled liquid phase at long-term annealing within the supercooled liquid region prior to crystallization (Fig. 4.5 (b)). Dark spots in the HRTEM image were analyzed by using EDX mapping and found to be Ag-rich. Nanocrystallization of Ag-rich particles takes place in these regions after separation at long-term annealing (Fig. 4.6). However, if the sample is heated quickly enough intensive eutectic-like crystallization of the studied alloy starts at a higher temperature.

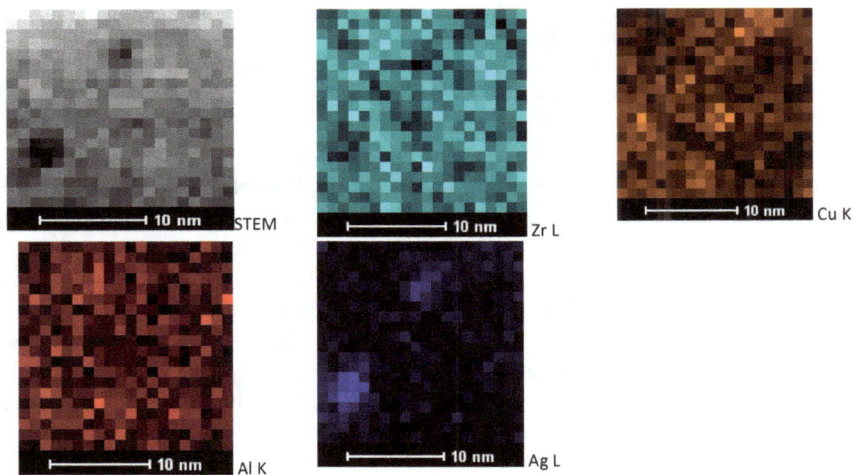

Fig. 4.6. STEM image of the $Cu_{36}Zr_{48}Al_8Ag_8$ glassy alloy sample annealed for 25 ks at 675 K as well as the corresponding EDX elemental maps as indicated. Reproduced from [27] with permission of Springer.

The $Zr_{62.5}Cu_{22.5}Al_{10}Fe_5$ alloy also undergoes nanoscale phase separation on heating prior to crystallization when kept at the supercooled liquid temperature owing to the repulsive Fe-Cu interaction (Fig. 4.7(a)) [28].

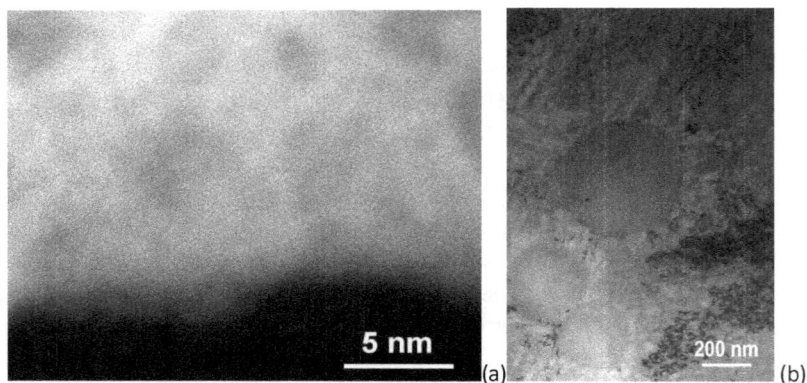

Fig. 4.7. (a) High-angle annular dark-field (HAADF) image of the $Zr_{62.5}Cu_{22.5}Al_{10}Fe_5$ sample annealed at 713 K for 300 s. Reproduced from [28] with permission of Taylor & Francis. (b) A residual globular glassy phase in the $Cu_{35}Zr_{45}Ag_{20}$ alloy annealed at 722 K for 1 ks. Reproduced from [29] with permission of Elsevier.

A small volume fraction (~10 vol.%) of a residual globular glassy phase was observed in the structure of the $Cu_{35}Zr_{45}Ag_{20}$ alloy (Fig. 4.7(b)) while no phase separation was found in the case of the as-solidified sample [29]. The sub-micrometer scale residual amorphous phase is depleted in Ag compared to the crystalline region. It contains only about 14 at.% Ag. This residual amorphous phase requires a higher temperature for crystallization. Such a phase separation in the Cu-Zr-Ag system can be expected according to a simple eutectic-type Cu-Ag phase diagram. It has been also observed in the $Cu_{43}Zr_{43}Al_7Ag_7$ [30] and $Ag_{20}Cu_{48}Zr_{32}$ [31] glassy alloys on a nanoscale while in the $Cu_{35}Zr_{45}Ag_{20}$ alloy it takes place on a submicron level.

4.3 Dissolution of a metastable crystalline phase in a supercooled liquid

Dissolution of a metastable crystalline phase produced by mechanical cycling was observed on heating up the Zr-based alloy to supercooled liquid (Fig. 4.3). The as-solidified $Cu_{50}Zr_{30}Ti_{10}Pd_{10}$ bulk glassy alloy was found to contain nanoparticles (Fig. 4.8 (a)) identified as CuZr cubic phase with a lattice parameter of 0.326 nm [32]. These particles nucleate but did not grow to the size more than 10 nm on cooling. However, dissolution of the CuZr nanoparticles was observed in the supercooled liquid (Fig 4.8 (b,c)). Dissolution of the CuZr

Materials Research Forum LLC
https://doi.org/10.21741/9781644901014

nanoparticles in a supercooled liquid of the $Cu_{50}Zr_{30}Ti_{10}Pd_{10}$ alloy before devitrification shows their instability in the temperature range of 750-800 K. In accordance with the Cu-Zr phase diagram, the CuZr phase underwents eutectoid transformation at 988 K which is above the supercooled liquid region. Thus, CuZr phase becomes thermodynamically unstable and dissolves in the supercooled liquid when atomic diffusivity is enhanced by temperature.

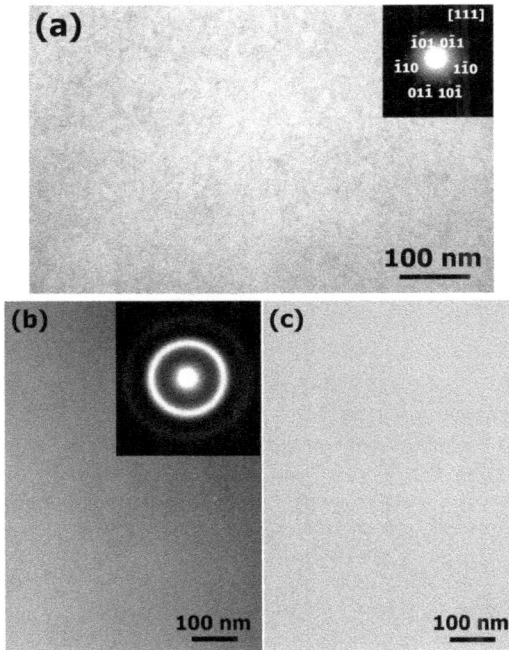

Fig. 4.8. $Cu_{50}Zr_{30}Ti_{10}Pd_{10}$, TEM. (a) bright-field image in as-solidified state. The insert – nanobeam diffraction pattern and its indexing according to CuZr phase. (b) bright-field and (c) dark-field images after heating up to T_g of 748 K and fast cooling. The insert in (b) – selected-area electron diffraction pattern. Reproduced from [32] with permission of Taylor & Francis.

4.4 Crystallization of glassy alloys.

4.4.1 General features.

Although it might be difficult to establish an intrinsic physical difference between amorphous and glassy alloys such a slightly arbitrary differentiation is useful. An alloy is called glassy if it is produced by rapid solidification of a liquid. Many of metallic glasses transform to a supercooled liquid before crystallization at high enough heating rate. The formation of a supercooled liquid has a significant influence on the devitrification process in metallic glasses [33] owing to the change of the local atomic structure in the supercooled liquid region due to higher atomic mobility compared to that in the glassy phase.

In general, glassy alloys exhibiting the supercooled liquid region on heating have higher relative density (density of the glassy alloy compared to its crystalline state) and better glass-forming ability compared to amorphous alloys. Marginal glass-formers and growth controlled BMGs possess pre-existing nuclei or even nanoparticles in the amorphous matrix. Growth of these nuclei or particles starts at a relatively low temperature without an incubation period (Figs. 2.15 (b) and 2.16).

The following phase transformations take place during crystallization of the glassy alloys and supercooled liquids: polymorphous (a product phase has the same composition as the glassy phase), primary (a product phase has a composition different from that of the glassy phase), eutectic (two or more phases nucleate and grow conjointly). Crystal growth velocities in these processes can range from very low values in case of primary crystallization (of the order of an atomic layer per second and even less) [32] up to as high as ~100 m/s in case of pure metals [34] and chalcogenide glasses [29,35]. Eutectic colonies exhibit intermediate growth rate values, for example ~100 nm/s in case of Pd-Ni-Si-P glassy alloys [36].

Below the liquidus temperature the liquid and glassy phases have a higher Gibbs free energy (G) than the competing crystalline phase, and thus have a driving force for crystallization. However, there is an energy barrier. Nucleation of a crystalline phase can be homogeneous and heterogeneous. The classical homogeneous nucleation theory states that transformation of a glass or a supercooled liquid to a crystal produces the Gibbs free energy change ($\Delta G_v < 0$ per unit volume) which for a spherical nuclei is $(4/3)\pi r^3 \Delta G_v$. However, nucleation process leads to the creation of an interface with energy ($\Delta G_s > 0$ per unit interface) area which for a spherical nuclei is $4\pi r^2 \Delta G_s$. The total Gibbs free energy change upon nucleation (ΔG) of a crystal with a radius r is:

$$\Delta G = (4/3)\pi r^3 \Delta G_v + 4\pi r^2 \Delta G_s \qquad (4.1)$$

In case of a solid-solid glass→crystal transformation there is one more term connected to the energy increase due to elastic deformation ($\Delta G_e > 0$) because the parent and daughter phases usually have different unit volume. This energy term does not have an influence on the shape of the ΔG curve but changes ΔG_c and r_c, in some cases significantly. At small r influence of $4\pi r^2 \Delta G_s$ is significant while at larger r the $(4/3)\pi r^3 \Delta G_v$ term start to dominate (Fig. 4.9). There are two critical points in the plot where the first and second derivatives are equal to zero. The first one corresponds to a critical nuclei size r_c which equals to $-2\Delta G_s/\Delta G_v$. If we insert this value in eq. (4.1) then the critical work will be $\Delta G_c = (16\pi/3)\Delta G_s^3/\Delta G_v^2$. An inflection point where the second derivative is zero corresponds to $r_i = -\Delta G_s/\Delta G_v = r_c/2$. In the case of heterogeneous nucleation the energy barrier against nucleation can be significantly reduced enhancing nucleation.

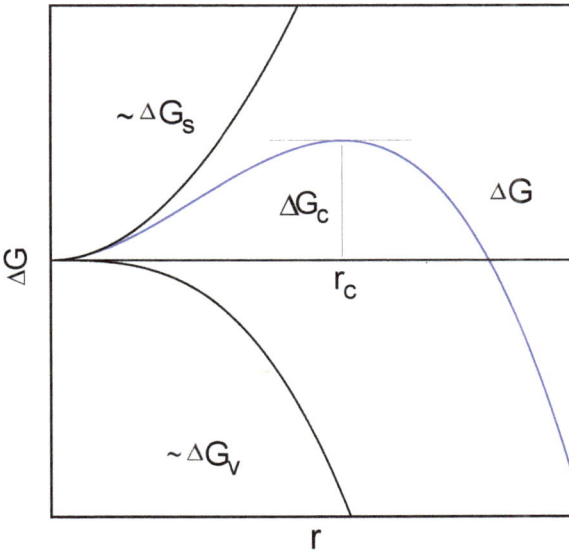

Fig. 4.9. Gibbs free energy difference between two phases as a function of nucleus radius (r) at constant temperature.

If crystallization occurs by nucleation and growth mechanism, then nucleation rate and growth rate have different temperature dependences. The number of critical-size nuclei per unit volume is proportional to $e^{-\Delta G_c/RT}$ while the frequency of atomic transfer through the interface is proportional to $e^{-Q_n/RT}$. Thus, the nucleation rate I is:

$$I = I_0 e^{-Q_n/RT} e^{-\Delta G_c/RT}, \qquad (4.2)$$

where $Q_n \sim 1/\Delta G^2$ is the activation energy for transfer of atoms across the surface of a nucleus and ΔG_c as a function of supercooling (ΔT) is the energy required to form a critical nucleus. The growth rate (u) is proportional to the frequencies of the transition from liquid/glass $v_{l\text{-}s}$ to solid and back $v_{s\text{-}l}$. $v_{l\text{-}s} = v \cdot \exp(-\Delta G/RT)$. Thus, in case of the interface controlled growth:

$$u = u_0 e^{-Q_g/RT} (1 - e^{-\Delta G/RT}), \qquad (4.3)$$

where Q_g is the activation energy for growth. The nucleation and growth rate dependencies as a function of supercooling are schematically shown in Fig. 4.10.

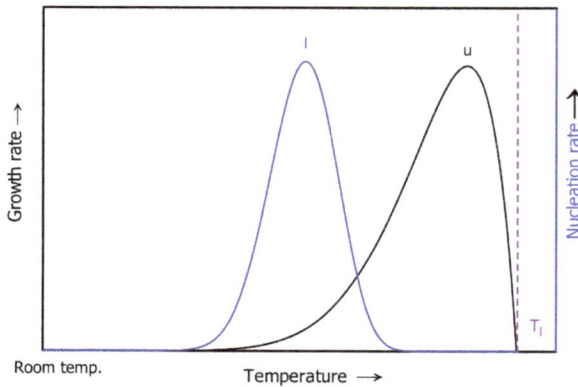

Fig. 4.10. Nucleation (I) and growth (u) rates as a function of supercooling.

As a result a schematic time-temperature-transformation (TTT) diagram (to be discussed below) can be plotted (Fig. 4.11).

Very high nucleation and growth rates are found for liquid and glassy pure metals. In molecular dynamics simulation [37] the population density of Ni precipitates forming below is the TTT diagram nose temperature is of the order of 10^{24} m^{-3}. High growth rate of 20–70 m/s corresponds is very typical for pure metals.

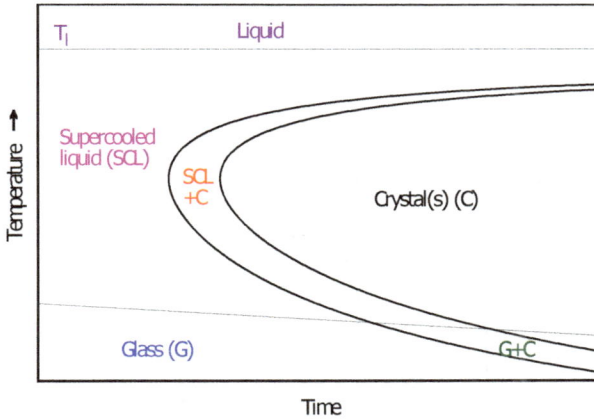

4.11. A schematic time-temperature-transformation (TTT) diagram.

Kinetics of the crystallization processes in metallic glasses was studied substantially [38,39]. If I and u are time independent and the reaction is interface-controlled then kinetics of the formation of spherical crystals in the glassy alloys can be analyzed by Kolmogorov [40] –Johnson-Mehl [41] –Avrami [42] general exponential equation for the fraction transformed x:

$$x=1-\exp(-(\pi/3)\cdot Iu^3t^n) \tag{4.4}$$

Nevertheless, this equation is also used for other reactions. Nevertheless, this equation is also used for other reactions. Depending on the Avrami exponent n different behavior can be observed (Fig. 4.12). The n values close to 2.5 correspond to nucleation and diffusion controlled growth while $n=4$ corresponds to nucleation and interface controlled growth [43].

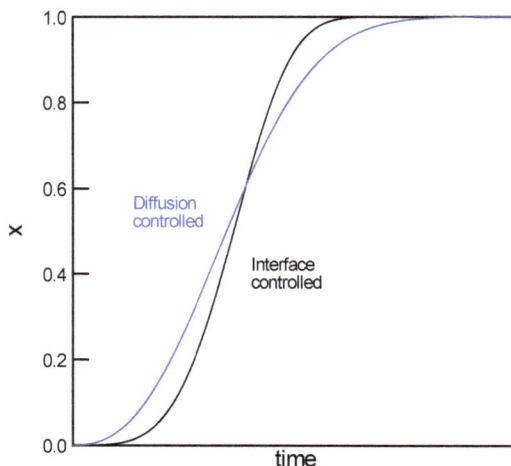

Fig. 4.12. Schematic representation of the fraction transformed (x) as a function of time in case of the interface and diffusion controlled growth. The kinetic constants (related to the nucleation and growth rates) are not equal.

For example, the $Cu_{50}Zr_{30}Ti_{10}Ni_{10}$ alloy [44] exhibits a completion of the transition to a liquid state on heating at about 740 K (Fig. 4.13 (a)). Crystallization takes place by nucleation and growth of the $(Cu,Ni)_{10}(Zr,Ti)_7$ phase. The isothermal calorimetry traces (Fig. 4.13 (b)) taken within the supercooled liquid region, glass-transition region and from the glassy phase shown were used to create the Avrami plot (Fig. 4.13 (c)) and the TTT-diagram (Fig. 4.13 (d)).

The glass-formation, crystallization kinetics and structure changes in the $Cu_{55-x}Zr_{45}Ag_x$ alloys on heating were studied and compared with that of the binary $Cu_{55}Zr_{45}$ alloy [24,45]. During the exothermic reactions the supercooled liquid crystallizes forming mainly oC68 $Cu_{10}Zr_7$ phase in the $Cu_{55}Zr_{45}$ alloy and a mixture of oC68 $(Cu,Ag)_{10}Zr_7$ and tP4 $(Ag,Cu)Zr$ solid solution phases in the Ag-bearing alloys. The crystallization started after a certain incubation period which indicates nucleation and growth transformation mechanism with a steady-state nucleation and 3-dimensional interface-controlled growth of nuclei (Fig. 4.14).

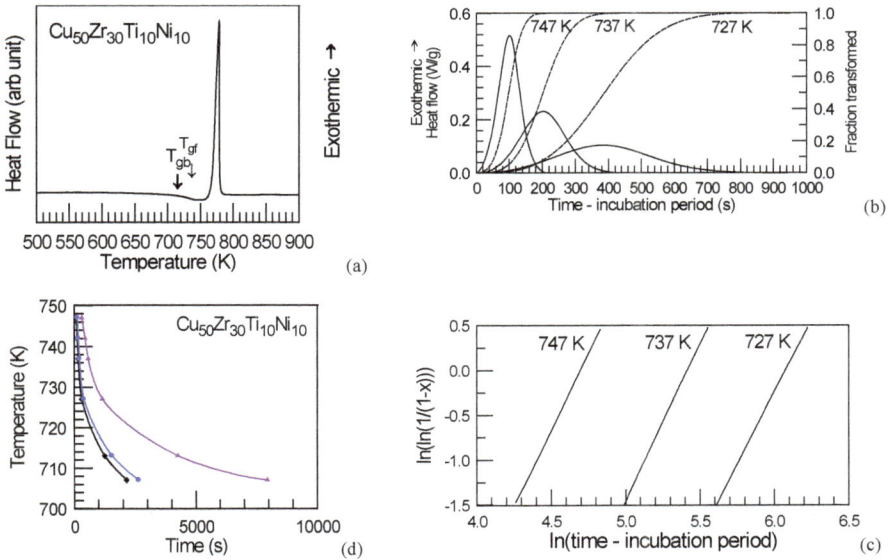

Fig. 4.13. DSC trace (a), isothermal calorimetry curves - solid lines and the fraction transformed - dotted lines (b), the Avrami plot (c) and a low-temperature part of the TTT diagram (diamond symbols and circles indicate the incubation period defined by using different methods and triangles indicate the completion of transformation). (d) of the $Cu_{50}Zr_{30}Ti_{10}Ni_{10}$ alloy. The data is partly taken from [45] with permission of Trans Tech Publications Ltd.

The XRD results indicate that the $(Cu,Ag)_{10}Zr_7$ phase is the main structure constituent in the $Cu_{45}Zr_{45}Ag_{10}$ alloy while the $Cu_{35}Zr_{45}Ag_{20}$ alloy contains a large fraction of $(Ag,Cu)Zr$ phase. A small volume fraction (~10 vol.%) of a residual globular glassy phase is observed in the structure of the $Cu_{35}Zr_{45}Ag_{20}$ alloy (Fig. 4.7b) while no phase separation was found in the case of the as-solidified sample and the sample annealed at T_g. According to EDX analysis (Table 4.1) the residual amorphous phase is depleted in Ag compared to the crystalline region. It contains only about 14 at.% Ag.

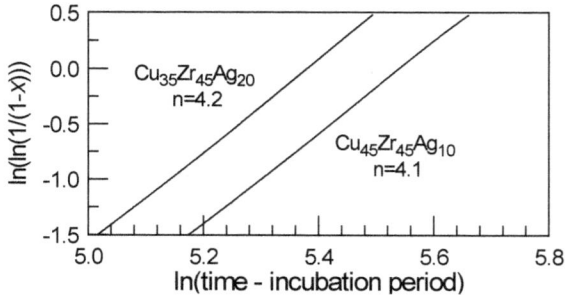

Fig. 4.14. The Avrami plots for the $Cu_{45}Zr_{45}Ag_{10}$ and $Cu_{35}Zr_{45}Ag_{20}$ glassy alloys constructed using the isothermal calorimetry. Reproduced from [29] with permission of Elsevier.

Table 4.1. Chemical composition of the thermodynamic phases observed in the $Cu_{35}Zr_{45}Ag_{20}$ alloy annealed at 722 K for 1 ks. Reproduced from [29] with permission of Elsevier.

Phases\Elements	Cu, at.%	Zr, at.%	Ag, at.%
oC68 $(Cu,Ag)_{10}Zr_7$	34	43	23
tP4 $(Ag,Cu)Zr$	20	54	26
Residual amorphous	34	52	14

Small changes in composition often lead to drastic changes in the crystallization behavior of metallic glasses. Devitrification of the $Cu_{57.5}Zr_{30}Ti_{10}Ni_{2.5}$ glassy alloy starts from formation of the intermediate nanoscale intermetallic compound. The $Cu_{55}Zr_{30}Ti_{10}Ni_5$ and $Cu_{52.5}Zr_{30}Ti_{10}Ni_{7.5}$ glassy alloys devitrify forming directly the equilibrium oC68 $Cu_{10}Zr_7$ phase. Partial replacement of Zr by Ti and Cu by Ni in $Cu_{10}Zr_7$ caused a decrease of the lattice parameters. The $(Cu,Ni)_{10}(Zr,Ti)_7$ phase is formed in the $Cu_{50}Zr_{30}Ti_{10}Ni_{10}$ glassy phase (Table 4.2, Fig. 4.15).

Table 4.2. Phase composition of the $Cu_{60-x}Zr_{30}Ti_{10}M_x$ glassy alloys after the initial, intermediate and final stages of the phase transformations corresponding to the DSC exothermic peaks. Reproduced from [45] with permission of Trans Tech Publications Ltd.

M element and its content	Primary stage	Intermediate	Fully crystallized
-	NIC* 1	NIC 2	$Cu_{10}Zr_7+Cu_3Ti_2+X$ **
2.5 at.% Ni	NIC 2	-	$Cu_{10}Zr_7+Cu_3Ti_2$
5 at.% Ni	$(Cu,Ni)_{10}(Zr,Ti)_7+NIC\ 3$	-	-***
7.5 at.% Ni	$(Cu,Ni)_{10}(Zr,Ti)_7+NIC\ 3$	-	-***
10 at.% Ni	$(Cu,Ni)_{10}(Zr,Ti)_7$	-	-***
5 at.% Co	NIC 2	NIC 2+X	$Cu_{10}Zr_7+X$

* NIC – nanoscale intermetallic compound; ** unidentified phase

*** at long enough annealing time the Cu_3Ti_2 phase starts to precipitate from $(Cu,Ni)_{10}(Zr,Ti)_7$

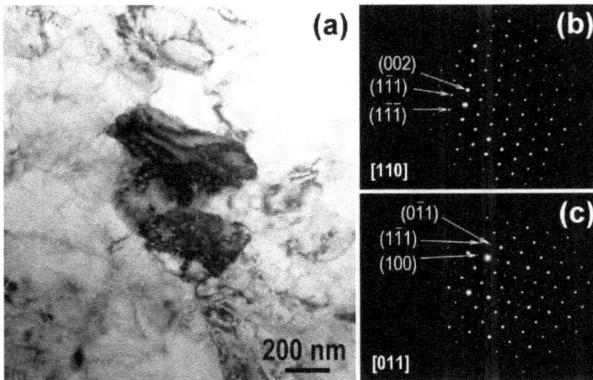

Fig. 4.15. Structure of the $Cu_{50}Zr_{30}Ti_{10}Ni_{10}$ alloy heat treated at 737 K for 0.9 ks, TEM. (a) Bright-field image. (b) and (c) SAED pattern and its indexing according to the Aba2 lattice. Zone axes are written in square brackets. Reproduced from [45] with permission of Trans Tech Publications Ltd.

The characteristic temperatures of the $Cu_{60-x}Zr_{30}Ti_{10}M_x$ glassy alloys are plotted in Fig. 4.16. Both T_x and T_g values at Ni content lower that 10 at.% change in an irregular way in consistent with the changes in the devitrification behavior observed (see Table 4.2).

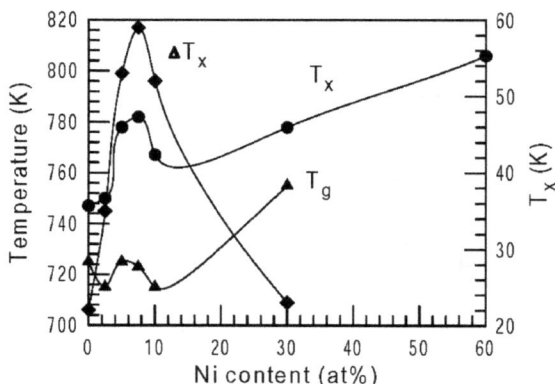

Fig. 4.16. T_g, T_x and ΔT_x of the $Cu_{60-x}Zr_{30}Ti_{10}Ni_x$ glassy alloys as a function of Ni content. Reproduced from [45] with permission of Trans Tech Publications Ltd.

Alloys containing 5, 7.5, 10, 30 and 60 at.% Ni (see Fig. 4.13 (a)) exhibit a single strong DSC exothermic peak. As one can see in Fig. 4.16 small additions of Ni up to 10 at.% extend the supercooled liquid region which attains the maximum value of 58 K at 7.5 at.% Ni. The supercooled liquid region in the glassy alloys containing more than 3.75 at.% and less than 20 at.% Ni is about two times larger than that in the $Cu_{60}Zr_{30}Ti_{10}$ glassy alloy.

Highly negative mixing enthalpy values -49 kJ/mol and -35 kJ/mol, are found for Ni-Zr and Ni-Ti atomic pairs, respectively [46]. These values are larger than those for the Cu-Zr and Cu-Ti atomic pairs of -23 and -9 kJ/mol, respectively. Negative mixing enthalpy is favorable for glass-formation and may be responsible for the widening of ΔT_x in Ni-bearing alloys compared to $Cu_{60}Zr_{30}Ti_{10}$.

Three types of the metastable nanoscale intermetallic compounds (NIC) were obtained in the studied alloys on devitrification (see Table 4.2). NIC 1 having a cubic lattice was obtained in the $Cu_{60}Zr_{30}Ti_{10}$ alloy only. Formation of the primary NIC 2 was found in the $Cu_{57.5}Zr_{30}Ti_{10}Ni_{2.5}$ and $Cu_{55}Zr_{30}Ti_{10}Co_5$ alloys. NIC 3 was found in the $Cu_{55}Zr_{30}Ti_{10}Ni_5$ and $Cu_{52.5}Zr_{30}Ti_{10}Ni_{7.5}$ alloy. The $(Cu,Ni)_{10}(Zr,Ti)_7$ compound only precipitates in the

$Cu_{60-10}Zr_{30}Ti_{10}Ni_{10-60}$ alloys (it does not contain Cu in the $Ni_{60}Zr_{30}Ti_{10}$ alloy) at the initial stage of devitrification.

The linearity of the Avrami plot of the $Cu_{50}Zr_{30}Ti_{10}Ni_{10}$ glassy alloy (Fig. 4.13(a)) indicates that this phase transformation obeys the kinetic law (4.4). The Arrhenius plot of the incubation time for devitrification as a function the testing temperature shows that the $Cu_{50}Zr_{30}Ti_{10}Ni_{10}$ glassy alloy has a high activation energy for nucleation for the $(Cu,Ni)_{10}(Zr,Ti)_7$ of 340 kJ/mol (Fig. 4.17 (a)). One should be also noted that notwithstanding on the structural differences in the as-cast state if a BMG sample is free of crystallites its crystallization kinetics is identical to that of ribbon samples of the same glassy alloy (Fig. 4.17 (b)) [47].

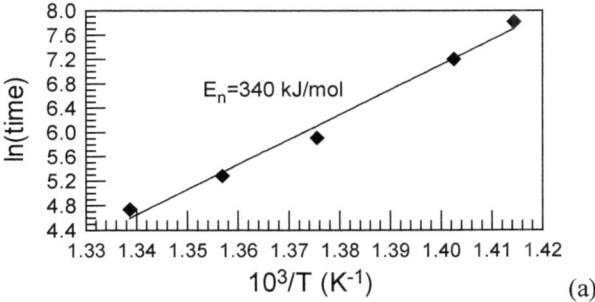

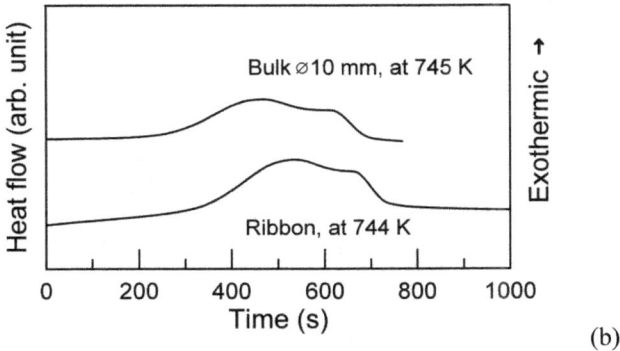

Fig. 4.17. (a) Arrhenius plot of the $Cu_{50}Zr_{30}Ti_{10}Ni_{10}$ glassy alloy. Reproduced from [45] with permission of Trans Tech Publications Ltd. (b) Differential isothermal calorimetry traces of the bulk and ribbon $Cu_{36}Zr_{48}Al_8Ag_8$ glassy alloy samples scanned at the temperatures below T_x. T_x at 40 K/min is about 790 K. Reproduced from [47] with permission of Trans Tech Publications Ltd.

A transition from classical nucleation and growth of the Zr_2Cu phase to quick crystallization of the $Cu_{10}Zr_7$ phase is found in a series of Zr-Cu-Al bulk metallic glasses [48]. The electron-beam irradiation causes primary nanocrystallization even in alloys that exhibit a eutectic transformation mechanism on heating [49,50].

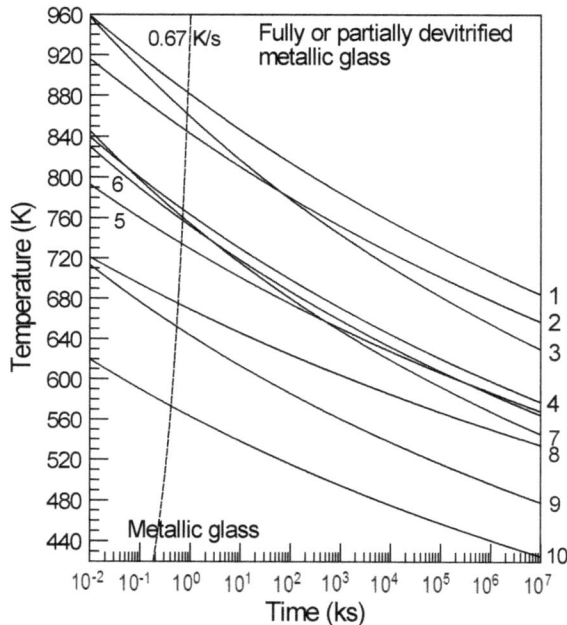

Fig. 4.18. Continuous heating transformation diagrams for different alloys numbered from 1 to 10. Corresponding alloy compositions are listed in Table 4.3. The curves correspond to T_p - peak temperature at which the transformation rate of the initial crystallization reaction attains the maximum. Reproduced from [56] with permission of Elsevier.

4.4.2 Phase transformation diagrams.

The time-temperature-transformation diagrams (Fig. 4.13 (d)) created in the isothermal mode or under continuous heating are useful for comparison of the thermal stabilities of different glasses against devitrification and for selection of the heat treatment regimes. Such diagrams have been created for various metallic glasses [51,52,53,54,55]. Comparison of the long-term thermal stabilities of different metallic glasses was performed using continuous heating transformation (CHT) diagrams (Fig. 4.18) [56]. CHT diagrams constructed by applying a corollary from the Kissinger analysis method using the DSC data at different heating rates allow estimating devitrification behavior of glassy alloys on a long-term scale. Early transition metals (ETMs): Ti-, Zr- and Hf-based metallic glasses were found to have a similar time-temperature gradient of the curves independently from their initial devitrification products (crystal or quasicrystal) and transformation mechanism. Metalloid-rich Si- and Ge-based glasses exhibit quite high long-term thermal stability and low average temperature gradient of the CHT curve. Late transition metals (LTMs): Cu- and Ni-based metallic glasses showed high thermal stability at a short time scale.

Table 4.3. Alloy numbers (Nr.), compositions of metallic glasses, initial crystallization products (IDP), crystallization mechanism (M) and the apparent activation energy (Q$_a$) for crystallization. Data from [56] with permission of Elsevier.

Nr	Composition	IDP[a]	M[b]	Q_a, kJ/mol
1	$Hf_{60}Pd_{30}Ni_{10}$	C	PR	400
2	$Hf_{65}Pd_{17.5}Ni_{10}Al_{7.5}$	QC	PR	400
3	$Ni_{66}Hf_{20}Al_{14}$	C	CX	320
4	$Ti_{45}Ni_{20}Cu_{25}Sn_5Zr_5$	C	PR	320
5	$Si_{55}Al_{20}Fe_{10}Ni_5Cr_5Zr_5$	C	PM	340
6	$Ti_{50}Ni_{25}Cu_{25}$	C	PM	300
7	$Cu_{60}Zr_{30}Ti_{10}$	C	S, PM	265
8	$Ge_{68}Cr_{14}Al_{10}Ce_4Sm_4$	C	CX	360
9	$Zr_{55}Cu_{20}Ti_{15}Ni_{10}$	QC	PR	250
10	$Ge_{60}Ni_{35}La_5$	C	PM	240

[a] C-crystal, QC-quasicrystal

[b] PR-primary, PM-polymorphous, CX-complex, i.e., more than 2 phases simultaneously including eutectic-type transformation, S-possible spinodal

CHT diagrams also can be recalculated from the isothermal ones [57] using a method close to that used for steels [58]. The recalculation method is based on the calculation and integration of the fraction of the incubation period ($F\tau$) passed on heating from room temperature at each step. The procedure is as follows: to sum all $F\tau=t_h/\tau$ values starting from 298 K with the step of 1 K where t_h is the time used to heat the specimen for 1 K at a given heating rate β. Then the total heating time t_{ht} up to T_x is $(T_x-298)/\beta$. This method was proved to provide a good correspondence between the data at least in the mid-temperature range (Fig. 4.19). The devitrification of metallic glasses is strongly temperature dependent since crystal nucleation and growth have different dependencies on temperature. However, as the incubation time representing an early stage of the reaction is mostly dominated by the nucleation process the present approach is useful.

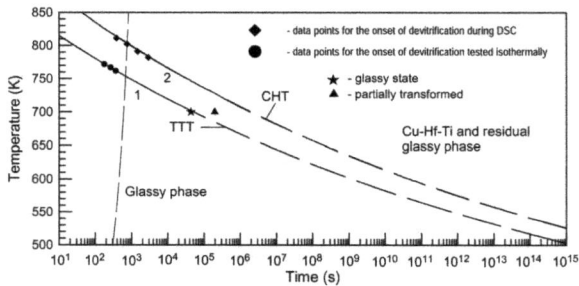

Fig. 4.19. The TTT (curve 1) and CHT (curve 2) diagrams for the primary precipitation of a metastable phase in the $Cu_{60}Hf_{25}Ti_{15}$ glassy alloy. The dashed vertical line indicates a heating rate of 0.67 K/s. Reproduced from [57] with permission of Elsevier.

Long-term room temperature stability of different types of metallic glasses and partially crystalline alloys stored for at least 15 years at ambient conditions was tested recently. Most of the alloys retained their initial structure and showed only a moderate decrease in the crystallization temperature (Table 4.4) [59]. It confirms high stability of metallic glasses at room temperature suggested from Fig. 4.18 and 4.19. However, some of Cu-Zr-system based alloys showed visible surface oxidation and partly transformed into a crystalline state forming micron-scale Cu particles in air at ambient conditions.

Table 4.4. Chemical compositions of the studied alloys in the as-prepared state (marked as initial) and after room-temperature aging for some time (marked as aged), the number of years passed after preparation (NY) and the related temperatures as well as temperature differences absolute (ΔT_g and ΔT_x), and relative (ΔT_r^x). Reproduced from [59] with permission of MDPI.

Alloy	NY	ΔT_g	ΔT_x	ΔT_r^x x 100	Initial T_g^i (K)	T_x^i (K)	Aged T_g^a (K)	T_x^a (K)
$Al_{85}Gd_8Ni_5Co_2$	19	4	3	0.52	554	575	550	572
$Al_{85}Sm_8Ni_5Co_2$	18	1	2	0.35	557	575	556	573
$Mg_{86}Mm_4Ni_{10}$	17	-	3	0.66	-	456	-	453
$Mg_{86}Y_2Mm_2Ni_{10}$	17	-	1	0.22	-	451	-	450
$Cu_{55}Zr_{30}Ti_{10}Au_5$	17	3	3	0.38	744	787	741	784
$Cu_{55}Zr_{30}Ti_{10}Pd_5$	17	-6	2	0.26	735	784	741	782
$Cu_{57.5}Ni_{2.5}Zr_{30}Ti_{10}$	18	3	1	0.13	705	749	702	748
$Ni_{55}Zr_{30}Ti_{10}Pd_5$	16	-4	0	0.00	787	811	791	811
$Ni_{55}Zr_{30}Ti_{10}Pt_5$	16	5	3	0.36	816	845	811	842
$Ni_{60}Zr_{30}V_{10}$	18	-2	4	0.47	812	852	814	848
$Ti_{50}Ni_{22}Cu_{25}Sn_3$	20	0	4	0.53	697	758	697	754
$Ti_{50}Ni_{20}Cu_{25}Sn_5$	20	6	4	0.52	702	767	696	763
$Ti_{50}Ni_{20}Cu_{20}Sn_{10}$	20	4	3	0.39	731	774	727	771
$Ti_{50}Ni_{22}Cu_{22}V_6$	19	-	3	0.41	-	733	-	730
$Zr_{65}Al_{7.5}Ni_{10}Cu_{12.5}Y_5$ *	19	-	1	0.15	-	665	-	664
$Si_{45}Al_{20}Fe_{10}Ge_{10}Ni_5Cr_5Zr_5$ **	22	-	2	0.27	-	752		750
$Al_{85}Ni_5Co_2Y_6Pd_2$ ***	19	-	0	0.00	-	523	-	523
$Cu_{60}Zr_{30}V_{10}$ ****	18	-	12	1.52	-	790	-	778
$Zr_{50}Cu_{50}$ ****	17	-	54	8	675	742	-	688

* amorphous+crystalline Zr; ** amorphous+nanoscale Ge; *** amorphous+nanoscale Al; **** initially amorphous but oxidized and partly crystallized forming micron scale Cu crystals.

4.4.3 Nanocrystallization.

Nanostructured materials can be defined as the substances having a very small grain size typically below 100 nm [60]. Invention of the age-hardenable Al-based alloys, with nanoscale precipitates was the first well known example of use of nanoparticles for practical purpose in physical metallurgy. A widely used method for the formation of nanomaterials is crystallization of the glassy alloys. Metallic glassy alloys crystallize at large supercooling when growth rate is relatively low while nucleation rate is usually high which with a small critical nucleus size lead to the formation of a nanostructure in various alloys. Some of the composites possess enhanced mechanical properties compared to single-phase glassy alloys.

Although there are some reports that, in particular cases, nanocrystalline structure can be obtained after eutectic [61] and polymorphous [62,63] crystallization of glassy alloys, the most common mechanism leading to formation of a nanostructure is primary crystallization. Nanostructured alloys are readily obtained on primary crystallization of glasses with a long-range diffusion controlled growth [64,65]. For example, the primary transformation reaction in the Al-rich amorphous alloys with high Al concentrations above 85 at.%, in many alloys, is related to the precipitation of a primary FCC-Al phase. Nanocrystallization was also observed in Fe-based glasses [66,67] and many others.

Different Al-RE-TM glasses show primary precipitation of the Al solid solution nanoparticles on heating [68] with extremely high nucleation rate exceeding 10^{20} m^{-3}s^{-1} [69,70]. The formation of Al at the primary stage is quite typical for Al-based amorphous alloys and glass-formers, for example: Al-Y-Ni-Co [71], Al-Fe-Y, Al-Sm [72], Al-Ni-Nd [73] Al-Fe-Nd [74] and so on. After reaching a certain size (typically few tens of nanometers) Al particles change their morphology from spherical to dendritic [75]. Pressure effects on crystallization metallic glasses were also studied [76]. The neutron diffraction study of the Al-Y-Ni alloys indicated that Y-rich glasses are homogeneous while Ni-rich glasses have compositional inhomogeneity [77].

FCC Al lattice parameter measurements and atom probe ion field microscopy investigation [78] showed very low concentration of the alloying elements in nanocrystalline Al in accordance with the phase diagrams of Al-RE and Al-TM systems. Segregation of the RE metal having low trace diffusivity in Al on the Al/amorphous phase interface is considered to be one of the most important reasons for the low growth rate of Al. By analogy with these results the extended X-ray absorption fine structure analysis of grain boundaries in the nanocrystalline $Fe_{85}Zr_7B_6Cu_2$ alloys showed very low Fe content in the grain boundaries between the BCC Fe nano grains. The primary crystallization requires a long-range diffusion that impedes crystal growth [79].

Another type of phase transformation in an amorphous solid leading to formation of a nanostructure is phase separation. For instance, it has been reported that in some Al–TM–RE glasses devitrification leading to formation of a nanostructure appears to be preceded by the amorphous phase separation [80,81].

In some alloys nanocrystallization only insignificantly affects flow ability of the supercooled liquid. The $Cu_{55}Zr_{30}Ti_{10}Co_5$ glassy sample exhibited a low viscosity at the temperatures above T_g till the end of the crystallization process owing to the equiaxed morphology of the nanoscale crystalline phase (Fig. 4.20). This alloy can be forged in the semi-solid state at low load and high speed, thus producing in-situ glassy/crystal composites [82].

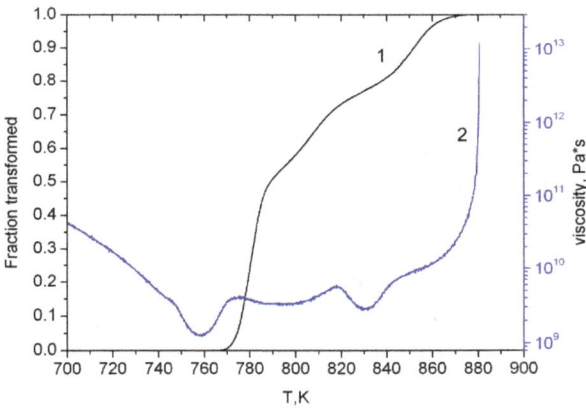

Fig. 4.20. Volume fraction of crystalline phases (1) as a function of temperature calculated from the portion of heat release at each phase transformation divided by the total heat release and the viscosity of the $Cu_{55}Zr_{30}Ti_{10}Co_5$ samples (2). Reproduced from [82] with permission of Elsevier.

On heating the $Fe_{48}Cr_{15}Mo_{14}C_{15}B_6Y_2$ [83] and $Fe_{48}Cr_{15}Mo_{14}C_{15}B_6Tm_2$ [84] glassy alloys formed nanoscale particles of the χ-$Fe_{36}Cr_{12}Mo_{10}$ phase. Its structure is cI58 (α-Mn type) and it has a nominal chemical composition of $Fe_{62}Cr_{21}Mo_{17}$. Larger particles of this phase were also found together with $(Fe,Cr)_{23}(C,B)_6$ carbide in a $Fe_{50}Cr_{15}Mo_{14}C_{15}B_6$ alloy after partial crystallization [85]. Figure 4.21 shows X-ray diffraction patterns of the $Fe_{48}Cr_{15}Mo_{14}C_{15}B_6Tm_2$ alloy in the as-solidified and annealed state. One can see the formation of a very broad (330) peak related to the crystalline χ-$Fe_{36}Cr_{12}Mo_{10}$ phase after annealing for 420 s at 893 K. Precipitation of the nanoscale χ-$Fe_{36}Cr_{12}Mo_{10}$ phase was also observed in the $Fe_{48}Cr_{15}Mo_{14}C_{15}B_6Tm_2$ alloy by TEM at an early crystallization stage [86].

Rather surprisingly this phase was also reported to form on crystallization in Cr- and Mo-free glassy alloys [87 , 88] though its size was significantly larger. After annealing $Fe_{48}Cr_{15}Mo_{14}C_{15}B_6Y_2$ at 973 K the (330) peak shows separation into two, owing to the formation of $Fe_{23}(B,C)_6$ phase (cF116 $Cr_{23}C_6$ type) and Fe_3Mo_3C (cF112 Fe_3W_3C type) likely by a quasi-peritectic reaction [83] (residual liquid$+\chi$-$Fe_{36}Cr_{12}Mo_{10}$ $\rightarrow$ $Fe_{23}(B,C)_6$ + Fe_3Mo_3C).

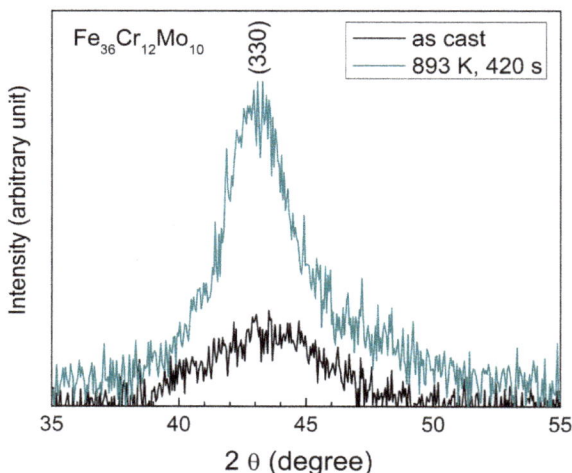

Fig. 4.21. X-ray diffraction patterns of the $Fe_{48}Cr_{15}Mo_{14}C_{15}B_6Tm_2$ alloy in as-solidified and annealed state.

The crystal size can be estimated from the main (330) X-ray diffraction peak of the primary nanocrystalline χ-$Fe_{36}Cr_{12}Mo_{10}$ phase of about 3 degrees of full width at half maximum (FWHM). According to the standard procedure after elimination of the instrumental broadening [89] and taking into account the effect of dispersed crystal blocks scattering only, direct application of the Scherrer formula suggests a coherent scattering crystallite size of about 3 nm diameter. Separating the broadening effects caused by small particle size and by inhomogeneous strain [90], and noting the crystal diameter of 10 nm provided by TEM observation, results in an estimate of the internal strain of 0.021. This is a very large value comparable to those obtained after intensive plastic deformation, for example ball milling [91]. This large inhomogeneous strain is likely to be related to the observed low growth rate of the particles compared to the relatively large particles of χ- phase found in the alloys containing no Cr, Mo and RE metals.

The large difference in the mass density between the metallic glassy and χ-$Fe_{36}Cr_{12}Mo_{10}$ phase, corresponding to large differences in the specific volume (about 4.2% and 6.5% for $Fe_{48}Cr_{15}Mo_{14}C_{15}B_6Tm_2$ and $Fe_{48}Cr_{15}Mo_{14}C_{15}B_6Y_2$ alloys, respectively), is likely to give a large hydrostatic tensile stress in the χ-$Fe_{36}Cr_{12}Mo_{10}$ phase. Such a stress may limit drastically the growth rate of this phase. On the other hand, these stresses in the crystals are not uniform. There is about one Y atom per unit cell of χ-$Fe_{36}Cr_{12}Mo_{10}$ phase of the $Fe_{48}Cr_{15}Mo_{14}C_{15}B_6Y_2$ alloy. Taking into account the Avogadro number and volume of a spherical nanoparticle having 10 nm in diameter such a particle should contain $4.7 \cdot 10^4$ atoms total and about 960 of Y atoms. The number of Y atoms in a single nanoparticle is considered to be large enough to ensure its low growth rate.

Schematic transformation diagrams for the $Fe_{48}Cr_{15}Mo_{14}C_{15}B_6RE_2$ (Fe-RE2) and $Fe_{50}Cr_{15}Mo_{14}C_{15}B_6$ (Fe-RE0) alloys are shown in Fig. 4.22. The cooling and heating regimes correspond to: 1) ribbon samples with the glassy phase and pre-existing nuclei of χ-phase in the $Fe_{48}Cr_{15}Mo_{14}C_{15}B_6RE_2$ alloys; 2) bulk samples of 2.7 mm thickness (glassy in the case of $Fe_{48}Cr_{15}Mo_{14}C_{15}B_6RE_2$ and crystalline in the case of $Fe_{50}Cr_{15}Mo_{14}C_{15}B_6$); 3) bulk samples of 3.5 mm thickness (formation of nanoscale χ-phase in $Fe_{48}Cr_{15}Mo_{14}C_{15}B_6Y_2$); and 4) bulk samples of 4.2 mm thickness leading to the formation of crystalline phases in $Fe_{48}Cr_{15}Mo_{14}C_{15}B_6Y_2$. The heating paths 5 and 6 correspond to the DSC and isothermal calorimetry studies, respectively.

In contrast to the $Fe_{50}Cr_{15}Mo_{14}C_{15}B_6$ glass in which an incubation period is observed, crystallization of the $Fe_{48}Cr_{15}Mo_{14}C_{15}B_6RE_2$ (RE=Y or Tm) alloys starts immediately after heating. This lack of an incubation period was found to apply for annealing of both RE-bearing alloys below T_x in both bulk and ribbon states. For example, differential isothermal calorimetry of the $Fe_{48}Cr_{15}Mo_{14}C_{15}B_6Y_2$ alloy was carried out at 882 and 898 K and precipitation of the χ-$Fe_{36}Cr_{12}Mo_{10}$ phase was observed. Similar results were obtained for $Fe_{48}Cr_{15}Mo_{14}C_{15}B_6Tm_2$ while $Fe_{50}Cr_{15}Mo_{14}C_{15}B_6$ exhibited a clear nucleation and growth reaction with an incubation period and no χ-$Fe_{36}Cr_{12}Mo_{10}$ phase.

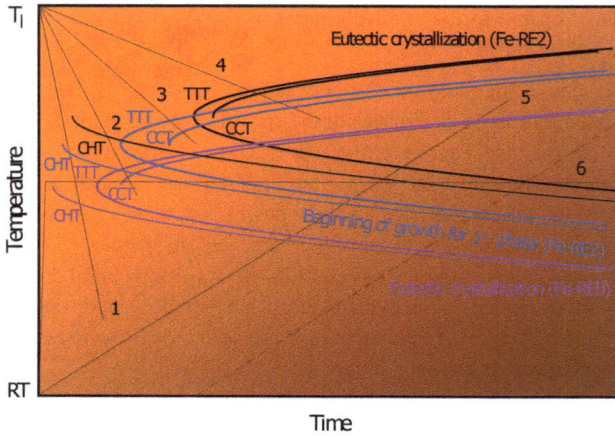

*Fig. 4.22. Schematic time-temperature-transformation diagram for the
$Fe_{48}Cr_{15}Mo_{14}C_{15}B_6RE_2$ (Fe-RE2) and $Fe_{50}Cr_{15}Mo_{14}C_{15}B_6$ (Fe-RE0) alloys. T_l and RT
indicate liquidus and room temperature respectively. The numbered dashed lines indicate
the cooling-heating pathways. TTT, CCT and CHT indicate
time-temperature-transformation (isothermal), continuous cooling transformation and
continuous heating transformation diagrams, respectively, marked according to the colors
used. Reproduced from [83] with permission of Elsevier.*

Thus, the $Fe_{50}Cr_{15}Mo_{14}C_{15}B_6$ glass is a typical BMG which GFA is limited by the rate of
crystal nucleation, while GFA of the $Fe_{48}Cr_{15}Mo_{14}C_{15}B_6RE_2$ alloys is limited by the rate of
crystal growth. The diffusion coefficients of Mo (D_{Mo}) at 873 and 903 K of $2.0 \cdot 10^{-20}$ and
$1.0 \cdot 10^{-19}$ m^2/s, respectively, are very close to those for self diffusion of Fe (D_{Fe}) of $1.9 \cdot 10^{-20}$
and $1.1 \cdot 10^{-19}$ m^2/s in crystalline alloys [83,84]. Chromium also has a similar D value in this
temperature region. On the other hand, Y (and likely other RE metals) has much lower D
value of about $4.2 \cdot 10^{-23}$ m^2/s at 900 K which corresponds to a diffusion length (L) (defined as
$2(D \cdot t)^{1/2}$ where t is the diffusion time) of 0.4 nm in 900 s. The self-diffusion coefficient of Fe
in the $Fe_{91}Zr_9$ glassy alloy at 900 K is about $2 \cdot 10^{-15}$ m^2/s [92] which is 4 orders in magnitude
higher than that in α-Fe. In order to be able to diffuse 10 nm in 900 s, yttrium should have a
diffusion coefficient of about 10^{-20} m^2/s which is a reasonable value compared to that in α-Fe.
Slow diffusion of Y explains its homogeneous distribution in the partially transformed
$Fe_{48}Cr_{15}Mo_{14}C_{15}B_6Y_2$ sample. Thus, the $Fe_{48}Cr_{15}Mo_{14}C_{15}B_6RE_2$ glassy alloys exhibit growth
of pre-existing nuclei of the χ-$Fe_{36}Cr_{12}Mo_{10}$ phase, though these nuclei are too small to be
clearly recognized even by high-resolution TEM. The χ-$Fe_{36}Cr_{12}Mo_{10}$ phase has the

structural type of α-Mn in which local atomic structure is found to be close to that of the glassy phase and describes well the first maximum of the radial distribution function. Y and Tm additions prevent the relatively fast eutectic crystallization observed in the $Fe_{48}Cr_{15}Mo_{14}C_{15}B_6$ alloy.

The structure of Fe-based soft magnetic alloys like: Finemet $Fe_{73.5}Cu_1Nb_3Si_{13.5}B_9$ [93] and Nanoperm $Fe_{84}Zr_{3.5}Nb_{3.5}B_8Cu_1$ [94] after annealing consists of BCC Fe nanocrystals below 20 nm in size finely dispersed in the amorphous matrix. Cu, Nb or Zr elements despite on their low contents are responsible for the formation of a nanostructure upon annealing. For example, devitrification of the $Fe_{73.5}Cu_1Nb_3Si_{13.5}B_9$ alloy, which initially has an isotropic amorphous structure, starts from the formation of Cu-enriched zones [95].

Atom probe field ion microscopy and high resolution transmission electron microscopy [96] indicated that Cu forms nano clusters in the $Fe_{73.5}Si_{13.5}B_9Nb_3Cu_1$ amorphous matrix which act as the sites for heterogeneous nucleation of the BCC Fe particles on devitrification [97]. X-ray absorption fine structure studies also showed that Cu clusters with near-FCC structure were present from very early stages of the crystallization process [98]. The density of the clusters estimated by three dimensional atom probe is in the order of 10^{24} m^{-3} at the average cluster size of about 2-3 nm [99].

Not only pure metals and their solid solutions but some intermetallic compound particles can also have a nanoscale size. For example, crystallization of the nanocrystal-forming Ti-based alloys, for example, the $Ti_{50}Ni_{20}Cu_{23}Sn_7$ alloy begins from the primary precipitation of a nanoscale equiaxed, almost spherical particles of cF96 Ti_2Ni solid solution phase (other alloying elements are partially dissolved in this phase) with a lattice parameter of 1.138 nm [100,101]. Formation of such a nanoscale cF96 phase having a large cubic unit cell was also observed in the Zr- and Hf-based alloys [102 , 103]. cF96 Ti_2Ni phase exhibits a diffusion-controlled growth.

An extremely small size and low growth rate of cF96 crystals were observed in the $Hf_{55}Co_{25}Al_{20}$ glassy alloy [104]. This alloy undergoes a double-stage devitrification forming the primary cubic cF96 Hf_2Co (Fig. 4.23) phase from the supercooled liquid by the steady state nucleation and diffusion-controlled growth of nuclei followed by the subsequent devitrification of the residual glassy matrix forming Hf_2Al and an unidentified phase. The cF96 nanoscale particles having a very small size are hardly distinguishable from the glassy matrix (Fig. 4.24) likely owing to local structural similarity between cF96 Hf_2Co phase and the glassy matrix. They also have a very low average growth rate of the order of 10^{-11} m/s.

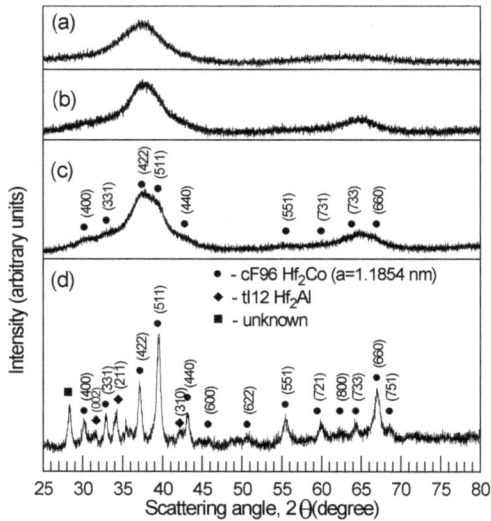

Fig. 4.23. XRD patterns of the $Hf_{60}Co_{25}Al_{20}$ glassy alloy (a) in as-solidified state and annealed (b) at 907 K for 0.9 ks, (c) at 907 K for 2.7 ks and using DSC at 0.67 K/s up to 1100 K. 907 K corresponds to the first exothermic peak temperature. Reproduced from [104] with permission of Elsevier.

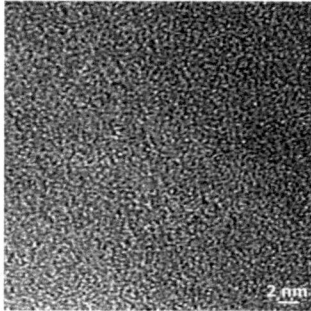

Fig. 4.24. HRTEM image of the $Hf_{60}Co_{25}Al_{20}$ glassy alloy annealed at 907 K for 0.9 ks.

In many cases redistribution of the alloying elements on a short scale precedes crystallization like it was observed in the Mg-Ni-Mm and Mg-Ni-Y-Mm glassy alloys [105]. Local structure of the $Mg_{86}Ni_{10}Y_2Mm_2$ and $Mg_{82}Ni_{14}Y_2Mm_2$ metallic glasses containing MRO zones changes prior to formation of the crystalline phases. Splitting of the amorphous halo peak occurs after heating the first exothermic peak and indicates redistribution of the alloying elements in the amorphous matrix and formation of the Mg-enriched zones (Fig. 4.25). This process occurs without an incubation period and leads to embrittlement of the samples. At the same time, $Mg_{86}Ni_{10}Mm_4$ exhibits formation of the primary HCP Mg crystals.

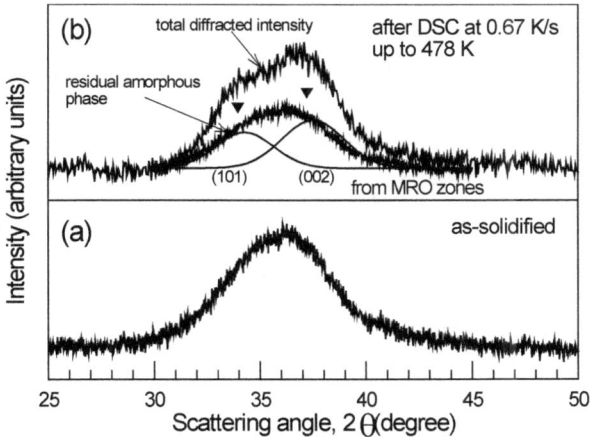

Fig. 4.25. X-ray diffraction patterns of the $Mg_{86}Ni_{10}Y_2Mm_2$ metallic glass. (a) as-solidified and (b) after DSC at 0.67 K/s up to 478 K. Triangles indicate positions of the (002) and (101) peaks of HCP Mg. Reproduced from [105] with permission of Taylor & Francis.

The amorphous Ge-Cr-Al-RE alloys also exhibit local order changes prior to nanoscale devitrification. For example, some structure changes occur in the $Ge_{70}Cr_{16}Al_{10}Nd_4$ alloy prior to devitrification. The position of the first peak of the radial distribution function (RDF) shifts to lower value after heating, and becomes closer to the values of pure amorphous and crystalline Ge which have tetrahedral configuration in the nearest neighbor region. The pronounced peak also appears at 0.41 nm in RDF for the heat treated sample. The ratio between the second and the first nearest neighbor distances (0.41 and 0.250 nm, respectively) is 1.64 which is almost identical with the value of 1.63 for a-Ge and c-Ge [106]. The results indicate that during heat treatment tetrahedral configuration of Ge atoms was formed. As a result primary Ge nanoparticles are also formed in many Ge-Al-Cr-RE alloys on further heating.

The as-solidified microstructure of the rapidly solidified Ge-Al-Cr-RE alloys examined by TEM showed the presence of finely dispersed zones of about 1 nm in size [107] homogeneously distributed in the amorphous matrix [108]. These zones are responsible for the split of the first diffraction maximum in the XRD patterns in Fig. 4.26. Such a structure is typical for different rapidly solidified Ge-Al-Cr-RE and $Ge_{60}Ni_{35}La_5$ amorphous alloys [109]. Upon annealing nanoscale particles of cF8 Ge are formed (Fig. 4.26 and 4.27).

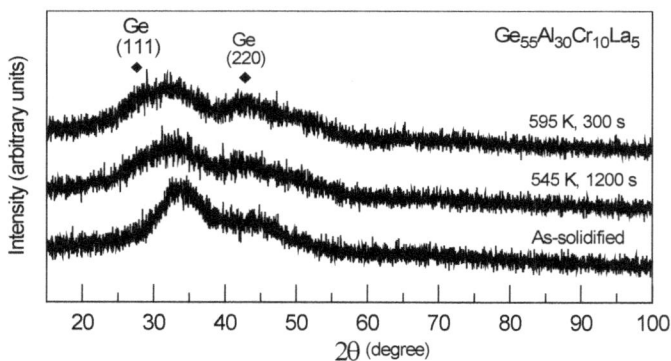

Fig. 4.26. XRD pattern of the ribbon sample of the $Ge_{55}Al_{30}Cr_{10}La_5$ alloy in as-solidified and annealed state as indicated (Cu Kα radiation). Reproduced from [107] with permission of Elsevier.

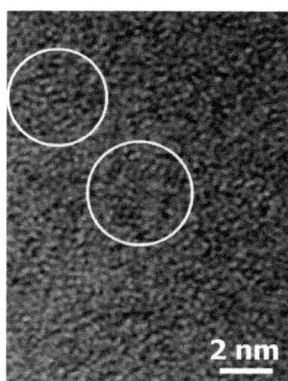

Fig. 4.27. High resolution TEM image of the $Ge_{55}Al_{30}Cr_{10}La_5$ alloy in the annealed at 545 K for 20 min. Ge clusters are encircled.

4.4.4 Difference in crystallization of metallic glasses above and below T_g

On heating glassy and amorphous $Al_{85}RE_8Ni_5Co_2$ alloys, (RE denotes rare-earth metal) namely, $Al_{85}Y_8Ni_5Co_2$, $Al_{85}Y_{8-x}Nd_xNi_5Co_2$ and $Al_{85}Gd_8Ni_5Co_2$ and others exhibit different crystallization behavior above and below T_g (Fig. 4.28 and Table. 4.5) [110]. The $Al_{85}RE_8Ni_5Co_2$ glassy alloys showed precipitation of the FCC Al nanoparticles after continuous heating using DSC at high enough heating rates (0.67 K/s and higher) or isothermal annealing at the temperature above T_g [111]. At the same time, Y-, Gd- and Dy-bearing metallic glasses [112] as well as the $Al_{85}Y_4Nd_4Ni_5Co_2$ [113] showed simultaneous formation of the intermetallic compound(s) and Al nanoparticles, or primary formation of the intermetallic compound after annealing below T_g. For example, the $Al_{85}Y_8Ni_5Co_2$ alloy shows formation of an intermetallic compound conjointly with Al nanoparticles after isothermal annealing up to the completion of the primary phase transformation below T_g. After the completion of the primary phase transformation a lot of relatively small intermetallic compound particles below 50 nm size exist in the structure of the $Al_{85}Y_8Ni_5Co_2$ alloy [114]. The intermetallic compounds are metastable and have a multicomponent composition. The volume fraction of the intermetallic compound is higher than that of Al and the fraction of Al depends on the annealing temperature below T_g.

Table 4.5. Isothermal crystallization of the $Al_{85}Y_4Nd_4Ni_5Co_2$ alloy in the vicinity of glass-transition region measured at 1.67 K/s is 561 K.

Transformation	Temperature, K	Crystalline Phases	n**
Primary, diffusion-controlled growth	568	nano-Al	2.6
	563	nano-Al+traces IMC*	3.1
Eutectic-like, interface-controlled growth	543	nano-Al+IMC	4
	538	nano-Al+IMC	4
	533	nano-Al+IMC	4

*IMC Intermetallic compound; ** the Avrami exponent

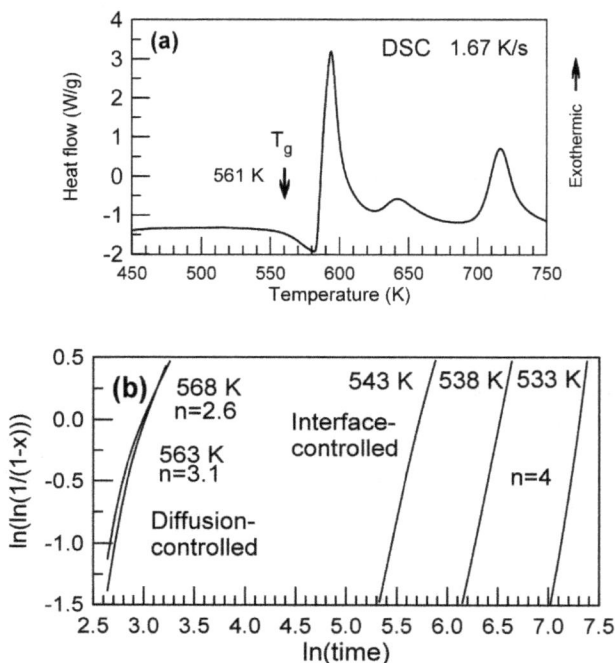

Fig. 4.28. (a) DSC curve of the $Al_{85}Ni_5Y_4Nd_4Co_2$ alloy obtained at a heating rate of 1.67 K/s, (b) Avrami plot for the isothermal calorimetry data at different temperatures above and below T_g of about 561 K. Reproduced from [110] with permission of Elsevier.

Below T_g most of the Al-based glassy alloys showed a single-stage transformation obeying the kinetic law (eq. 4.4) for the fraction transformed (x) as a function of time (t) with the Avrami exponent (n) close to 4 while n values on the primary crystallization of nanoscale Al in the $Al_{85}Dy_8Ni_5Co_2$, $Al_{85}Y_8Ni_5Co_2$, $Al_{85}Y_4Nd_4Ni_5Co_2$ and other metallic glasses above T_g varies not only with temperature but with time as well exhibiting non-steady state nucleation (Fig. 4.28 (b)). Al nanocrystals only are formed on rapid enough heating and annealing above T_g. The transition from the interface-controlled conjoint growth of Al and intermetallic compound (IM) to the diffusion-controlled growth of Al takes place within the glass-transition region. This behavior was also observed in the $Al_{85}Ni_7Gd_8$ alloy [115].

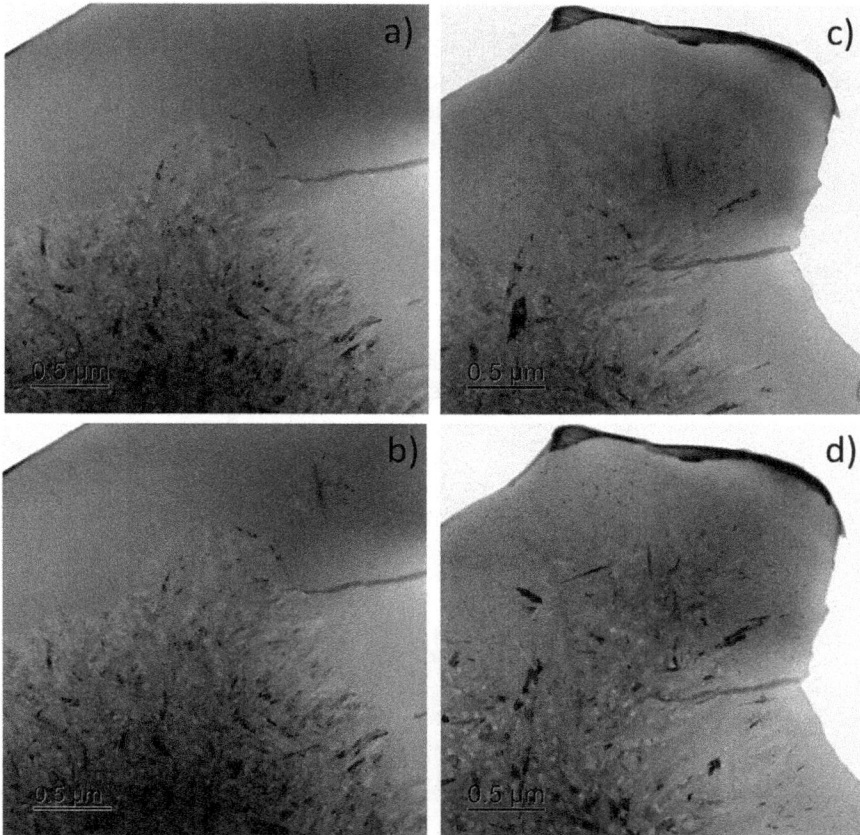

Fig. 4.29. Bright-field TEM images indicating growth of a single colony towards the residual liquid phase observed in $Al_{85}Y_8Ni_5Co_2$ at: (a) 603 K, (b) 608 K, (c) 613 K and (d) 633 K. Reproduced from [116] with permission of Elsevier.

The detailed study of the eutectic-like reaction observed below T_g in $Al_{85}Y_8Ni_5Co_2$ showed that it is not exactly an eutectic one but consists of two stages: precipitation of a primary intermetallic compound immediately followed by crystallization of nanoscale Al around it (Fig. 4.29) [116]. In parallel with growth of the existing colonies triggered by an intermetallic compound, the formation of new needle-like intermetallic compound particles takes place in

Metallic Glasses and Their Composites – 2nd Edition
Materials Research Foundations **85** (2021)

Materials Research Forum LLC
https://doi.org/10.21741/9781644901014

the glassy matrix. It initiates the formation and growth of new eutectic colonies as shown in Fig. 4.29 (a,b) (see a needle-like precipitate formed in the upper right corner). As it is schematically shown in (Fig. 4.30) these colonies are abnormal, namely, there is often no boundary between the aluminum nanoparticles and the intermetallic compound which are separated by the glassy matrix. On the other hand, this transformation takes place statistically as a single-step one with a nearly constant Avrami exponent. At a relatively high heating rate when temperature is raised quickly above T_g the crystallization process changes to the primary crystallization of a solid solution of Al followed by the formation of intermetallic compounds by subsequent reactions separated by time or temperature.

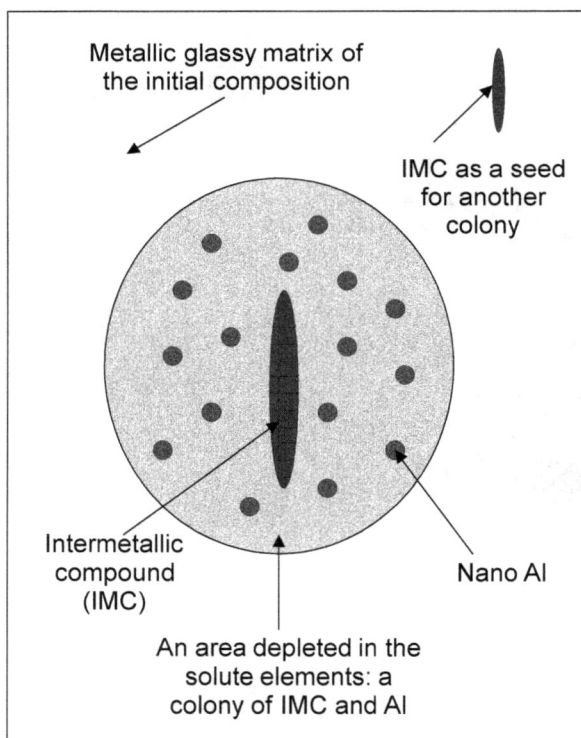

Fig. 4.30. A scheme of a growing colony in the $Al_{85}Y_8Ni_5Co_2$ glassy alloy.

The above-mentioned may indicate that the devitrification products in such alloys inherit the local structure (short-range order) of the glassy (amorphous) solid phase while the glasses devitrifying from the supercooled liquid state may inherit a short range order of the liquid. The origin of such behavior may also be connected with the difference in the structure, physical and thermal properties between the glassy phase and the supercooled liquid.

The crystallization of various Al-based glassy and amorphous alloys can be described by one of the three following types of mechanisms:

- If an alloy does not show T_g on heating prior to devitrification/crystallization and exhibits nucleation and growth transformation mechanism then it forms an intermetallic compound(s) (IM) or IM+nanoscale Al from the amorphous matrix.

- If an alloy does not show glass-transition on heating prior to devitrification and has pre-existing nuclei formed on cooling at high temperature then it forms nanoscale primary Al.

- If an alloy shows glass-transition on heating and exhibits the nucleation and growth transformation mechanism then it forms nanoscale Al above T_g and IM+Al or only IM below T_g.

These crystallization mechanisms can be associated with a new topological empirical criterion which was proposed for the design of multicomponent Al-rich amorphous alloys [117]. Such a criterion, is similar to the topological λ_t instability criterion [118], as the ratio of solute atoms radius (r_i) of concentration C_i to Al atomic radius (r_{Al}) defined by equation:

$$\lambda_t = \sum_{i=B}^{Z} C_i \cdot \left| \frac{r_i^3}{r_{Al}^3} - 1 \right|$$

$$(4.5)$$

The λ_t parameter can successfully predict the thermal behavior of the Al-based alloys. It indicates compositions which exhibit the supercooled liquid region ($\lambda_t > 0.1$), nanocrystalline behavior ($\lambda_t < 0.1$) or alloys with an intermediate, nano-glassy behavior, where nanocrystallization is preceded by the supercooled liquid region ($\lambda_t \approx 0.1$). For example, the $Al_{84}Y_6Ni_4Co_2Sc_4$ alloy has an empirical topological criterion $\lambda_t = 0.1$ while the $Al_{85}Y_6Ni_5Co_2Zr_2$ one has $\lambda_t = 0.091$ close to 0.1.

$Pd_{43}Ni_{10}Cu_{27}P_{20}$ [119] and $Zr_{41.2}Ti_{13.8}Cu_{12.5}Ni_{10}Be_{22.3}$ [120] BMG alloys also show some differences in the crystallization behavior above and below the TTT nose temperature and a transition from nucleation control to growth control has been suggested.

CHT diagrams of the amorphous alloys (Fig. 4.31) [121] which has no supercooled liquid prior to crystallization like the $Al_{84}Nd_8Ni_4Co_2$ one correspond well to the TTT diagrams while less correlation is found in the case of two other glassy alloys which transform to a supercooled liquid and have different crystallization behavior below and above T_g. This fact again is in line with the difference in their crystallization behavior above and below T_g.

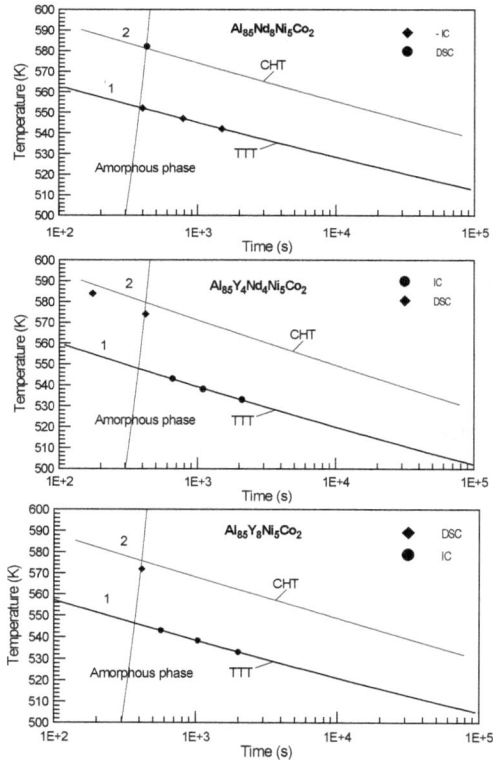

Fig. 4.31. The TTT (curve 1) and CHT (curve 2) diagrams for the initial phase transformations in the glassy alloys. The dashed vertical line indicates a heating rate of 0.67 K/s. Reproduced from [121] with permission of the Society of Glass Technology.

Heat treatment of the $Al_{85}RE_8Ni_5Co_2$ glassy alloys (exhibiting glass-transition prior to crystallization) below T_g leaded to formation of at least two phases one of which is Al, another – metastable intermetallic compound while these glassy alloys after continuous heating using DSC at high enough heating rate of 0.67 K/s, and higher, or annealing at the temperature above T_g showed precipitation of the Al nanoparticles only (Table 4.5). An exceptional case was in the case of the $Al_{85}Y_4Sm_4Ni_5Co_2$ alloy which showed formation of the only primary Al nanoparticles even at isothermal annealing below T_g. On the other hand amorphous alloys which crystallize directly and do not exhibit glass-transition on heating did not show any temperature dependence of their crystallization products. The alloys exhibiting no supercooled liquid region, for example La-, Mm- (Mischmetal) and Nd- bearing $Al_{85}RE_8Ni_5Co_2$ alloys ($Al_{85}La_8Ni_5Co_2$, $Al_{85}Nd_8Ni_5Co_2$, and $Al_{85}Mm_8Ni_5Co_2$), being annealed isothermally at various temperatures showed formation of the same crystallization products per alloy. It indicates an important role of the supercooled liquid in crystallization of the Al-based glasses.

Crystallization of a metallic glass forming a mixture of IM phase and Al looks similar to an irregular eutectic type. However, Al and IM phase in many cases do not have a common interface and likely nucleate separately from the glassy phase. Below T_g most of the glassy alloys showed a single-stage transformation obeying the kinetic law (4.4) with the Avrami exponent close to 4 while n found for the primary phase transformation of the $Al_{85}Dy_8Ni_5Co_2$, $Al_{85}Y_8Ni_5Co_2$, $Al_{85}Y_4Nd_4Ni_5Co_2$ (Table 4.6) and other metallic glasses above T_g is lower and varies not only with temperature but with time as well.

Table 4.6. Compositions, glass transition (T_g) and crystallization temperature (T_x) of the $Al_{85}RE_8Ni_5Co_2$ glassy alloys crystallizing by nucleation and growth mechanism during continuous heating in a DSC cell at high enough heating rate of 0.67 K/s.

Composition	T_g, K	T_x, K	Initial crystallization
$Al_{85}Y_8Ni_5Co_2$	544	573	Nanoscale Al
$Al_{85}La_8Ni_5Co_2$	-	582	$Al_4LaCo+X^*$
$Al_{85}Sm_8Ni_5Co_2$	559	574	Nanoscale Al
$Al_{85}Dy_8Ni_5Co_2$	549	574	Nanoscale Al
$Al_{85}Nd_8Ni_5Co_2$	-	578	X
$Al_{85}Mm_8Ni_5Co_2$	-	584	$Al_{11}La_3+Al_{11}Ce_3+X$
$Al_{85}Y_4Sm_4Ni_5Co_2$	545	568	Nanoscale Al
$Al_{85}Y_4Mm_4Ni_5Co_2$	555	581	Nanoscale Al
$Al_{85}Y_6Zr_2Ni_5Co_2$	567	597	Nanoscale Al

X-unidentified phase(s)

Difference in the crystallization mechanism has been also observed in the $Al_{92}Sm_8$ amorphous alloy prepared by rapid solidification of the melt and by solid-state processing (cold rolling) at ambient temperature. No Al nanocrystals were formed in the samples produced by cold rolling while a large number density of Al was obtained by devitrification of the melt-spun samples [122].

4.4.5 Crystallization upon deformation

4.4.5.1 Crystallization upon plastic deformation

Plastic deformation can cause nanocrystallization of a glassy phase in some alloys. For example, deformation of some Al-RE-TM amorphous alloys at room temperature causes precipitation of deformation-induced Al particles of 7-10 nm in diameter within the shear bands on bending [123] or nano-indentation [124]. Under the compressive testing, the nanoparticles were formed in shear bands while under tensile test they were found on the fracture surface only [125]. It was also shown that a high-energy ball milling also produced nanocrystalline particles from amorphous phase [126]. It has been suggested that a local temperature rise can play a role in mechanically-induced devitrification [127]. However, further studies on nano-indentation of glassy alloys at low strain rates also showed the formation of the nanoparticles when local rise of the temperature due to adiabatic heating is not supposed to take place [128]. This postulate has been confirmed by bending test at low temperature using dry ice [129].

At the same time some Al- and Zr-based glassy alloys do not show formation of nanoparticles on deformation [130]. The deformation-induced crystallization was studied in the glassy $Al_{85}Y_8Ni_5Co_2$ and $Al_{85}Gd_8Ni_5Co_2$ as well as amorphous $Al_{85}Nd_8Ni_5Co_2$ and $Al_{85}Mm_8Ni_5Co_2$ ribbon samples subjected to cold rolling with 33 % reduction in thickness [131]. The glassy $Al_{85}Y_8Ni_5Co_2$ and $Al_{85}Gd_8Ni_5Co_2$ alloys exhibiting the supercooled liquid region and the $Al_{85}Nd_8Ni_5Co_2$ and $Al_{85}Mm_8Ni_5Co_2$ alloys exhibiting just very beginning of the glass-transition phenomenon were chosen for the study. These alloys have close values of the crystallization temperatures and crystallize by the nucleation and growth reaction.

It is found that contrary to crystallization on heating, no intermetallic compounds but only a very small volume fraction of Al nano particles (<1 vol.%) was formed during the deformation-induced crystallization of the $Al_{85}Y_8Ni_5Co_2$ and $Al_{85}Nd_8Ni_5Co_2$ glassy alloys (Fig. 4.32), while the $Al_{85}Gd_8Ni_5Co_2$ and $Al_{85}Mm_8Ni_5Co_2$ samples remained fully disordered (Fig. 4.33). The most important factor responsible for the suppression of precipitation of the intermetallic compounds is believed to be the effect of pressure during cold rolling as FCC structure of Al is supposed to have a high atomic packing density of the FCC lattice. The applied pressure favors precipitation of the phases with high packing density. The effect of rising temperature within the shear bands also favors crystallization and enhances the diffusion-controlled growth. The higher growth rate of Al particles in the cold-rolled $Al_{85}Nd_8Ni_5Co_2$ amorphous alloy compared to the cold-rolled $Al_{85}Y_8Ni_5Co_2$ glassy alloy may indicate significantly higher deformation-induced diffusion rate of the alloying elements in the former alloy. As these alloys have a nucleation and growth crystallization behavior on heating they do not exhibit significant deformation induced crystallization compared to the Al-RE-Ni alloys studied earlier which can contain pre-existing nuclei and exhibit just growth behavior.

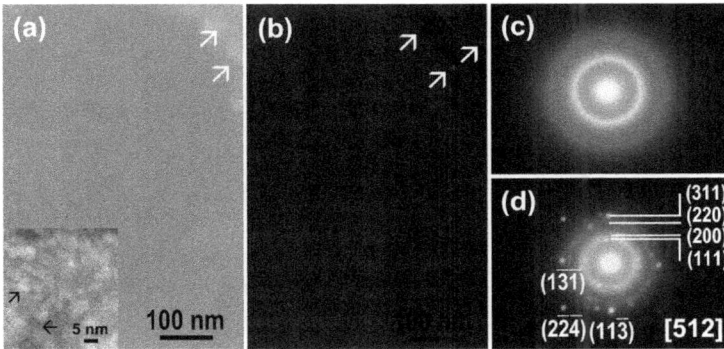

Fig. 4.32. The structure of the $Al_{85}Y_8Ni_5Co_2$ glassy alloy after cold-rolling, TEM. (a) bright-field, (b) dark-field image, (c) SAED pattern, (d) nanobeam diffraction pattern including the diffraction rings from a group of particles and a diffraction pattern from a single particle superimposed including their indexing according to cubic Al lattice. Dark-field image (b) was formed with the amorphous halo. The arrows indicate nanoscale Al particles. The insert in (a) is a high-resolution TEM image of the local region with high volume fraction of Al nanoparticles. Reproduced from [131] with permission of Elsevier.

It has been also shown that applied pressure enhances the precipitation of the Al from the amorphous phase in the $Al_{89}La_6Ni_5$ alloy by decrease in the crystallization temperature [132,133]. As these alloys have a nucleation and growth crystallization behavior on heating they do not exhibit significant deformation induced crystallization compared to the Al-RE-Ni alloys [134] studied earlier which can contain pre-existing nuclei and exhibit just growth behavior.

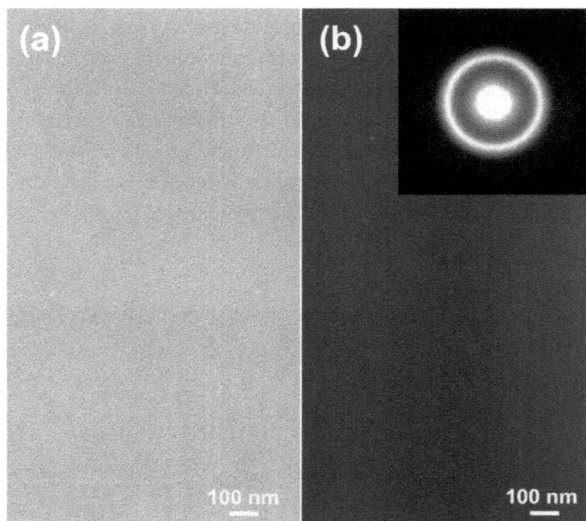

Fig. 4.33. The structure of the $Al_{85}Gd_8Ni_5Co_2$ glassy alloy after cold-rolling, TEM. (a) bright-field, (b) dark-field image, the insert SAED pattern. Reproduced from [131] with permission of Elsevier.

4.4.5.2 Crystallization upon elastic deformation

As has been shown in 4.1.1 not only plastic but even elastic deformation, mostly on a cycling loading with a large enough number of cycles, can also cause nanocrystallization of a glassy phase in some alloys [6]. Even same or very similar in composition bulk metallic glasses can give different nanocrystallization products depending on the sample geometry, applied stress and stress amplitude.

4.4.6 Formation of quasicrystals upon devitrification

An icosahedral phase is a quasicrystal having a long-range quasiperiodic translational order and an icosahedral orientational order (Fig. 4.34).

Fig. 4.34. HRTEM image of a single icosahedral quasicrystal formed in a bulk glassy $Zr_{65}Al_{7.5}Ni_{10}Pd_{17.5}$ alloy. The insert is the Fourier Transform image indicating 3-fold symmetry with irrational lattice spacing.

However, it has no three-dimensional translational periodicity. It was initially discovered in Al-Mn alloys [135,136] and later in the other binary and ternary Al-TM-Cr [137] alloys. Thermodynamically stable icosahedral quasicrystals were found in binary Cd-Ca and Cd-Yb [138]; ternary Al-Li-Cu, Al-Cu-Fe, Al-Cu-Ru, etc…; as well as quaternary systems [139]. A dodecagonal quasicrystalline structure can be made of the perovskite barium titanate $BaTiO_3$, which forms long-range ordered dodecagonal films when grown on a Pt(111) substrate [140]. Oxide quasicrystals may possess useful optical and electrical properties which could be very important for engineering applications.

Different indexing methods of the diffraction patterns of quasicrystals are used [141, 142, 143]. The Al-TM base icosahedral structure is presumed to consist of Mackay icosahedral clusters [144] while another type of icosahedral quasicrystals rather consists of

Materials Research Forum LLC
https://doi.org/10.21741/9781644901014

Bergman-type clusters [145] and others [146]. The Mendeleev number is also used as a parameter for choosing the alloys forming a quasicrystalline phase [147]. The application of Pettifor maps [148] finds the middle path between purely empirical rules and detailed quantum mechanical calculations.

Synchrotron-beam X-ray diffraction has shed some light on the cluster structure of the $Cd_{5.7}Yb$ binary icosahedral phase [149]. A large difference in X-ray atomic scattering factors of Cd and Yb atoms and a series of approximants with close chemical composition in this system helped to produce some model of a real structure of this quasicrystal. The structure is described as a collection of the connected icosahedral YbCd clusters linked by ytterbium atoms located between the clusters.

Quasicrystals can be quasiperiodic in one, two or three dimensions: One-dimensional quasicrystals: are quasiperiodic in one dimension [150]. Two-dimensional quasicrystals: octagonal quasicrystals [151] (Cr-Ni-Si, Mn-Fe-Si, etc.), decagonal quasicrystals (Al-Ni-Co, Al-Cu-Co, etc.) dodecagonal quasicrystals (Cr-Ni, V-Ni) are quasiperiodic in two dimensions [152]. Three-dimensional quasicrystals: icosahedral quasicrystals (Al-Cu-Fe, Al-Li-Cu, etc.) are quasiperiodic in three dimensions. The symmetry of an icosahedral structure can be described by a six-dimensional space group.

Rational approximants are crystals similar to quasicrystals which exhibit translational periodicity [153]. They have periodic structures with relatively large unit cells compared to ordinary crystals. Quasicrystalline-type short-range order can be found in metallic glassy alloys [154]. The existence of the chemical short range order clusters similar to those of a quasicrystal was found in the glassy state [155]. The icosahedral order exists in the liquid phase of Ti-Zr-Ni alloy [156]. The reduced supercooling (undercooling) before crystallization from the melt was found to be the smallest for quasicrystals, larger for crystal approximants and the largest for crystal phases [157]. A low energy barrier for nucleation of the icosahedral phase may explain the fact that only growth of the pre-existed icosahedral nuclei was observed in the $Zr_{55}Ni_{10}Al_{7.5}Cu_{7.5}Ti_{10}Ta_{10}$ alloy [158].

In the rapidly solidified $Ti_xZr_yHf_zNi_{20}$ system alloys the icosahedral phase is formed in the composition ranges close to the cI2 β solid solution phase and complex cF96 phase formation ranges (Fig. 4.35) [159].

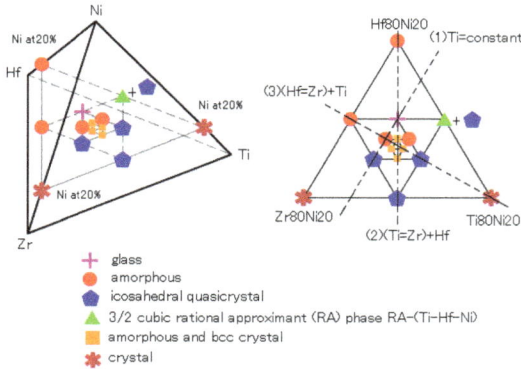

Fig. 4.35. The phase composition of the melt-spun $(Ti_xZr_yHf_z)_{80}Ni_{20}$ alloys. Reprinted from [159] with permission of Elsevier.

Hf-based alloys have a higher tendency to form a cubic cF96 phase compared to Zr-based ones. The alloys in the systems in which an equilibrium Hf-based cF96 phase exists do not show the formation of the icosahedral phase from the amorphous matrix. The metastable cF96 phase and the icosahedral phase are formed by primary crystallization from the amorphous phase inheriting the structure of the icosahedral clusters [160]. cF96 phase nanoparticles, as well as those of the icosahedral phase, formed on crystallization were found to have similar equiaxed, close to spherical morphology.

4.4.7 Formation of nanoscale quasicrystals

Nanoscale quasicrystals are formed on devitrification in various metallic glassy alloys. The icosahedral phase was found in the devitrified Zr-Cu-Al, Zr-Al-Ni-Cu [161] and Zr-Ti-Ni-Cu-Al [162] glassy alloys containing an impurity of oxygen above about 1800 mass ppm, although no icosahedral phase is formed if oxygen content is lower than 1700 mass ppm. Later the nanoscale icosahedral phase was obtained in devitrified Zr-Al-Ni-Cu-Ag [163], Zr-Al-Ni-Cu-Pd [164], and various other alloys including binary Zr-Pd [165] and Zr-Pt [166,167] system alloys at low (typically about 800 mass ppm) oxygen content. It was also produced in the noble metal-free Zr–Cu–Ti–Ni [168]. The icosahedral phase in Zr-based alloys is often formed in cooperation with cF96 phase and with cI2 βZr solid solution phase [169].

A nanoscale icosahedral quasicrystalline phase has been also produced upon heating glassy Hf-based alloys such as $Hf_{65}Pd_{17.5}Ni_{10}Al_{7.5}$ [170], $Hf_{65}Al_{7.5}Ni_{10}Cu_{12.5}Pd_5$ [171], and $Hf_{65}Au_{17.5}(Ni,Cu)_{10}Al_{7.5}$ [172] alloys. According to the quasilattice constant value, the icosahedral phase in these alloys consists of 137-atom Bergmann rhombic triacontahedra. The activation energy for such a transformation was found to be 400 and 450 kJ/mol for Pd and Au-bearing alloys, respectively. No glassy phase was formed in the $Hf_{65}Ag_{17.5}Cu_{10}Al_{7.5}$ alloy. Also no icosahedral phase is formed during annealing of the glassy $Hf_{65}Pd_{17.5}Cu_{10}Al_{7.5}$ alloy.

Formation of the nanoscale (3-10 nm in size) Cu-based icosahedral phase was observed in the Cu-Zr-Ti-NM (NM-noble metal) alloys containing Pd [173] and Au [174] after heat treatment of the glassy phase. Although, due to the small particle size, the diffraction peaks of the icosahedral phase are broad, five strongest peaks were identified and indexed (Fig. 4.36 (a)). Calculation gives the quasilattice constants a_q of 0.495 and 0.496 nm for the $Cu_{55}Zr_{30}Ti_{10}Pd_5$ and $Cu_{55}Zr_{30}Ti_{10}Au_5$ alloys, respectively. At the next exothermic peak the icosahedral phase transforms to an intermediate intermetallic crystalline phase, which subsequently transforms to oC68 $(Cu,NM)_{10}(Zr,Ti)_7$ phase (solid solution of Ti and Pd or Au elements in $Cu_{10}Zr_7$) at higher temperature. This solid solution partially decomposes during long-time annealing forming other equilibrium phases.

Pd- and Au-bearing alloys have a wide supercooled liquid region (Fig. 4.36 (b)) while Ag and Pt-bearing alloys having narrow supercooled liquid do not form the icosahedral phase on heating [175]. The supercooled liquid region for $Cu_{55}Zr_{30}Ti_{10}Pd_5$ and $Cu_{55}Zr_{30}Ti_{10}Au_5$ reached about 50 K and 45 K, respectively (see Fig. 4.36 (b)). These values are very close to those of the $Cu_{55}Zr_{30}Ti_{10}Ni_5$ and $Cu_{55}Zr_{30}Ti_{10}Co_5$ alloys while the $Cu_{55}Zr_{30}Ti_{10}Ag_5$ and $Cu_{55}Zr_{30}Ti_{10}Pt_5$ glassy alloys exhibited narrower supercooled liquid regions.

The $Cu_{55}Zr_{30}Ti_{10}Pd_5$ glassy alloy shows a rather high Poisson's ratio exceeding 0.42 which is maintained after the structural relaxation and primary crystallization. The Young's and Shear moduli decrease upon primary crystallization while the bulk modulus exhibits a maximum after the structural relaxation (Fig. 4.37) [176].

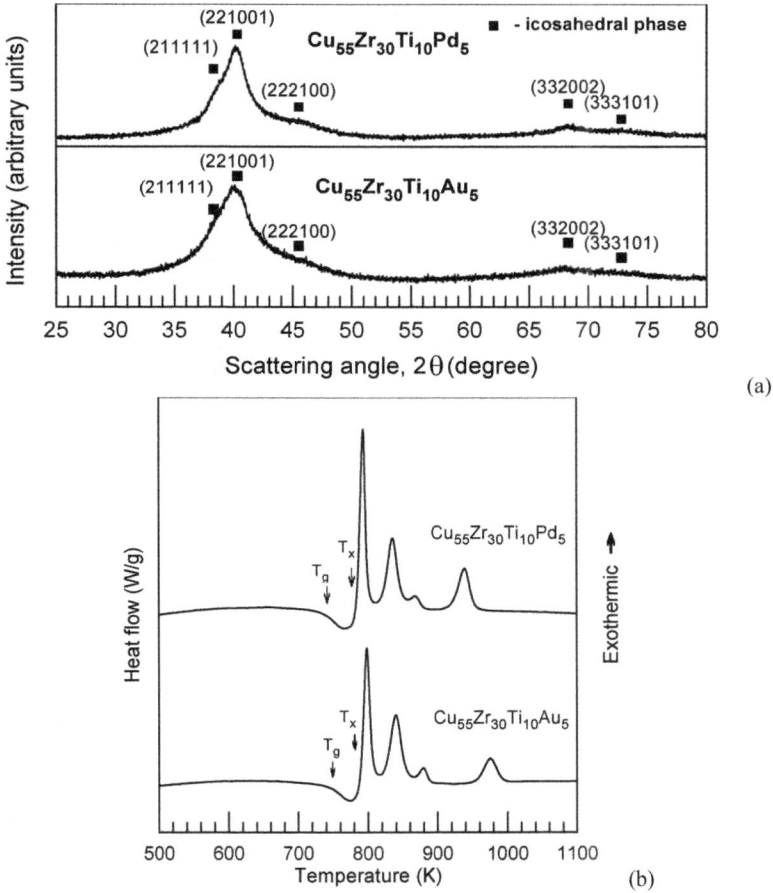

Fig. 4.36. (a) X-ray diffraction patterns of the $Cu_{55}Zr_{30}Ti_{10}Pd_5$ and $Cu_{55}Zr_{30}Ti_{10}Au_5$ glassy alloys annealed for 600 s at 785 K and 420 s at 797 K, respectively. (b) DSC traces of these alloys. The heating rate 0.67 K/s. The data are used from [174,175] with permission of Elsevier.

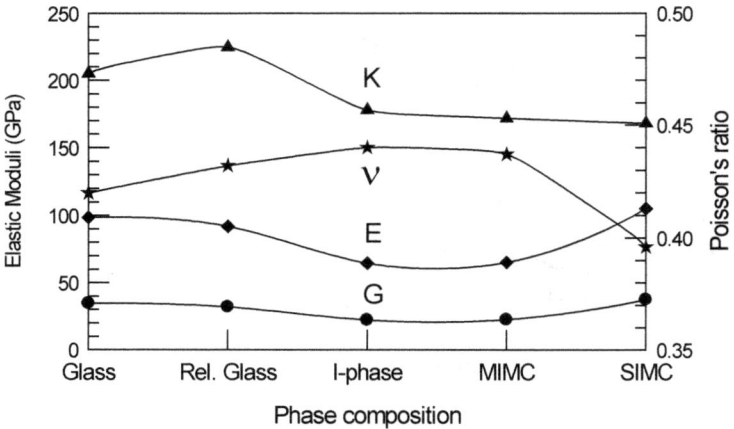

Fig. 4.37. Some elastic moduli and the Poisson's ratio of the $Cu_{55}Zr_{30}Ti_{10}Pd_5$ alloy as a function of temperature. The studied materials were as-cast glass (Glass), glass relaxed at 740 K for 120 s (Rel. Glass), the glass containing the nanoscale icosahedral phase (I-phase), metastable intermetallic compound (MIMC) and stable intermetallic compound(s) (SIMC) formed at the first, second and third exothermic peaks of Fig. 4.36 (b), respectively. Reproduced from [176] with permission of Elsevier.

The structure of the $Cu_{55}Zr_{30}Ti_{10}Pd_5$ sample annealed for 0.6 ks at 785 K is shown in Fig. 4.38 (a,b). The average size of the quasicrystalline particles ranged from 3 to 10 nm. The sharp rings in the selected-area electron diffraction pattern correspond to the X-ray diffraction peaks. The nanobeam diffraction patterns shown in Fig. 4.38 (c,d) represent the five- and two-fold symmetry of the icosahedral phase, respectively. One should point out that some works impartially state that nanoscale icosahedral particles may belong to a high-order approximants. High symmetry of the observed nanobeam electron diffraction patterns indicated that the ordered phase in the $Cu_{55}Zr_{30}Ti_{10}Pd_5$ alloy has a high degree of the icosahedral symmetry, although, due to a small size it does not have all features of a quasicrystal.

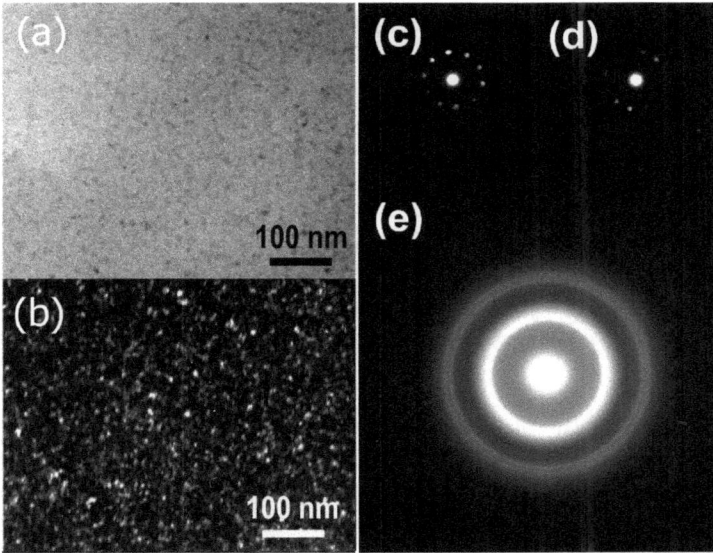

Fig. 4.38. (a) bright- and (b) dark-field TEM images of the $Cu_{55}Zr_{30}Ti_{10}Pd_5$ alloy annealed for 0.6 ks at 785 K. (c) and (d) nanobeam diffraction pattern of five- and two-fold symmetry, respectively. (e) selected-area electron diffraction (SAED) pattern.

Fig. 4.39 shows 3D atom probe (3DAP) elemental maps within the analyzed volume of approximately 4 nm × 4 nm × 130 nm of the $Cu_{55}Zr_{30}Ti_{10}Pd_5$ sample annealed at 785 K for 600 s [177]. Fig. 4.40 shows the atom probe concentration depth profiles of Cu, Zr, Ti, and Pd calculated from the volume of 2 nm × 2 nm × 130 nm that was selected from the volume analyzed in Fig. 4.40 and the calculated concentration frequency distribution diagrams. The data indicate that the concentration frequency diagrams of Cu, Zr and Ti show strong deviation from those expected from a homogeneous solid solution. This indicates that the nanoicosahedral phase in the $Cu_{55}Zr_{30}Ti_{10}Pd_5$ glassy alloy has slightly different composition from the composition of the glass as that of $Zr_{65}Al_{7.5}Ni_{10}Cu_{7.5}Ag_{10}$ [178]. Cu concentration is larger than 50 at.% in all regions indicating that the icosahedral phase in this alloy contains more than 50 at.% of Cu.

Metallic Glasses and Their Composites – 2nd Edition   Materials Research Forum LLC

Materials Research Foundations **85** (2021)   https://doi.org/10.21741/9781644901014

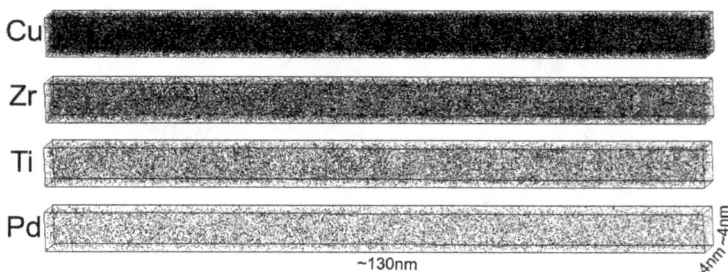

Fig. 4.39. 3 D atom map for the $Cu_{55}Zr_{30}Ti_{10}Pd_5$ alloy sample after annealing at 785 K for 600 s. Reproduced from [177] with permission of the American Institute of Physics.

3DAP results (Fig. 4.39 and 4.40) indicate that the composition of the icosahedral phase in the $Cu_{55}Zr_{30}Ti_{10}Pd_5$ alloy is close to that of the residual glassy matrix, though it is definitely different from that of the glassy matrix as the concentration fluctuation higher than the statistical error was detected. This is in line with the previously reported diffusion-controlled growth mechanism of the icosahedral phase. An alternative mechanism limiting growth rate of the icosahedral phase may be connected with incompatibility of 5-fold symmetry with 3-dimensional translational periodicity which may initiate internal elastic stresses. This idea is supported by the fact that the largest icosahedral phase particles in the $Cu_{55}Zr_{30}Ti_{10}Pd_5$ alloy hardly exceed 10 nm in size and transform to a metastable intermetallic compound of further heating.

The isothermal calorimetry trace for the $Cu_{55}Zr_{30}Ti_{10}Pd_5$ and $Cu_{55}Zr_{30}Ti_{10}Au_5$ alloys exhibited an exothermic peak after the certain incubation period which is required for nucleation of the icosahedral phase. Primary precipitation of the icosahedral phase obeys the kinetic law (eq. 4.4). The values of the Avrami exponent n=2.3 and 2.8 obtained for the $Cu_{55}Zr_{30}Ti_{10}Pd_5$ and $Cu_{55}Zr_{30}Ti_{10}Au_5$ alloys, respectively, are close to 2.5 which corresponds to 3-dimensional diffusion-controlled growth of nuclei at a constant nucleation rate. Bulk glassy samples of 2 mm diameter were formed $Cu_{55}Zr_{30}Ti_{10}Pd_5$ and $Cu_{55}Zr_{30}Ti_{10}Ag_5$ ingots while the $Cu_{55}Zr_{30}Ti_{10}Au_5$ bulk alloy sample has had a mixed glassy and crystalline structure. No glassy phase was formed in the $Cu_{55}Zr_{30}Ti_{10}Pt_5$ bulk alloy sample whereas rapidly solidified ribbon samples with a glassy structure were obtained for all the $Cu_{55}Zr_{30}Ti_{10}NM_5$ alloys. Pd and Au additions significantly expand the supercooled liquid region of the Cu-Zr-Ti glassy alloy.

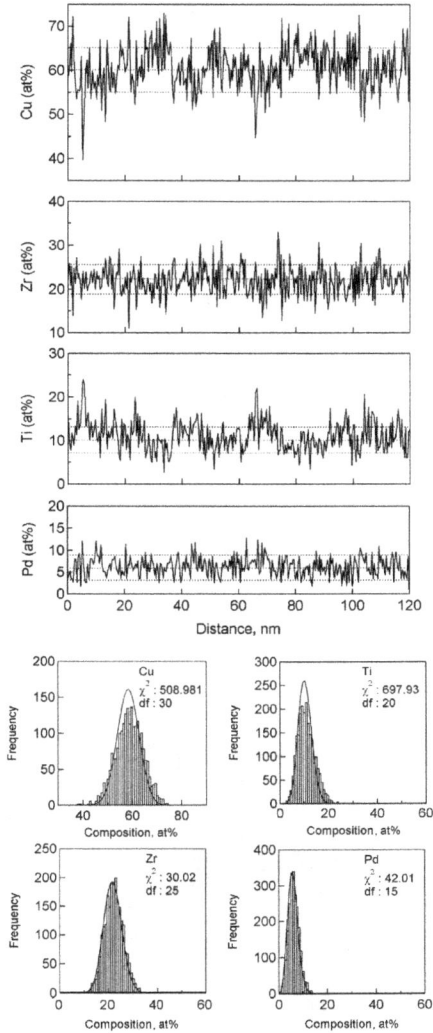

Fig. 4.40. 3D atom probe concentration depth profiles for different alloying elements obtained from the $Cu_{55}Zr_{30}Ti_{10}Pd_5$ sample after annealing at 785 K for 600 s and the concentration frequency distribution diagrams. Reproduced from [177] with permission of the American Institute of Physics.

In a later work the icosahedral phase was formed only in the $Cu_{55}Zr_{30}Ti_{10}Pd_5$ alloy while the $Cu_{55}Zr_{40}Pd_5$, $Cu_{55}Zr_{20}Ti_{20}Pd_5$, $Cu_{45}Zr_{40}Ti_{10}Pd_5$, $Cu_{60}Zr_{25}Ti_{10}Pd_5$ and $Cu_{65}Zr_{20}Ti_{10}Pd_5$ alloys did not form the icosahedral phase [179]. Notwithstanding on a limited homogeneity area of the icosahedral phase observed, according to the Avrami analysis this phase exhibited nucleation and 3-dimentional diffusion-controlled growth below and above the glass-transition region. The above mentioned also indicates (in addition to 3DAP results) that the composition of these two the glassy and icosahedral phases is not exactly equal.

Compared to the Zr-Cu-Pd alloys studied earlier the $Cu_{55}Zr_{40}Pd_5$ alloy exhibited formation of the equilibrium $(Cu,Pd)_{10}Zr_7$ phase from the supercooled liquid likely by polymorphous mechanism. This fact rules out possibility for the Cu-Zr-Ti-Pd icosahedral phase of being a Cu-rich part of the compositional homogeneity area of the Zr-Cu-Pd one. Thus, the filling sequence of the coordination shells in the Zr-Cu-Pd and Cu-Zr-Ti-Pd icosahedral phases must be different.

The $Zr_{60}Cu_{30-x}Co_xAl_{10}$ (x=0, 2.5, 5, 7.5, 10, 15, 20, 25, 30) metallic glasses were studied recently [180]. An ultrahigh grain number density (10^{25} m^{-3}) of a nanoscale metastable Co-rich icosahedral-like phase, with a size as small as about 1-2 nm, was obtained by partial crystallization of the $Zr_{60}Co_{30}Al_{10}$ metallic glass (Fig. 4.41). The nanoscale icosahedral phase remains stable over quite a large annealing temperature range and time. Nanoscale chemical and structural heterogeneity with Co rich areas were observed (Fig. 4.42). The interface energy, σ_{i-a} between the nano icosahedral phase and the amorphous matrix is about 14 mJ/m^2, indicating a low nucleation barrier. The calculated nucleation rate values are of the order of 10^{22}-10^{23} $m^{-3}s^{-1}$.

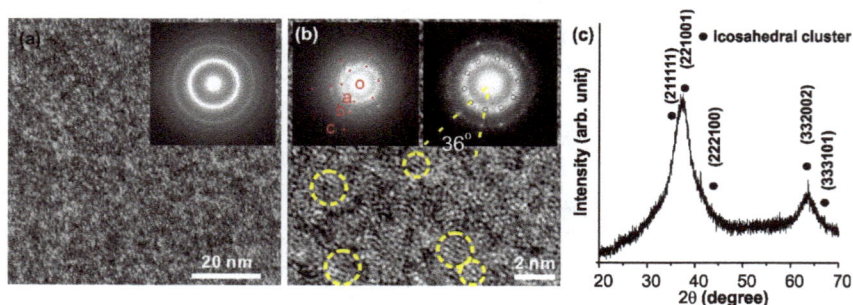

Fig. 4.41. (a) Dark-field TEM image, (b) HRTEM image of the $Zr_{60}Co_{30}Al_{10}$ alloy annealed at 823 K. The insert in (a) is the selected-area electron diffraction patterns; the inserts in (b) are nanobeam diffraction pattern. (c) XRD pattern of the $Zr_{60}Co_{30}Al_{10}$ alloy heated to 823 K. Reproduced from [180] with permission of Elsevier.

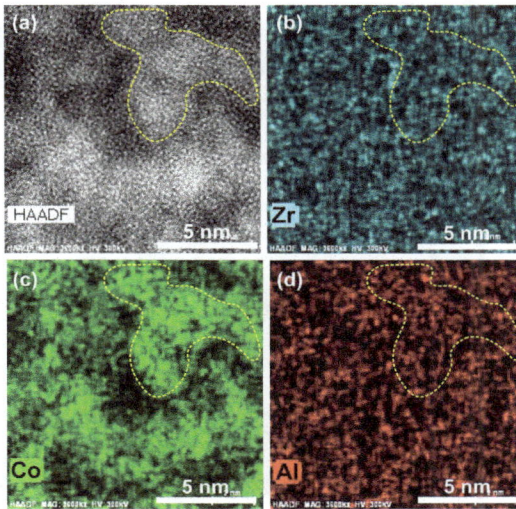

Fig. 4.42. The HAADF image(a) and EDX mapping results (b-d) of the Zr60Co30Al10 alloy glassy ribbon up to the end of the first DSC crystallization peak. Reproduced from [180] with permission of Elsevier.

4.4.8 Crystallization of the amorphous alloys containing pre-existing nuclei and nanoparticles, formation of nanocomposites

As it has been mentioned in Chapter 2 the $Fe_{48}Cr_{15}Mo_{14}C_{15}B_6Tm_2$ and $Fe_{48}Cr_{15}Mo_{14}C_{15}B_6Y_2$ alloys are the example of growth limited BMGs. Fig. 2.14 showed the differential isothermal calorimetry (DIC) trace of $Fe_{48}Cr_{15}Mo_{14}C_{15}B_6Tm_2$ at 868 K which did not show an incubation period.

A nano-dispersed structure can be obtained directly from the melt upon rapid solidification at proper alloying. It is also found that the addition of Pd to Al-Y-Ni-Co alloys substituting Y caused disappearance of the supercooled liquid region as well as the formation of the highly dispersed primary FCC-Al nanoparticles about 3-7 nm in size homogeneously embedded in the glassy matrix upon solidification [181]. The value of the activation energy for the growth of Al nanocrystals of 146 kJ/mol obtained using the Kissinger analysis method is very close to the activation energy for self diffusion of pure Al. An extremely high density of precipitates of the order of 10^{24} m^{-3} is obtained (Fig. 4.43). This is the highest precipitation density observed so far in Al-based metallic glasses.

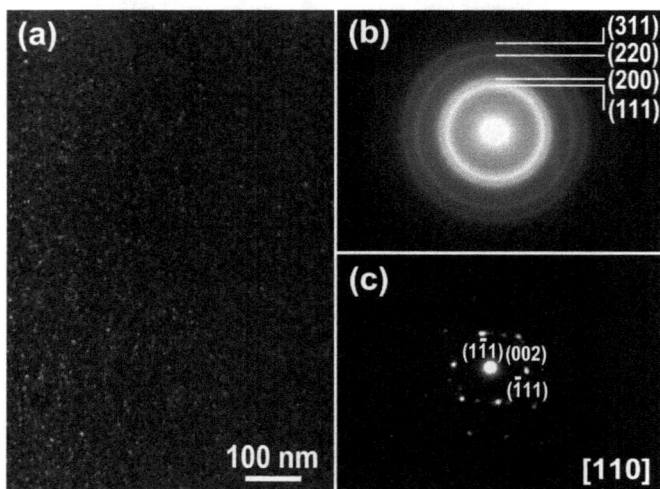

Fig. 4.43. The structure of the $Al_{85}Y_4Ni_5Co_2Pd_4$ glassy alloy in as-solidified state, TEM. (a) Dark-field image, (b) SAED and (c) nanobeam diffraction pattern. Reproduced from [181] with permission of Elsevier.

The broad X-ray diffraction peaks in the XRD pattern of the $Al_{85}Y_4Ni_5Co_2Pd_4$ glassy alloy (Fig. 4.44) were fitted with the Gaussian function. The resulted d-spacings for (111) and (200) of 0.2338 nm and 0.2030 nm, respectively, correspond very well to that of pure Al. The broad diffraction peak from the amorphous phase contributes to the intensity of (111).

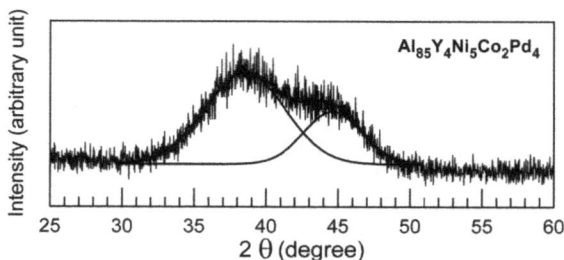

Fig. 4.44. The X-ray diffraction pattern of the $Al_{85}Y_4Ni_5Co_2Pd_4$ as-rapidly solidified alloy and the peaks fitting with the Gaussian function. Reproduced from [181] with permission of Elsevier.

As indicated by the high-resolution TEM (HRTEM) (Fig. 4.45), in some areas where nuclei of Al nano-particles were located close enough to each other, the particles got in touch with each other upon growth and the crystal lattice of Al nanoparticles is heavily distorted. Compared to Al nanocrystals produced by the annealing of the glassy matrix which are usually found to be free of linear defects Al nanocrystals in the as-solidified Al-Y-Ni-Co-Pd sample contain microstrain and linear defects like dislocations. The source of the dislocations and the distortions of the crystalline lattice is impingement of the growing particles.

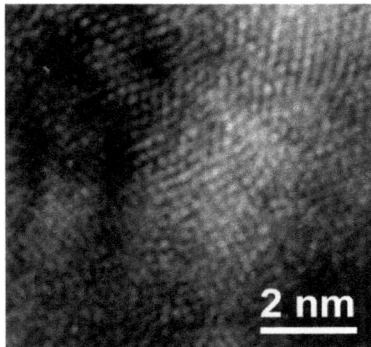

Fig. 4.45. HRTEM image of the $Al_{85}Y_4Ni_5Co_2Pd_4$ glassy alloy in as-solidified state.

As the diffraction peaks were significantly broadened it was possible to estimate the coherent scattering area (D_{hkl}) size (crystallite size). The resulted D_{hkl} values for two strong diffraction peaks (111) and (200) were 2.5 and 3.5 nm, respectively. The latter value is more reasonable as broad diffraction peak from the residual glassy phase contributes to the intensity of (111) peak. High degree of micro-strain within the Al lattice (see Fig. 4.45) is a reason for the slight discrepancy between the particles size calculated by using the XRD peaks and observed by TEM.

The alloys with low RE/TM content ratio in general crystallize by growth of pre-existed nuclei formed on solidification which is reflected in a weak crystallization pre-peak seen in Fig. 4.36. Also low glass-forming ability of Al-based alloys is connected with high liquidus temperature (Fig. 4.46) caused by RE metals which form refractory intermetallic compounds such as Al_3RE. The enthalpies of crystallization and melting found on heating are 120 and 297 J/g, respectively.

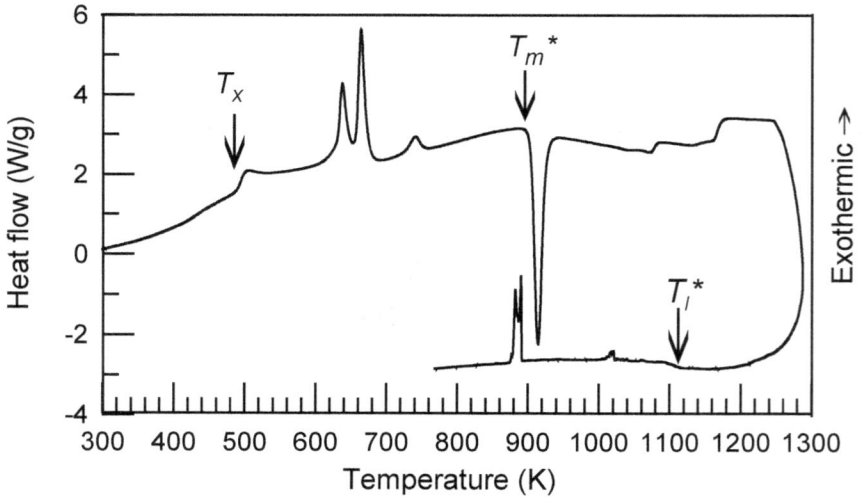

Fig. 4.46. DSC curve of the $Al_{85}Y_4Ni_5Co_2Pd_4$ glassy alloy on heating at 0.67 K/s and cooling at 0.17 K/s. T_l^* indicates an approximate liquidus temperature with some undercooling.

The $Al_{85}Y_8Ni_{5-3}Co_2Cu_{0-2}$ metallic glasses also exhibited a change from the nucleation and growth of a crystalline phase(s) in the $Al_{85}Y_8Ni_5Co_2$ one to growth of pre-existing nanoparticles (of 2-5 nm size) in $Al_{85}Y_8Ni_3Co_2Cu_2$ (Fig. 4.47) [182]. The addition of Cu also caused disappearance of the supercooled liquid region (Fig. 4.48). The heat released during primary exothermic heat effect (A) decreases from 42 J/g in Cu-free alloy up to 27 J/g and 23 J/g in the alloys containing 1 and 2 at.% Cu, respectively.

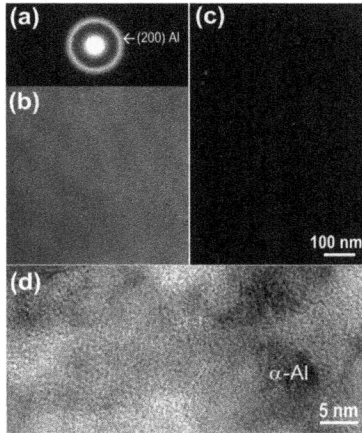

Fig. 4.47. (a) selected-area electron diffraction pattern, (b) bright-field (c) dark-field and (d) high-resolution image TEM of the $Al_{85}Y_8Ni_3Co_2Cu_2$ metallic glass in as-solidified state. Reproduced from [182] with permission of the Materials Research Society.

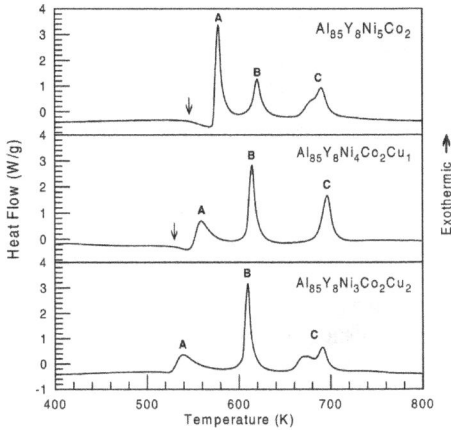

Fig. 4.48. DSC traces of the $Al_{85}Y_8Ni_{5-3}Co_2Cu_{0-2}$ metallic glasses obtained at 0.67 K/s. The arrows indicates increment of C_p related to glass-transition. Reproduced from [182] with permission of the Materials Research Society.

The melt-spun $(Fe_{0.75}Pt_{0.25})_{75-70}B_{25-30}$ ribbons were found to be X-ray amorphous [183]. These alloys were also found to possess good hard magnetic properties including a high intrinsic coercivity values up to 400 kA/m in the nano-crystallized state [184]. These alloys are promising to be applied as nanocomposite permanent magnets. Although only broad diffraction peaks are observed in the XRD pattern of the rapidly-solidified $(Fe_{0.75}Pt_{0.25})_{75}B_{25}$ alloy some nanoparticles were observed by high-resolution TEM (Fig. 4.49). These particles do not produce the detectable X-ray diffraction peaks due to their small size and limited volume fraction.

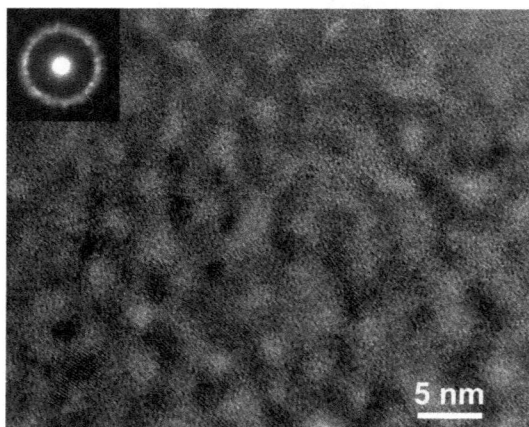

Fig. 4.49. High-resolution TEM image of the rapidly-solidified $(Fe_{0.75}Pt_{0.25})_{75}B_{25}$ alloy in the as-solidified state. The insert: NBD pattern from the amorphous phase. Reproduced from [185] with permission of Elsevier.

No exothermic peak but just a variation of the baseline is observed during isothermal calorimetry trace taken at 647 K, which indicates absence of the incubation period (Fig. 4.50) [185].

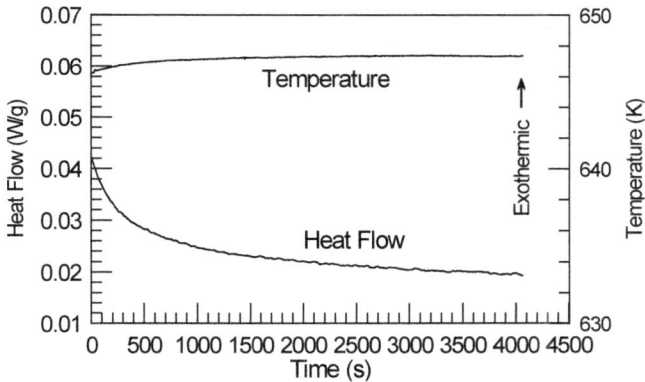

Fig. 4.50. Differential isothermal calorimetry (DIC) trace obtained at 647 K. Reproduced from [185] with permission of Elsevier.

Structure of the rapidly solidified $(Fe_{0.75}Pt_{0.25})_{75}B_{25}$ alloy contains a limited volume fraction of the pre-existing nuclei and nanoscale cubic cF4 Fe(Pt) solid solution particles embedded in the amorphous matrix. Nanoparticles of the cF4 Fe(Pt) phase start growing at the elevated temperature and then undergo ordering forming the tP4 FePt compound of about 15 nm in size which is followed by the formation of the tI12 Fe_2B phase from the residual amorphous matrix. Ordering of the cF4 phase takes place after the incubation period, existence of which allows anticipating nucleation and growth mechanism. This process takes place conjointly with further growth of the tP4 FePt particles in expense of the residual amorphous matrix. Formation of the tI12 Fe_2B phase from the residual amorphous matrix begins when the residual amorphous matrix is enriched in B up to a certain value. The observed low growth rate of tP4 FePt phase leading to the formation of a nanostructure is due to the diffusion-controlled growth mechanism and very low calculated diffusion coefficients of Fe and Pt in cF4 Fe at the temperature range of transformation (795-860 K).

It should be also mentioned that recently developed nanostructured metallic glasses exhibited high stability versus crystallization [186]. The Au-based nanoglass is kinetically ultrastable compared to its melt-spun ribbon counterpart. The nanoglass shows a 20 K higher glass transition temperature, a 32 K higher crystallization peak temperature, larger activation energy, slower crystal growth rate, and lower enthalpy compared to the ribbon sample consisting of the entire glassy phase.

Also microstructure of the $Si_{48}Al_{20}Fe_{10}Ge_7Ni_5Cr_5Zr_5$ alloy having an amorphous type X-ray diffraction pattern is inhomogeneous and contains Ge particles of less than 5 nm in size embedded in an amorphous matrix [187]. Ge having a lower mixing enthalpy with the other alloying elements than Si is rejected from the amorphous matrix and precipitates forming fine nanoparticles, though it has an unlimited solubility in crystalline Si. Similar results were obtained for Al-Si-Ge-TM alloys [188].

4.4.9 Homogeneous nucleation and the nature of incubation period

The $Ti_{50}Ni_{23}Cu_{22}Sn_5$ bulk glass-forming alloy (initially with a homogeneously glassy structure (Fig. 4.51(a)) exhibits a nucleation and growth transformation mechanism. According to thermodynamic calculations phase separation is unlikely in this alloy. Although the structure of the alloy at the early crystallization stage remains structurally disordered within the incubation period, it becomes chemically ordered on the nanoscale. Formation of the nanoscale Ti-enriched zones and the regions enriched in the solute late 3d transition metals (Ni and Cu) is observed by HRTEM and elemental mapping (Fig. 4.51(b)) [189]. MD simulation also illustrated the structural changes within the incubation period prior to nanocrystallization. It is suggested that the formation of such a chemically ordered structure gradually reduces the energy barrier for nucleation (W_c) of the cF96 crystalline phase with time until it becomes comparable with the thermal energy fluctuations. The cF96 phase begins to precipitate in the entire volume of the sample with a high particle density only after the incubation period while the crystallization process is further completed within a similar time period. Calculations based on the classical nucleation theory at 753 K indicate W_c of about 12.4 RT. W_c is greater than the thermal energy as it should be for nucleation and growth type transformation with an incubation period.

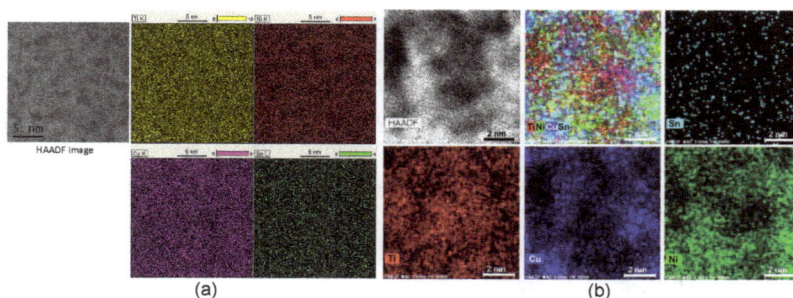

Fig. 4.51. HAADF and STEM-EDX mapping of the as-cast $Ti_{50}Ni_{23}Cu_{22}Sn_5$ alloy sample (a) and the sample annealed at 753 K for 40 s as indicated (b). Reproduced from [189] with permission of Elsevier.

Although the structure of the $Ti_{50}Ni_{23}Cu_{22}Sn_5$ alloy remains spatially disordered within the incubation period (Fig. 4.52a), the formation of the nanoscale Ti-enriched regions still takes place (Fig. 4.51). The chemical ordering towards $Ti_2(Ni,Cu)$ composition and possible topological changes caused by that likely reduce the barrier for nucleation of the cF96 crystalline phase which starts to precipitate after 145 s with high nucleation density (Fig. 4.52(b)). The process has some similarities with the formation of Cu-rich Guinier–Preston type I (GP-I) zones in Al-based alloys when only compositional fluctuation without a crystal symmetry changes is observed: Cu-rich layer in the FCC Al lattice and Ti-rich glassy phase in the $Ti_{50}Ni_{23}Cu_{22}Sn_5$ liquid alloy, respectively. This process is known to precede formation of the intermetallic phases.

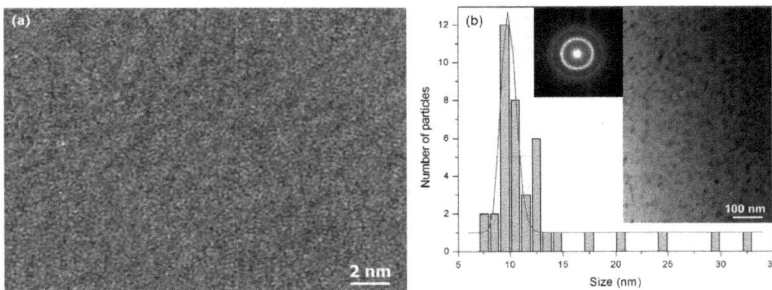

Fig. 4.52. (a) High resolution TEM image of the sample annealed for 80 s at 753 K. (b) Particle size distribution after 280 s at 753 K from the bright-field TEM image shown in the insert (right inserted image) together with SAED (left inserted image). Reproduced from [189] with permission of Elsevier.

4.5 Peritectic like reactions involving glassy phase

Such phase transformation involving a glassy/amorphous/supercooled liquid phase was initially observed in the rapidly-solidified $Ge_{60}Al_{25}La_{15}$ alloy in which an amorphous and a crystalline phase produce another crystalline phase [190]. Another rapidly solidified $Zr_{65}Ni_{10}Al_{7.5}Cu_{7.5}Ti_5Nb_5$ alloy has a mixed structure containing glassy and a submicron β-Zr solid solution phase. The glassy+BCC β-Zr solid solution structure transforms to a mixture consisting of residual glassy and icosahedral phases after the completion of the first exothermic reaction (Fig. 4.53-4.55), and this reaction is also a single-type reaction (Fig. 4.56) [191]. The existence of the residual glassy phase and the average Avrami exponent value of 2.7 indicate that the transformation "glassy+BCC β-Zr solid solution→residual glassy+icosahedral phase" is diffusion-controlled and takes place at nearly constant nucleation rate. The difference in chemical composition between β-Zr and the icosahedral phase also supports the diffusion control of the reaction.

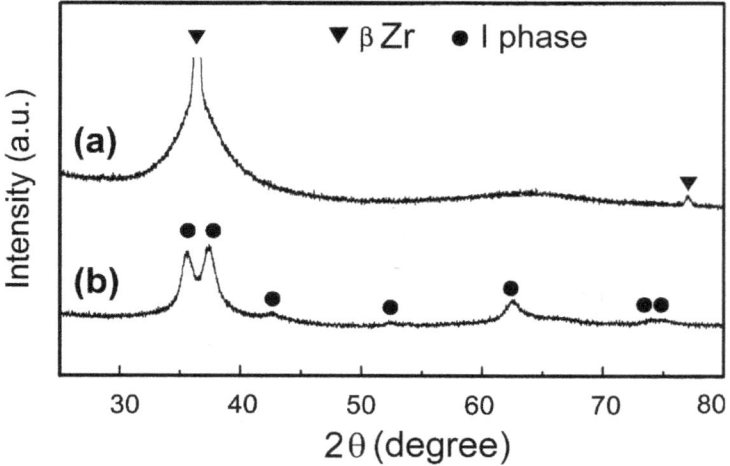

Fig. 4.53. X-ray diffraction patterns of the $Zr_{65}Ni_{10}Al_{7.5}Cu_{7.5}Ti_5Nb_5$ amorphous alloy: (a) in as solidified state and (b) annealed for 300 s at 723 K.

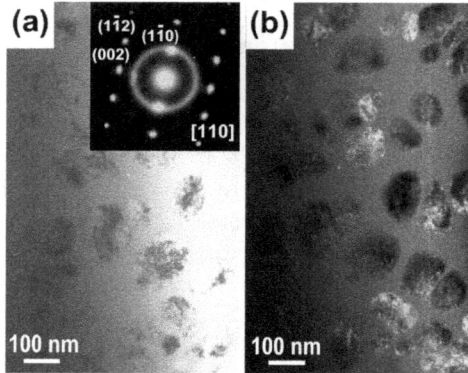

Fig. 4.54. (a) bright-field and (b) dark-field TEM images of the studied alloy as solidified state. The insert - selected-area electron diffraction pattern. Reproduced from [191] with permission of Elsevier.

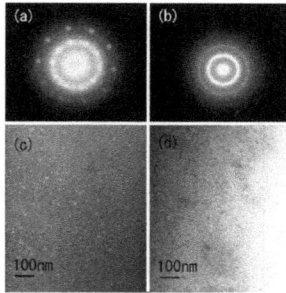

Fig. 4.55. (a) Nanobeam diffraction patterns, (b) selected-area electron diffraction pattern (c) dark-field and (d) bright-field TEM images of the $Zr_{65}Ni_{10}Al_{7.5}Cu_{7.5}Ti_5Nb_5$ alloy annealed for 300 s at 723 K. Reproduced from [191] with permission of Elsevier.

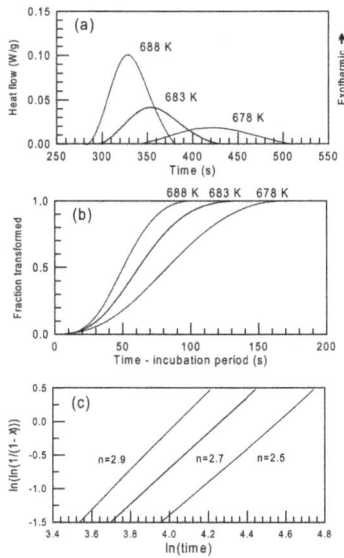

Fig. 4.56. (a) isothermal differential calorimetry curves, (b) fraction transformed as a function of time curves and (c) the Avrami plot of the $Zr_{65}Ni_{10}Al_{7.5}Cu_{7.5}Ti_5Nb_5$ alloy annealed isothermally at different temperature. Reproduced from [191] with permission of Elsevier.

This phase transformation seems to be, in general, close to a peritectic one. However, one can see a large size difference between the crystalline precipitates in the as-solidified sample (Fig. 4.54) compared to the annealed one (Fig. 4.55). No inherited structure is found after phase transformation: submicron-size grains of β-Zr solid solution disappear in the beginning of the phase transformation followed by the formation of fine nanoscale precipitates of the icosahedral phase through an exothermic reaction (see Fig. 4.56). After the completion of all phase transformations on heating the structure was found to consist of the equilibrium Zr_2Ni, Zr_2Cu and the unidentified phases.

A common feature of these phase transformations in the $Ge_{60}Al_{25}La_{15}$ and $Zr_{65}Ni_{10}Al_{7.5}Cu_{7.5}Ti_5Nb_5$ alloys is that they are both diffusion-controlled. The reactions described above look, in general, similar to the peritectic one though absence of the diffusion transfer through the product phase, lack of a common interface and absence of an inherited structure differentiate such type of reactions from a classical type peritectic reaction. A reaction "glassy+nanoscale I-phase → cF96 Hf_2Fe phase" observed in the melt-spun $Ti_{40}Zr_{20}Hf_{20}Fe_{20}$ alloy is closer to typical peritectic one [192].

References

[1] H.B. Yu, W.H. Wang, H.Y. Bai, K. Samwer, The β relaxation in metallic glasses, Natl. Sci. Rev., 1 (2014) 429-461. https://doi.org/10.1093/nsr/nwu018

[2] H.B. Yu, W.H. Wang and K. Samwer, The β relaxation in metallic glasses: an overview, Materials Today, 16 (2013) 183. https://doi.org/10.1016/j.mattod.2013.05.002

[3] M. S. Blanter, I. S. Golovin, H. Neuhäuser, and H.-R. Sinning, Internal Friction in Metallic Materials. A Handbook, Springer Verlag, (2007) p. 540. https://doi.org/10.1007/978-3-540-68758-0

[4] D. P. Wang, J. C. Qiao and C. T. Liu, Relating structural heterogeneity to β-relaxation processes in metallic glasses, Mater. Res. Lett., 7 (2019) 305–311. https://doi.org/10.1080/21663831.2019.1604441

[5] J.M. Pelletier, D. V. Louzguine-Luzgin, S. Li and A. Inoue, Elastic and viscoelastic properties of glassy, quasicrystalline and crystalline phases in $Zr_{65}Cu_5Ni_{10}Al_{7.5}Pd_{12.5}$ alloys, Acta Materialia, 59 (2011) 2797-2806. https://doi.org/10.1016/j.actamat.2011.01.018

[6] D.V. Louzguine-Luzgin, M.Yu. Zadorozhnyy, S.V. Ketov, J. Jiang, I.S. Golovin, A.S. Aronin, Influence of cyclic loading on the structure and double-stage structure relaxation behavior of a Zr-Cu-Fe-Al metallic glass, Materials Science and Engineering: A, 742 (2019) 526-531. https://doi.org/10.1016/j.msea.2018.11.031

[7] A.L. Greer, Atomic transport and structural relaxation in metallic glasses, Journal of Non-Crystalline Solids, 61-62 (1984) 737-748.
https://doi.org/10.1016/0022-3093(84)90633-1

[8] F.O. Méar, B. Lenk, Y. Zhang and A.L. Greer, Structural relaxation in a heavily cold-worked metallic glass, Scripta Materialia, 59 (2008) 1243–1246.
https://doi.org/10.1016/j.scriptamat.2008.08.023

[9] D.V. Louzguine-Luzgin, M. Ito, S.V. Ketov, A.S. Trifonov, J. Jiang, C.L. Chen, K. Nakajima, Exceptionally high nanoscale wear resistance of a $Cu_{47}Zr_{45}Al_8$ metallic glass with native and artificially grown oxide, Intermetallics, 93 (2018) 312-317.
https://doi.org/10.1016/j.intermet.2017.10.011

[10] D. V. Louzguine, A. R. Yavari, K. Ota, G. Vaughan and A. Inoue, Synchrotron X-ray radiation diffraction studies of thermal expansion, free volume change and glass transition phenomenon in Cu-based glassy and nanocomposite alloys on heating, Journal of Non-Crystalline Solids, 351 (2005) 1639.
https://doi.org/10.1016/j.jnoncrysol.2005.04.054

[11] Q. Hu, J.M. Wang, Y.H. Yan, J.Z. Zou, X.R. Zeng, Invar effect of Fe-based bulk metallic glasses, Intermetallics, 93 (2018) 318-322.
https://doi.org/10.1016/j.intermet.2017.10.012

[12] D. V. Louzguine-Luzgin, A. Inoue, A. R. Yavari, and G. Vaughan, Thermal expansion of a glassy alloy studied using a real-space pair distribution function, Appl. Phys. Lett., 88 (2006) 121926. https://doi.org/10.1063/1.2187955

[13] D. V. Louzguine-Luzgin, A. R. Yavari, M. Fukuhara, K. Ota, G. Xie, G. Vaughan and A. Inoue, Free volume and elastic properties changes in Cu-Zr-Ti-Pd bulk glassy alloy on heating, Journal of Alloys and Compounds, 431 (2007) 136.
https://doi.org/10.1016/j.jallcom.2006.05.069

[14] O. Haruyama, M. Kohda, N. Nishiyama and T. Egami, Volume relaxation in ternary bulk Pd- and Pt-P based metallic glasses, Journal of Physics: Conference Series, 144 (2009) 012050. https://doi.org/10.1088/1742-6596/144/1/012050

[15] A. Ishii, F. Hori, A. Iwase, Y. Fukumoto, Y. Yokoyama and T. J. Konno, Relaxation of free Volume in $Zr_{50}Cu_{40}Al_{10}$ bulk metallic glasses studied by positron annihilation measurements, Materials Transactions, 49 (2008) 1975 - 1978.
https://doi.org/10.2320/matertrans.MAW200826

[16] D. J. Magagnosc, G. Kumar, J. Schroers, P. Felfer, J. M. Cairney and D. S. Gianola, Effect of ion irradiation on tensile ductility, strength and fictive temperature in metallic

glass nanowires, Acta Mater., 74 (2014) 165–182.
https://doi.org/10.1016/j.actamat.2014.04.002

[17] D.V. Louzguine-Luzgin, V. Yu. Zadorozhnyy, S.V. Ketov, Z. Wang, A.A. Tsarkov and A.L. Greer, On room-temperature quasi-elastic mechanical behaviour of bulk metallic glasses, Acta Materialia, 129 (2017) 343–351.
https://doi.org/10.1016/j.actamat.2017.02.049

[18] J. Saida, R. Yamada, M. Wakeda and S. Ogata, Thermal rejuvenation in metallic glasses, Science and Technology of Advanced Materials, 18 (2017) 152-162.
https://doi.org/10.1080/14686996.2017.1280369

[19] S. V. Ketov, Y. H. Sun, S. Nachum, Z. Lu, A. Checchi, A. R. Beraldin, H. Y. Bai, W. H. Wang, D. V. Louzguine-Luzgin, M. A. Carpenter and A. L. Greer, Rejuvenation of metallic glasses by non-affine thermal strain, Nature 524 (2015) 200–203.
https://doi.org/10.1038/nature14674

[20] J. W. Cahn and R. J. Charles, The initial stages of phase separation in glasses. physics and ghemistry of glasses, 6 (1965) 181-191

[21] A. Inoue, S. Chen and T. Masumoto, Zr-Y base amorphous alloys with two glass transitions and two supercooled liquid regions, Mater. Sci. Eng. A, 179/180 (1994) 346.
https://doi.org/10.1016/0921-5093(94)90224-0

[22] A. Kündig, M. Ohnuma, D. Ping, T. Ohkubo, and K. Hono, In situ formed two-phase metallic glass with surface fractal microstructure, Acta Mater., 52 (2004) 2441.
https://doi.org/10.1016/j.actamat.2004.01.036

[23] T. Wada, D. V. Louzguine-Luzgin, and A. Inoue, Preparation of Zr-based metallic glass nanowires and nanoparticles by selective etching, Scripta Materialia, 57 (2007) 901.
https://doi.org/10.1016/j.scriptamat.2007.07.032

[24] G.E. Abrosimova, and A.S. Aronin, Evolution of the amorphous-phase structure in metal–metal type metallic glasses, J. Synch. Investig., 9 (2015) 887.
https://doi.org/10.1134/S1027451015050031

[25] G.E. Abrosimova and A.S. Aronin, Two-phase amorphous alloys of Al-Ni-Y system, J. Synch. Investig., 9 (2015) 134. https://doi.org/10.1134/S102745101501022X

[26] D. V. Louzguine-Luzgin, T. Wada, H. Kato, J. Perepezko and A. Inoue, In situ phase separation and flow behavior in the glass transition region, Intermetallics, 18 (2010) 1235-1239. https://doi.org/10.1016/j.intermet.2010.03.018

[27] D. V. Louzguine-Luzgin, I. Seki, T. Wada, and A. Inoue, Structural relaxation, glass transition, viscous formability, and crystallization of Zr-Cu-Based bulk metallic glasses on

heating, Metallurgical and Materials Transactions A, 43 (2012) 2642-2648.
https://doi.org/10.1007/s11661-011-1005-4

[28] D. V. Louzguine-Luzgin, G. Xie, Q. Zhang and A. Inoue, Effect of Fe on the glass-forming ability, structure and devitrification behavior of Zr-Cu-Al bulk glass-forming alloys, Philosophical Magazine, 90 (2010) 1955–1968.
https://doi.org/10.1080/14786430903571495

[29] D. V. Louzguine-Luzgin, G. Xie, W. Zhang, and A. Inoue, Devitrification behavior and glass-forming ability of Cu–Zr–Ag alloys, Materials Science and Engineering: A, 465 (2007) 146–152. https://doi.org/10.1016/j.msea.2007.02.039

[30] J.C. Oh, T. Ohkubo, Y.C. Kim, E. Fleury, and K. Hono, Phase separation in $Cu_{43}Zr_{43}Al_7Ag_7$ bulk metallic glass, Scripta Mater., 53 (2005) 165.
https://doi.org/10.1016/j.scriptamat.2005.03.046

[31] A. A. Kundig, M. Ohnuma, T. Ohkubo, T. Abe, and K. Hono, Glass formation and phase separation in the Ag–Cu–Zr system, Scripta Materialia, 55 (2006) 449.
https://doi.org/10.1016/j.scriptamat.2006.05.012

[32] D. V. Louzguine, A. R. Yavari and A. Inoue, Devitrification behaviour of Cu-Zr-Ti-Pd bulk glassy alloys, Philosophical Magazine, 83 (2003) 2989–3003.
https://doi.org/10.1080/1478643031000151394

[33] D. V. Louzguine and A. Inoue, Influence of a supercooled liquid on devitrification of Cu-, Hf- and Ni- based metallic glasses, Mater. Sci. and Eng. A, 375-377 (2004) 346.
https://doi.org/10.1016/j.msea.2003.10.140

[34] J. Orava, A. L. Greer, B. Gholipour, D. W. Hewak, C. E. Smith, Characterization of supercooled liquid $Ge_2Sb_2Te_5$ and its crystallization by ultrafast-heating calorimetry, Nat. Mater., 11 (2012) 279–283. https://doi.org/10.1038/nmat3275

[35] P. Zalden, A. von Hoegen, P. Landreman, M. Wuttig, and A. M. Lindenberg, How supercooled liquid phase-change materials crystallize: snapshots after femtosecond optical excitation, Chem. Mater., 27 (2015) 5641–5646.
https://doi.org/10.1021/acs.chemmater.5b02011

[36] N. Chen, L. Gu, G. Q. Xie, D. V. Louzguine-Luzgin, A. R. Yavari, G. Vaughan, S. D. Imhoff, J. H. Perepezko, T. Abe and A. Inoue, Flux-induced structural modification and phase transformations in a $Pd_{40}Ni_{40}Si_4P_{16}$ bulk-glassy alloy, Acta Materialia, 58 (2010) 5886-5897. https://doi.org/10.1016/j.actamat.2010.07.003

[37] D. V. Louzguine-Luzgin, M. Miyama, K. Nishio, A. A. Tsarkov and A. L. Greer, Vitrification and nanocrystallization of pure liquid Ni studied using molecular-dynamics simulation, J. Chem. Phys. 151, (2019) 124502. https://doi.org/10.1063/1.5119307

[38] J. H. Perepezko, R. J. Hebert, W. S. Tong, J. Hamann, H. R. Rösner, and G. Wilde, Nanocrystallization reactions in amorphous aluminum alloys, Mater. Trans., 44 (2003) 1982. https://doi.org/10.2320/matertrans.44.1982

[39] J. H. Perepezko, R. J. Hebert , and W. S. Tong, Amorphization and nanostructure synthesis in Al alloys, Intermetallics, 10 (2002) 1079-1088. https://doi.org/10.1016/S0966-9795(02)00144-9

[40] A. N. Kolmogorov, A statistical theory for the recrystallisation of metals, Isz. Akad. Nauk. USSR Ser. Matem., 3 (1937) 355-359.

[41] M. W. A. Johnson and K. F. Mehl, Reaction kinetics in processes of nucleation and growth, Transactions of the American Institute of Mining and Metallurgical Eng .Trans. Am. Inst. Mining. Met. Eng., 135 (1939) 416-442.

[42] M. Avrami, Granulation, phase change, and microstructure kinetics of phase change. III, J. Chem. Phys,. 9 (1941) 177-184. https://doi.org/10.1063/1.1750872

[43] J. W. Christian, The theory of transformations in metals and alloys, Pergamon Press Ltd., Oxford, 1975, p. 369.

[44] D. V. Louzguine and A. Inoue, Effect of Ni on stabilization of the supercooled liquid and devitrification of Cu-Zr-Ti bulk glassy alloys, J. Non-Cryst. Sol., 325 (2003) 187. https://doi.org/10.1016/S0022-3093(03)00338-7

[45] D. V. Louzguine and A. Inoue, Influence of Ni and Co additions on supercooled liquid region, devitrification behaviour and mechanical properties of Cu-Zr-Ti bulk metallic glass, Journal of Metastable & Nanocrystalline Materials, 15-16, (2003) 31-36. https://doi.org/10.4028/www.scientific.net/JMNM.15-16.31

[46] A. Takeuchi and A. Inoue, Calculations of mixing enthalpy and mismatch entropy for ternary amorphous alloys, Mater. Trans. JIM, 41 (2000) 1372-1378. https://doi.org/10.2320/matertrans1989.41.1372

[47] D. V. Louzguine-Luzgin, G. Xie, S. Li, Q. S. Zhang, W. Zhang, C. Suryanarayana and A. Inoue, Glass-forming ability and differences in the crystallization behavior of ribbons and rods of $Cu_{36}Zr_{48}Al_8Ag_8$ bulk glass-forming alloy, Journal of Materials Research, 24 (2009) 1886-1895. https://doi.org/10.1557/jmr.2009.0219

[48] S. Lan, Z.D. Wu, X.O. Wei, J, Zhou, Z.P. Lu, J. Neuefeind, X.L. Wang, Structure origin of a transition of classic-to-avalanche nucleation in Zr-Cu-Al bulk metallic glasses, Acta Materialia, 149 (2018) 108-118. https://doi.org/10.1016/j.actamat.2018.02.028

[49] T. Nagase, Y. Umakoshi, Thermal crystallization and electron irradiation induced phase transformation behavior in $Zr_{66.7}Cu_{33.3}$ metallic glass, Mater. Trans., 46 (2005) 616-621. https://doi.org/10.2320/matertrans.46.616

[50] G. Xie, Q. Zhang, D.V. Louzguine-Luzgin, W. Zhang and A. Inoue, Nanocrystallization of $Cu_{50}Zr_{45}Ti_5$ Metallic Glass Induced by Electron Irradiation, Mater. Trans., 47 (2006) 1930-1933. https://doi.org/10.2320/matertrans.47.1930

[51] A. Peker and W. L. Johnson, Time-temperature-transformation diagram of a highly processable metallic glass, Mater. Sci. Eng. A., 179-180 (1994) 173-175. https://doi.org/10.1016/0921-5093(94)90187-2

[52] N. Nishiyama and A. Inoue, Supercooling investigation and critical cooling rate for glass formation in Pd–Cu–Ni–P alloy, Acta Mater., 47 (1999) 1487-1495. https://doi.org/10.1016/S1359-6454(99)00030-0

[53] J. Loeffler, J. Schroers and W.L. Johnson, Time-temperature-transformation diagram and microstructures of bulk glass forming $Pd_{40}Cu_{30}Ni_{10}P_{20}$, Appl. Phys. Lett., 77 (2000) 681-683. https://doi.org/10.1063/1.127084

[54] R. Janlewing, and U. Köster, Nucleation in crystallization of Zr–Cu–Ni–Al metallic glasses Mater. Sci. Eng. A, 304-306 (2001) 833-838. https://doi.org/10.1016/S0921-5093(00)01574-4

[55] J. H. Kim, S. G. Kim, and A. Inoue, In situ observation of solidification behavior in undercooled Pd-Cu-Ni-P alloy by using a confocal scanning laser microscope, Acta Mater., 49 (2001) 615-622. https://doi.org/10.1016/S1359-6454(00)00353-0

[56] D. V. Louzguine and A. Inoue, Comparison of the long-term thermal stability of various metallic glasses under continuous heating, Scripta Materialia, 47 (2002) 887-891. https://doi.org/10.1016/S1359-6462(02)00268-3

[57] D. V. Louzguine-Luzgin and A. Inoue, Relation between time–temperature transformation and continuous heating transformation diagrams of metallic glassy alloys, Physica B: Condensed Matter, 358 (2005) 174-180. https://doi.org/10.1016/j.physb.2005.01.141

[58] R. A. Grange, and J.M. Kiefer, Transformation of austenite on continuous cooling and its realation to transformation at constant temperature, Trans. ASM, 29 (1941) 85-144.

[59] D. V. Louzguine-Luzgin and J. Jiang, On long-term stability of metallic glasses, Metals, 9 (2019) 1076. https://doi.org/10.3390/met9101076

[60] H. Gleiter, Nanocrystalline materials, Progress in Materials Science, 33 (1989) 223-315. https://doi.org/10.1016/0079-6425(89)90001-7

[61] L. Battezzati, C. Antonione, G. Riontino, F. Marino, and H.R. Sinning, The crystallization of a $Pd_{60}Ti_{20}Si_{20}$ metallic glass, Acta Metall. Mater., 39 (1991) 2107-2115. https://doi.org/10.1016/0956-7151(91)90181-Y

[62] W. N. Myung, L. Battezzati, M. Baricco, K. Oaki, A. Inoue, and T. Masumoto, Kinetic and thermodynamic aspects of crystallization in Cu-Ti-Ni and Cu-Ti-Al metallic glasses, Mater. Sci. Eng. A, 179-180 (1994) 371. https://doi.org/10.1016/0921-5093(94)90229-1

[63] U. Koster, and J. Meinhardt, Crystallization of highly undercooled metallic melts and metallic glasses around the glass transition temperature, Mater. Sci. Eng. A, 178 (1994) 271-278. https://doi.org/10.1016/0921-5093(94)90553-3

[64] A. R. Yavari and D. Negri, Effect of concentration gradients on nanostructure development during primary crystallization of soft-magnetic iron-based amorphous alloys and its modelling, Nanostructured Materials, 8 (1997) 969-986. https://doi.org/10.1016/S0965-9773(98)00047-6

[65] D. V. Louzguine-Luzgin, Aluminum-base amorphous and nanocrystalline materials, Metal Science and Heat Treatment, 53 (2012) 472–477. https://doi.org/10.1007/s11041-012-9417-3

[66] A. R. Yavari and O. Drbohlav, Thermodynamics and kinetics of nanostructure formation in soft-magnetic nanocrystalline alloys, Mater. Trans. JIM, 36 (1995) 896. https://doi.org/10.2320/matertrans1989.36.896

[67] A. Mitra, H.Y. Kim, D.V. Louzguine, N. Nishiyama, B. Shen, and A. Inoue, Structure and magnetic properties of amorphous and nanocrystalline $Fe_{40}Co_{40}Cu_{0.5}Zr_9Al_2Si_4B_{4.5}$ alloys, Journal of Magnetism and Magnetic Materials, 278 (2004) 299-305. https://doi.org/10.1016/j.jmmm.2003.12.1315

[68] M. Gogebakan, P. J. Warren, and B. Cantor, Crystallization behaviour of amorphous $Al_{85}Y_{11}Ni_4$ alloy, Mater. Sci. Eng., A 226-228 (1997) 168-172. https://doi.org/10.1016/S0921-5093(96)10611-0

[69] J. C. Foley, D. R. Allen and J. H. Perepezko, Analysis of nanocrystal development in Al-Y-Fe and Al-Sm glasses, Scripta Mater., 35 (1996) 655-660. https://doi.org/10.1016/1359-6462(96)00196-0

[70] M. Calin, A. Rudiger and U. Koester, Primary crystallization of Al-based metallic glasses, J. Metastable and Nanocryst. Mater., 8 (2000) 359. https://doi.org/10.4028/www.scientific.net/JMNM.8.359

[71] N. Bassim, C.S. Kiminami, and M.J. Kaufman, Phases formed during crystallization of amorphous $Al_{84}Y_9Ni_5Co_2$ alloy, J. Non-Cryst. Solids, 273 (2000) 271-276. https://doi.org/10.1016/S0022-3093(00)00135-6

[72] R. I. Wu, G. Wilde, and J. H. Perepezko, Glass formation and primary nanocrystallization in Al-base metallic glasses, Materials Science and Engineering A, 301 (2001) 12-17. https://doi.org/10.1016/S0921-5093(00)01390-3

[73] A. Inoue, Amorphous, nanoquasicrystalline and nanocrystalline alloys in Al-based systems, Prog. Mater. Sci., 43 (1998) 365-520. https://doi.org/10.1016/S0079-6425(98)00005-X

[74] A. R. Yavari, W. J. Botta, C.A.D. Rodrigues, C. Cardoso, and R.Z. Valiev, Nanostructured bulk $Al_{90}Fe_5Nd_5$ prepared by cold consolidation of gas atomised powder using severe plastic deformation, Scr. Mater., 46 (2002) 711-716. https://doi.org/10.1016/S1359-6462(02)00057-X

[75] M. C. Gao and G. J. Shiflet, Devitrification phase transformations in amorphous $Al_{85}Ni_7Gd_8$ alloy, Intermetallics, 10 (2002) 1131-1139. https://doi.org/10.1016/S0966-9795(02)00139-5

[76] Y. X. Zhuang, J. Z. Jiang, T. J. Zhou, H. Rasmussen, L. Gerward, M. Mezouar, W. Crichton, and A. Inoue, Crystallization of $Pd_{40}Cu_{30}Ni_{10}P_{20}$ bulk glass under pressure, Appl. Phys. Lett., 77 (2000) 4133. https://doi.org/10.1063/1.1332409

[77] Z. Altounian, S. Saini, J. Mainville and R. Bellissent, A neutron diffraction study of Al-rich glasses, Physica B: Condensed Matter, 241-243 (1997) 915-917. https://doi.org/10.1016/S0921-4526(97)00752-7

[78] K. Hono, Y. Zhang, A. P. Tsai, A. Inoue and T. Sakurai, Solute partitioning in partially crystallized Al-Ni-Ce(-Cu) metallic glasses, Scripta Mater., 32 (1995) 191. https://doi.org/10.1016/S0956-716X(99)80035-1

[79] A. R. Yavari and D. Negri, Effect of concentration gradients on nanostructure development during primary crystallization of soft-magnetic iron-based amorphous alloys and its modeling, Nanostr. Mater., 8 (1997) 969. https://doi.org/10.1016/S0965-9773(98)00047-6

[80] A. K. Gangopadhyay, T. K. Croat and K. F. Kelton, The effect of phase separation on subsequent crystallization in Al-Gd-Ni, Acta Mater., 48 (2000) 4035-4043. https://doi.org/10.1016/S1359-6454(00)00196-8

[81] K. F. Kelton, T. K. Croat, A. K. Gangopadhyay, L. Q. Xing, A. L. Greer, M. Weyland, X. Li and K. Rajan, Mechanisms for nanocrystal formation in metallic glasses, J. Non-Cryst. Sol., 317 (2003) 71-77. https://doi.org/10.1016/S0022-3093(02)02004-5

[82] S.V. Ketov, A. Inoue, H. Kato, D.V. Louzguine-Luzgin, Viscous flow of $Cu_{55}Zr_{30}Ti_{10}Co_5$ bulk metallic glass in glass-transition and semi-solid regions, Scripta Materialia, 68 (2013) 219–222. https://doi.org/10.1016/j.scriptamat.2012.10.037

[83] D. V. Louzguine-Luzgin, A.I. Bazlov, S.V. Ketov, A.L. Greer and A. Inoue, Crystal growth limitation as a critical factor for formation of Fe-based bulk metallic glasses, Acta Materialia, 82 (2015) 396–402. https://doi.org/10.1016/j.actamat.2014.09.025

[84] D.V. Louzguine-Luzgin, A.I. Bazlov, S.V. Ketov and A. Inoue, Crystallization behavior of Fe- and Co-based bulk metallic glasses and their glass-forming ability, Materials Chemistry and Physics, 162 (2015) 197–206. https://doi.org/10.1016/j.matchemphys.2015.05.058

[85] M. J. Duarte, A. Kostka, J.A. Jimenez, P. Choi, J. Klemm, D. Crespo, D. Raabe, F.U. Renner, Crystallization, phase evolution and corrosion of Fe-based metallic glasses: An atomic-scale structural and chemical characterization study, Acta Materialia 71 (2014) 20–30.

[86] A. Hirata, Y. Hirotsu, K. Amiya and A. Inoue, Crystallization process and glass stability of an $Fe_{48}Cr_{15}Mo_{14}C_{15}B_6Tm_2$ bulk metallic glass, Physical Review B, 78 (2008) 144205. https://doi.org/10.1103/PhysRevB.78.144205

[87] I. V. Lyasotskii, N. B. Dyakonova, E. N. Vlasova, D. L. Dyakonov and M. Yu. Yazvitskii, Metastable and quasiperiodic phases in rapidly quenched Fe–B–Si–Nb(Cu) alloys, Phys. Stat. Sol. A, 203 (2006) 259–270. https://doi.org/10.1002/pssa.200521126

[88] I. V. Lyasotsky, N. B. Dyakonova, and D. L. Dyakonov, Metastable primary precipitation phases in multicomponent glass forming Fe-base alloys with metalloids, Journal of Alloys and Compounds, 586 (2014) 20–23. https://doi.org/10.1016/j.jallcom.2013.03.112

[89] H. H. Tian and M. M. Atzmon, Comparison of X-ray analysis methods used to determine the grain size and strain in nanocrystalline materials, Philos. Mag. A, 79 (1999) 1769-1776. https://doi.org/10.1080/014186199251698

[90] E. J. Mittemeijer, and U. Welzel, The "state of the art" of the diffraction analysis of crystallite size and lattice strain, Z. Kristallogr., 223 (2008) 552–560. https://doi.org/10.1524/zkri.2008.1213

[91] E. Purushotham, and N. Krishna, X-ray determination of crystallite size and effect of lattice strain on Debye–Waller factors of platinum nano powders, Bull. Mater. Sci., 36 (2013) 973-976. https://doi.org/10.1007/s12034-013-0553-1

[92] F. Faupel, W. Frank, M.-P. Macht, H. Mehrer, V. Naundorf, K. Rätzke, H. R. Schober, S. K. Sharma and H. Teichler, Diffusion in metallic glasses and supercooled melts, Rev. Mod. Phys., 75 (2003) 237-250. https://doi.org/10.1103/RevModPhys.75.237

[93] H. Yoshizawa, K. Yamauchi, T. Yamane and H. Sugihara, Common mode choke cores using the new Fe-based alloys composed of ultrafine grain structure, J. Appl. Phys., 64 (1988) 6047. https://doi.org/10.1063/1.342150

[94] K. Suzuki, N. Kataoka, A. Inoue, A. Makino and T. Masumoto, High saturation magnetization and soft magnetic properties of BCC Fe-Zr-B alloys with ultrafine grain structure, Mater. Trans. JIM, 31 (1990) 743-746. https://doi.org/10.2320/matertrans1989.31.743

[95] K. Hono, Nanoscale microstructural analysis of metallic materials by atom probe field ion microscopy, Progress in Materials Science, 47 (2002) 621-729. https://doi.org/10.1016/S0079-6425(01)00007-X

[96] K. Hono, K. Hiraga, Q. Wang, A. Inoue and T. Sakurai, The microstructure evolution of a $Fe_{73.5}Si_{13.5}B_9Nb_3Cu_1$ nanocrystalline soft magnetic material, Acta Metall. et Mater., 40 (1992) 2137-2147. https://doi.org/10.1016/0956-7151(92)90131-W

[97] W. J. Botta F., D. Negri and A. R. Yavari, Crystallization of Fe-based amorphous alloys, J. Non-Cryst. Sol., 247 (1999) 19-25. https://doi.org/10.1016/S0022-3093(99)00024-1

[98] J. D. Ayers, V. G. Harris, J. A. Sprague, W. T. Elam and H. N. Jones, On the formation of nanocrystals in the soft magnetic alloy $Fe_{73.5}Nb_3Cu_1Si_{13.5}B_9$, Acta Materialia, 46 (1998) 1861-1874. https://doi.org/10.1016/S1359-6454(97)00436-9

[99] M. Ohnuma, K. Hono, H. Onodera, J. S. Pedersen and S. Linderoth, Cu clustering stage before the crystallization in Fe–Si–B–Nb–Cu amorphous alloys, Nanostructured Materials, 12 (1999) 693-696. https://doi.org/10.1016/S0965-9773(99)00219-6

[100] D. V. Louzguine and A. Inoue, Nanocrystallization of Ti-Ni-Cu-Sn amorphous alloy, Scripta Materialia, 43 (2000) 371. https://doi.org/10.1016/S1359-6462(00)00425-5

[101] G. He, J. Eckert and W. Loser, Stability, phase transformation and deformation behavior of Ti- base mettalic glass and composite, Acta Materialia, 51 (2003) 1621-1631. https://doi.org/10.1016/S1359-6454(02)00563-3

[102] Z. Altounian, E. Batalla, J.O. Strom-Olsen, and J.L. Walter, The influence of oxygen and other impurities on the crystallization of NiZr$_2$ and related metallic glasses, J. Appl Phys, 61 (1987) 149-155. https://doi.org/10.1063/1.338847

[103] L. Q. Xing, J. Eckert, W. Loser, L. Schultz and D. M. Herlach, Crystallization behaviour and nanocrystalline microstructure evolution of a Zr$_{57}$Cu$_{20}$Al$_{10}$Ni$_8$Ti$_5$ bulk amorphous alloy, Phil. Mag. A, 79 (1999) 1095-1108. https://doi.org/10.1080/01418619908210349

[104] D. V. Louzguine, H. Kato, H. S. Kim and A. Inoue, Formation of 2–5 nm size pre-precipitates of cF96 phase in a Hf-Co-Al glassy alloy, J. Alloys and Comp., 359 (2003) 198. https://doi.org/10.1016/S0925-8388(03)00292-5

[105] D. V. Louzguine, L. V. Louzguina, and A. Inoue, Multistage devitrification of Mg-Ni-Mm and Mg-Ni-Y-Mm metallic glasses (Mm=mischmetal), Philosophical Magazine, 83 (2003) 203. https://doi.org/10.1080/0141861021000032687

[106] D. V. Louzguine, A. Inoue, M. Saito and Y. Waseda, Structural relaxation in Ge-Cr-Al-Nd amorphous alloy, Scripta Materialia, 42 (2000) 289. https://doi.org/10.1016/S1359-6462(99)00348-6

[107] D. V. Louzguine and A. Inoue, Ge-Al-Cr-La amorphous alloys containing crystalline-like zones, NanoStructured Materials, 11 (1999) 115. https://doi.org/10.1016/S0965-9773(99)00024-0

[108] D. V. Louzguine and A. Inoue, Influence of rare earth metals (RE) on formation range and structure of amorphous phase in Ge-Al-Cr-RE system, Mater. Trans. JIM, 40 (1999) 485. https://doi.org/10.2320/matertrans1989.40.485

[109] D. V. Louzguine, A. Takeuchi and A. Inoue, Structure and crystallization behavior of Al-free Ge-based amorphous alloys produced by rapid solidification of the melt, Journal of Non-Crystalline Solids, 289 (2001) 196. https://doi.org/10.1016/S0022-3093(01)00702-5

[110] D. V. Louzguine and A. Inoue, Crystallization behaviour of Al-based metallic glasses below and above the glass-transition temperature, Journal of Non-Crystalline Solids, 311 (2002) 281-293. https://doi.org/10.1016/S0022-3093(02)01375-3

[111] K. Pekala, P. Jaskiewicz, J. Latuch and A. Kokoszkiewicz, Kinetics of crystallization processes in Al−Y−Ni−Co studied by electrical resistivity, J. Non-Cryst. Solids, 211 (1997) 72-76. https://doi.org/10.1016/S0022-3093(96)00627-8

[112] D. V. Louzguine and A. Inoue, Comparative analysis of crystallization of Al$_{85}$RE$_8$Ni$_5$Co$_2$ (RE-Rare Earth Metals) metallic glasses with and without supercooled

liquid region, Materials Science Forum, 386-388 (2002) 117-122.
https://doi.org/10.4028/www.scientific.net/MSF.386-388.117

[113] D. V. Louzguine and A. Inoue, Strong influence of supercooled liquid on crystallization of the $Al_{85}Ni_5Y_4Nd_4Co_2$ metallic glass, Appl. Phys. Lett., 78 (2001) 3061. https://doi.org/10.1063/1.1371795

[114] D. V. Louzguine and A. Inoue, Influence of a supercooled liquid on crystallization behaviour of Al-Y-Ni-Co metallic glass, Materials Letters, 54 (2002) 75-80. https://doi.org/10.1016/S0167-577X(01)00542-0

[115] F. Q. Guo, S. J Poon, and G. J. Shiflet, in Supercooled liquids, glass transition, and bulk metallic glasses, edited by, T. Egami, A.L. Greer, A. Inoue, S. Ranganathan, MRS Symposium Proceedings, 754 (2003) CC11.6.

[116] A.I. Bazlov, N. Yu Tabachkova, V.S. Zolotorevsky, D.V. Louzguine-Luzgin, Unusual crystallization of $Al_{85}Y_8Ni_5Co_2$ metallic glass observed in situ in TEM at different heating rates, Intermetallics, 94 (2018) 192–199. https://doi.org/10.1016/j.intermet.2017.12.024

[117] R. D. Sá Lisboa, C. Bolfarini, W. J. Botta F. and C. S. Kiminami, Topological instability as a criterion for design and selection of aluminum-based glass-former alloys, Appl. Phys. Lett., 86 (2005) 211904-211906. https://doi.org/10.1063/1.1931047

[118] T. Egami, and Y. Waseda, Atomic size effect on the formability of metallic glasses, J. Non-Crystall. Sol., 64 (1984) 113-134. https://doi.org/10.1016/0022-3093(84)90210-2

[119] J. Schroers, Y. Wu, R. Busch and W. L. Johnson, Transition from nucleation controlled to growth controlled crystallization in $Pd_{43}Ni_{10}Cu_{27}P_{20}$ melts, Acta Mater., 49 (2001) 2773-2779. https://doi.org/10.1016/S1359-6454(01)00159-8

[120] W. L. Johnson, G. Kaltenboeck, M. D. Demetriou, J. P. Schramm, X. Liu, K. Samwer, C. P. Kim and D. C. Hofmann, Beating crystallization in glass-forming metals by millisecond heating and processing, Science, 332 (2011) 828-833. https://doi.org/10.1126/science.1201362

[121] D. V. Louzguine-Luzgin and A. Inoue, The outline of glass transition phenomenon derived from the viewpoint of devitrification process, Physics and Chemistry of Glasses - European Journal of Glass Science and Technology Part B, 50 (2009) 27-30.

[122] G. Wilde, H. Sieber and J.H. Perepezko, Glass formation in Al-rich Al–Sm alloys during solid state processing at ambient temperature, J. Non-Cryst. Solids, 250-252 (1999) 621-625. https://doi.org/10.1016/S0022-3093(99)00147-7

[123] H. Chen, Y. He, G. J. Shiflet, and S. J. Poon, Deformation-induced nanocrystal formation in shear bands of amorphous alloys, Nature (London), 367 (1994) 541-543. https://doi.org/10.1038/367541a0

[124] W. H. Jiang, F. E. Pinkerton, and M. Atzmon, Effect of strain rate on the formation of nanocrystallites in an Al-based amorphous alloy during nanoindentation, J. Appl. Phys., 93 (2003) 9287-9290. https://doi.org/10.1063/1.1571234

[125] W. H. Jiang and M. Atzmon, The effect of compression and tension on shear-band structure and nanocrystallization in amorphous $Al_{90}Fe_5Gd_5$: a high-resolution transmission electron microscopy study, Acta Mater., 51 (2003) 4095-4105. https://doi.org/10.1016/S1359-6454(03)00229-5

[126] Y. He, G. J. Shiflet, and S. J. Poon, Ball milling-induced nanocrystal formation in aluminum-based metallic glasses, Acta Metall. Mater., 43 (1995) 83-91. https://doi.org/10.1016/0956-7151(95)90264-3

[127] A. A. Csontos and G. J. Shiflet, Formation and chemistry of nanocrystalline phases formed during deformation in aluminum-rich metallic glasses, Nanostruct. Mater., 9 (1997) 281-289. https://doi.org/10.1016/S0965-9773(97)90068-4

[128] J. J. Kim, Y. Choi, S. Suresh, and A. S. Argon, Nanocrystallization during nanoindentation of a bulk amorphous metal alloy at room temperature, Science, 295 (2002) 654-700.

[129] W. H. Jiang, F. E. Pinkerton, and M. Atzmon, Deformation-induced nanocrystallization in an Al-based amorphous alloy at a subambient temperature, Scr. Mater., 48 (2003) 1195-1200. https://doi.org/10.1016/S1359-6462(02)00568-7

[130] J. Li, Z.L. Wang, and T.C. Hufnagel, Characterization of nanometer-scale defects in metallic glasses by quantitative high-resolution transmission electron microscopy, Phys. Rev. B, 65 (2002) 144. https://doi.org/10.1103/PhysRevB.65.144201

[131] D. V. Louzguine-Luzgin and A. Inoue, Comparative study of the effect of cold rolling on the structure of Al-RE-Ni-Co (RE = rare-earth metals) amorphous and glassy alloys, Journal of Non-Crystalline Solids, 352 (2006) 3903-3909. https://doi.org/10.1016/j.jnoncrysol.2006.06.022

[132] F. Ye and K. Lu, Pressure effect on crystallization kinetics of an Al–La–Ni amorphous alloy, Acta Materialia, 47 (1999) 2449-2454. https://doi.org/10.1016/S1359-6454(99)00104-4

[133] Y. X. Zhuang, J. Z. Jiang, T. J. Zhou, H. Rasmussen, L. Gerward, M. Mezouar, W. Crichton, and A. Inoue, Pressure effects on $Al_{89}La_6Ni_5$ amorphous alloy crystallization, Appl. Phys. Lett., 77 (2000) 4133. https://doi.org/10.1063/1.1332409

[134] H. Chen, Y. He, G. J. Shiflet and S. J. Poon, Deformation-induced nanocrystal formation in shear bands of amorphous alloys, Nature, 367 (1994) 541-543. https://doi.org/10.1038/367541a0

[135] D. Shechtman, I.A. Blech, D. Gratias, and J.W. Cahn, Metallic phase with long-range orientational order and no translational symmetry, Phys. Rev. Lett., 53 (1984) 1951. https://doi.org/10.1103/PhysRevLett.53.1951

[136] D. Levine and R. J. Steinhardt, Quasicrystals: a new class of ordered structures, Phys. Rew. Lett., 53 (1984) 2477. https://doi.org/10.1103/PhysRevLett.53.2477

[137] A. Inoue, H. M. Kimura, and T. Masumoto, Formation, thermal stability and electrical resistivity of quasicrystalline phase in rapidly quenched Al-Cr alloys, J. Mater. Sci., 22 (1987) 1758. https://doi.org/10.1007/BF01132404

[138] A.P. Tsai, J.Q. Guo, E. Abe, H. Takakura, T.J. Sato, A stable binary quasicrystal, Nature, 408 (2000) 537. https://doi.org/10.1038/35046202

[139] D. V. Louzguine-Luzgin and A. Inoue, Formation and properties of quasicrystals, Annual Review of Materials Research, 38 (2008) 403-423. https://doi.org/10.1146/annurev.matsci.38.060407.130318

[140] S. Förster, K. Meinel, R. Hammer, and W. Widdra, Quasicrystalline structure formation in a classical crystalline thin-film system, Nature, 502 (2013) 215. https://doi.org/10.1038/nature12514

[141] V. Elser, Indexing problems in quasicrystal diffraction, Phys. Rev. B, 32 (1985) 4892. https://doi.org/10.1103/PhysRevB.32.4892

[142] P.A. Bancel, P.A. Heiney, P.W. Stephens, A.I. Goldman, and P.M. Horn, Structure of rapidly quenched Al-Mn, Phys. Rev. Lett., 54 (1985) 2422. https://doi.org/10.1103/PhysRevLett.54.2422

[143] J.W. Cahn, D. Shechtman and D. Gratias, Metallic phase with long-range orientational order and no translational symmetry, J. Mater. Res., 1 (1986) 13.

[144] A.L. Mackay, Dense non-crystallographic packing of equal spheres, Acta Crystallogr., 15 (1962) 916. https://doi.org/10.1107/S0365110X6200239X

[145] G. Bergman, J.L.T. Waugh, and L. Pauling, The crystal structure of the metallic phase $Mg_{32}(Al, Zn)_{49}$, Acta Crystalogr., 10 (1957) 254. https://doi.org/10.1107/S0365110X57000808

[146] S. Ranganathan and A. Inoue, An application of Pettifor structure maps for the identification of pseudo-binary quasicrystalline intermetallics, Acta Mater., 54 (2006) 3647. https://doi.org/10.1016/j.actamat.2006.01.041

[147] H.S. Jeevan and S. Ranganathan, A new basis for the classification of quasicrystals, J. Non-Cryst. Solids, 334 (2004) 184. https://doi.org/10.1016/j.jnoncrysol.2003.11.035

[148] D.G. Pettifor, A chemical scale for crystal-structure maps, Solid State Commun., 51 (1984) 31. https://doi.org/10.1016/0038-1098(84)90765-8

[149] H. Takakura, C.P. Gómez, A. Yamamoto, M. de Boissieu, and A.P. Tsai, Atomic structure of the binary icosahedral Yb–Cd quasicrystal, Nature Mater., 6 (2007) 58. https://doi.org/10.1038/nmat1799

[150] R. Merlin, K. Bajema, R. Clarke, F.Y. Juang and P.K. Bhattacharya, Quasiperiodic GaAs-AlAs heterostructures, Phys. Rev. Lett., 55 (1985) 1768. https://doi.org/10.1103/PhysRevLett.55.1768

[151] N. Wang, H. Chen, and K.H. Kuo, Two-dimensional quasicrystal with eightfold rotational symmetry, Phys. Rev. Lett., 59 (1987) 1010. https://doi.org/10.1103/PhysRevLett.59.1010

[152] S. Ranganathan, K. Chattopadhyay, A. Singh and K.F. Kelton, Decagonal quasicrystals, Progr. Mater. Sci., 41 (1997) 195. https://doi.org/10.1016/S0079-6425(97)00028-5

[153] A.K. Srivastava and S. Ranganathan, Quasicrystals, crystals and multiple twins in rapidly solidified AlCrSi, AlMnSi and AlMnCrSi alloys, Acta. Mater., 44 (1996) 2935. https://doi.org/10.1016/1359-6454(95)00353-3

[154] W.W. Warren Jr, H.S. Chen, and J.J. Hauser, NMR spin-echo spectra of icosahedral quasicrystals, Phys. Rev., 32 (1985) 7614. https://doi.org/10.1103/PhysRevB.32.7614

[155] E. Matsubara, M. Sakurai, T. Nakamura, M. Imafuku, S. Sato, J. Saida and A. Inoue, Environmental structural studies in amorphous and quasicrystalline $Zr_{70}Al_6Ni_{10}Pt_{14}$ alloys, Scripta Mater., 44 (2001) 2297. https://doi.org/10.1016/S1359-6462(01)00895-8

[156] K. F. Kelton, G.W. Lee, A.K. Gangopadhyay, R.W. Hyers, T.J. Rathz, J.R. Rogers, M.B. Robinson and D.S. Robinson, First X-Ray Scattering studies on electrostatically levitated metallic liquids: demonstrated influence of local icosahedral order on the nucleation barrier, Phys. Rev. Lett., 90 (2003) 195504. https://doi.org/10.1103/PhysRevLett.90.195504

[157] K. F. Kelton, Crystallization of liquids and glasses to quasicrystals, J. Non-Cryst. Sol., 334-335 (2004) 253. https://doi.org/10.1016/j.jnoncrysol.2003.11.052

[158] Lj. Ouyang, D. V. Louzguine, H. M. Kimura, T. Ohsuna, S. Ranganathan and A. Inoue, Influence of icosahedral clusters on crystallization of $Zr_{55}Ni_{10}Al_{7.5}Cu_{7.5}Ti_{10}Ta_{10}$ glassy alloy, Journal of Metastable and Nanocrystalline Materials, 18 (2003) 37. https://doi.org/10.4028/www.scientific.net/JMNM.18.37

[159] N. Chen, D.V. Louzguine, S. Ranganathan, and A. Inoue, Formation ranges of icosahedral, amorphous and crystalline phases in rapidly solidified Ti–Zr–Hf–Ni alloys, Acta Materialia, 53 (2005) 759. https://doi.org/10.1016/j.actamat.2004.10.027

[160] D. V. Louzguine and A. Inoue, Nanoscale cF96 cubic versus icosahedral phase in devitrified Hf-based metallic glasses, Annales de Chimie - Science des Matériaux, 27 (2002) 91. https://doi.org/10.1016/S0151-9107(02)80049-6

[161] U. Koster, J. Meinhardt, S. Roos, and H. Liebertz, Formation of quasicrystals in bulk glass forming Zr–Cu–Ni–Al alloys, Appl. Phys. Lett., 69 (1996) 179. https://doi.org/10.1063/1.117364

[162] L. Q. Xing, J. Eckert, W. Loser, and L. Schultz, Effect of cooling rate on the precipitation of quasicrystals from the Zr–Cu–Al–Ni–Ti amorphous alloy, Appl. Phys. Lett., 73 (1998) 2110. https://doi.org/10.1063/1.122394

[163] M. W. Chen, T. Zhang, A. Inoue, A. Sakai, and T. Sakurai, Quasicrystals in a partially devitrified $Zr_{65}Al_{7.5}Ni_{10}Cu_{12.5}Ag_5$ bulk metallic glass, Appl. Phys. Lett., 75 (1999) 1697. https://doi.org/10.1063/1.124793

[164] A. Inoue, J. Saida, M. Matsushita and T. Sakurai, Formation of an icosahedral quasicrystalline phase in $Zr_{65}Al_{7.5}Ni_{10}M_{17.5}$ (M=Pd, Au or Pt) alloys, Mater. Trans. JIM, 41 (2000) 362. https://doi.org/10.2320/matertrans1989.41.362

[165] J. Saida, M. Matsushita, and A. Inoue, Transformation kinetics of nanoicosahedral phase from a supercooled liquid region in $Zr_{70}Pd_{30}$ binary glassy alloy, J. Appl. Phys., 88 (2000) 6081. https://doi.org/10.1063/1.1322377

[166] B. S. Murty, D. H. Ping, M. Ohnuma and K. Hono, Nanoquasicrystalline phase formation in binary Zr–Pd and Zr–Pt alloys, Acta Mater., 49 (2001) 3453. https://doi.org/10.1016/S1359-6454(01)00254-3

[167] J. Saida, M. Matsushita and A. Inoue, Nano icosahedral phase in Zr–Pd and Zr–Pt binary alloys, Journal of Alloys and Compounds, 342 (2002) 18. https://doi.org/10.1016/S0925-8388(02)00117-2

[168] D. V. Louzguine and A. Inoue, Formation of a nanoquasicrystalline phase in Zr-Cu-Ti-Ni metallic glass, Applied Physics Letters, 78 (2001) 1841. https://doi.org/10.1063/1.1358362

[169] X. Y. Yang, M. J. Kramer, E. A. Rozhkova , and D. J. Sordelet, Coincident lattice sites between cubic β-Zr(Pt) and an isochemical icosahedral phase in rapidly solidified $Zr_{80}Pt_{20}$ alloys, Scripta Materialia, 49 (2003) 885-890. https://doi.org/10.1016/S1359-6462(03)00440-8

[170] D. V. Louzguine, M. S. Ko and A. Inoue, Nanoquasicrystalline phase produced by devitrification of Hf-Pd-Ni-Al metallic glass, Appl. Phys. Lett., 76 (2000) 3424-3426. https://doi.org/10.1063/1.126667

[171] C. Li, J. Saida, M. Matsushita, and A. Inoue, Precipitation of icosahedral quasicrystalline phase in $Hf_{65}Al_{7.5}Ni_{10}Cu_{12.5}Pd_5$ metallic glass, Appl. Phys. Lett., 77 (2000) 528-530. https://doi.org/10.1063/1.127033

[172] D. V. Louzguine, M. S. Ko, and A. Inoue, Nanoscale icosahedral phase produced by devitrification of Hf-Au-Ni-Al and Hf-Au-Cu-Al metallic glasses, Scripta Mater., 44 (2001) 637-642. https://doi.org/10.1016/S1359-6462(00)00606-0

[173] D. V. Louzguine and A. Inoue, Nanoparticles with icosahedral symmetry in Cu-based bulk glass former induced by Pd addition, Scripta Mater., 48 (2003) 1325-1329. https://doi.org/10.1016/S1359-6462(03)00018-6

[174] D. V. Louzguine and A. Inoue, Gold as an alloying element promoting formation of a nanoicosahedral phase in a Cu-based alloy, J. Alloys Comp., 361 (2003) 153-156. https://doi.org/10.1016/S0925-8388(03)00409-2

[175] D. V. Louzguine, H. Kato, and A. Inoue, Investigation of mechanical properties and devitrification of Cu-based bulk glass formers alloyed with noble metals, Sci. Tech. Adv. Mater., 4 (2003) 327-331. https://doi.org/10.1016/j.stam.2003.09.003

[176] D. V. Louzguine-Luzgin, A. R. Yavari, M. Fukuhara, K. Ota, G. Xie, G. Vaughan , and A. Inoue, Free volume and elastic properties changes in Cu-Zr-Ti-Pd bulk glassy alloy on heating, Journal of Alloys and Compounds, 431 (2007) 136-140. https://doi.org/10.1016/j.jallcom.2006.05.069

[177] D. V. Louzguine-Luzgin, A. Inoue, D. Nagahama, and K. Hono, Composition and structure of Cu-based nanoicosahedral phase in Cu-Zr-Ti-Pd alloy, Applied Physics Letters, 87 (2005) 211918. https://doi.org/10.1063/1.2135143

[178] J. Z. Jiang, A. R. Rasmussen, C. H. Jensen, Y. Lin, and P. L. Hansen, Change of quasilattice constant during amorphous-to-quasicrystalline phase transformation in $Zr_{65}Al_{7.5}Ni_{10}Cu_{7.5}Ag_{10}$ metallic glass, Appl. Phys. Lett., 80 (2002) 2090. https://doi.org/10.1063/1.1463207

[179] D. V. Louzguine-Luzgin and A Yu. Churyumov, Dual-phase glassy/nanoscale icosahedral phase materials in Cu–Zr–Ti–Pd system alloys, Materials Characterization, 96 (2014) 6-12. https://doi.org/10.1016/j.matchar.2014.07.014

[180] Z. Wang, S.V. Ketov, C.L. Chen, Y. Shen, Y. Ikuhara, A.A. Tsarkov, D.V. Louzguine-Luzgin, and J.H. Perepezko, Nucleation and thermal stability of an icosahedral nanophase during the early crystallization stage in Zr-Co-Cu-Al metallic glasses, Acta Materialia, 132 (2017) 298-306. https://doi.org/10.1016/j.actamat.2017.04.044

[181] D. V. Louzguine-Luzgin and A. Inoue, Structure and transformation behaviour of a rapidly solidified Al–Y–Ni–Co–Pd alloy, J. Alloys and Comp., 399 (2005) 78-85. https://doi.org/10.1016/j.jallcom.2005.02.018

[182] D. V. Louzguine, and A. Inoue, Investigation of structure and properties of the Al-Y-Ni-Co-Cu metallic glasses, Journal of Materials Research, 17 (2002) 1014-1018. https://doi.org/10.1557/JMR.2002.0149

[183] A. Inoue , and W. Zhang, Nanocrystalline Fe-Pt-B base hard magnets with high coercive force obtained from amorphous precursor, J. Appl. Phys., 97 (2005) 10H308. https://doi.org/10.1063/1.1854252

[184] W. Zhang, D. V. Louzguine and A. Inoue, Synthesis and magnetic properties of Fe-Pt-B nanocomposite permanent magnets with low Pt concentrations, Applied Physics Letters, 85 (2004) 4998. https://doi.org/10.1063/1.1824172

[185] D. V. Louzguine-Luzgin, W. Zhang and A. Inoue, Nanoscale precipitates and phase transformations in a rapidly-solidified Fe-Pt-B amorphous alloy, Journal of Alloys and Compounds, 402 (2005) 78-83. https://doi.org/10.1016/j.jallcom.2005.03.089

[186] J.Q. Wang, N. Chen, P. Liu, Z. Wang, D.V. Louzguine-Luzgin, M.W. Chen, and J.H. Perepezko, The ultrastable kinetic behavior of an Au-based nanoglass, Acta Materialia, 79 (2014) 30–36. https://doi.org/10.1016/j.actamat.2014.07.015

[187] D. V. Louzguine and A. Inoue, Effect of Ge addition to $Si_{50-55}Al_{25-20}Fe_{10}Ni_5Cr_5Zr_5$ alloys obtained by melt spinning, NanoStructured Materials, 8 (1997) 1007-1013. https://doi.org/10.1016/S0965-9773(98)00043-9

[188] D. V. Louzguine and A. Inoue, Precipitation of nanogranular Ge particles in rapidly solidified Al-Si-Fe-Cr-Ge alloys, Materials Transactions JIM, 39 (1998) 504-507. https://doi.org/10.2320/matertrans1989.39.504

[189] Z. Wang, C.L. Chen, S. V. Ketov, K. Akagi, A. A. Tsarkov, Y. Ikuhara, D. V. Louzguine-Luzgin, Local chemical ordering within the incubation period as a trigger for

nanocrystallization of a highly supercooled Ti-based liquid, Materials & Design, 156 (2018) 504-513. https://doi.org/10.1016/j.matdes.2018.07.013

[190] D. V. Louzguine and A. Inoue, The structure and phase transformation behaviour of rapidly solidified alloys in the Ge-Al-La system, Mater. Res. Bull., 34 (1999) 1991-2001. https://doi.org/10.1016/S0025-5408(99)00200-7

[191] D. V. Louzguine, Lj. Ouyang, H. M. Kimura and A. Inoue, Transformation from glassy + beta-Zr to glassy+icosahedral structure in Zr-based alloy, Scripta Mater., 50 (2004) 973-976. https://doi.org/10.1016/j.scriptamat.2004.01.005

[192] N. Chen, D. V. Louzguine-Luzgin, S. Ranganathan and A. Inoue, Glassy and icosahedral phases in rapidly solidified Ti-Zr-Hf-(Fe, Co or Ni) alloys, J. Non-Cryst. Sol., 351 (2005) 2547–2551. https://doi.org/10.1016/j.jnoncrysol.2005.06.041

CHAPTER 5

Mechanical Properties and Deformation Behavior

Owing to the absence of a crystalline lattice and dislocations, a unique deformation mechanism is realized in bulk glassy alloys, which thus, exhibit high strength, high hardness, good wear resistance and large elastic deformation. In the present chapter the mechanical properties and deformation behavior of BMGs at different temperatures are discussed.

Contents

5.1   Mechanical properties of bulk metallic glasses and their deformation behavior at room temperature

5.1.1   Elastic properties

The moduli of elasticity define how a material responds to external loading until plastic deformation begins. The bulk modulus (K) is related to the hydrostatic pressure required to change the average interatomic distances in a body. The Young's modulus (E) is a physical value characterizing the properties of the material to resist tensile, compressive or bending elastic deformation. The shear modulus (G) characterizes the ability of a material to resist shear deformation. The moduli of elasticity and the Poisson's ratio (v) are mutually related. For example, in a homogeneous isotropic material (like a BMG at a long enough length scale) the shear modulus is related to the Young's modulus by the Poisson's ratio:

$$G = \frac{E}{2(1+v)} \tag{5.1}$$

The elastic properties of bulk metallic glasses are shown in Table 5.1 [1,2]. θ_D is the Debye temperature.

As one can see in Fig. 5.1 although BMGs have rather reduced E modulus values compared to crystal their stress-strain curves exhibit good linearity. A typical coefficient of determination for the linear fit R^2 can be as high as 0.9999.

Table 5.1. Elastic properties of bulk metallic glasses rounded to integers and the Debye temperature from two works marked as "Ref.".

Alloys	K (GPa)	G (GPa)	E (GPa)	ν	θ_D (K)	Ref.
$Zr_{55}Al_{10}Ni_{10}Cu_{15}Be_{10}$	112	34	94	0.36	297	1
$Zr_{65}Al_{10}Ni_{10}Cu_{15}$	112	36	97	0.36	293	1
$Zr_{61.88}Al_{10}Ni_{10.12}Cu_{18}$	108	29	80	0.38	263	1
$Zr_{62.325}Cu_{17.55}Ni_{10.125}Al_{10}$	108	30	82	0.37	266	1
$Zr_{61}Cu_{18.3}Ni_{12.8}Al_{7.9}$	101	29	79	0.37	260	1
$Zr_{62}Al_{10}Ni_{12.6}Cu_{15.4}$	109	29	80	0.38	262	1
$Pd_{40}Ni_{10}Cu_{30}P_{20}$	173	36	100	0.40	279	1
$Pd_{40}Ni_{40}P_{20}$	175	37	105	0.40	292	1
$Pt_{57.5}Cu_{14.7}Ni_{5.3}P_{22.5}$	199	33	95	0.42	206	1
$Fe_{48}Cr_{15}Mo_{14}C_{15}B_6Er_2$	192	86	224	0.31	489	1
$Fe_{41}Co_7Cr_{15}Mo_{14}C_{15}B_6Y_2$	193	84	220	0.31	488	1
$Mg_{60}Cu_{25}Gd_{15}$	47	20	52	0.31	261	1
$Mg_{55}Cu_{25}Ag_{10}Gd_{10}$	54	24	62	0.31	274	1
$Mg_{58.5}Cu_{30.5}Y_{11}$	49	20	54	0.32	293	1
$Ca_{50}Mg_{20}Cu_{30}$	29	13	33	0.31	–	1
$Sr_{50}Mg_{20}Zn_{20}Cu_{10}$	17	9	23	0.28	169	1
$Sr_{60}Mg_{20}Zn_{15}Cu_5$	15	8	20	0.28	157	1
$Ce_{60}Al_{15}Ni_{15}Cu_{10}$	37	19	48	0.28	146	1
$Ce_{65}Al_{10}Ni_{10}Cu_{10}Nb_5$	30	12	31	0.33	145	1
$Nd_{60}Al_{10}Fe_{20}Co_{10}$	47	21	54	0.31	159	1
$Nd_{60}Al_{10}Ni_{10}Cu_{20}$	43	14	37	0.36	189	1
$Er_{55}Al_{25}Co_{20}$	61	27	71	0.31	229	1
$La_{66}Al_{14}Cu_{10}Ni_{10}$	35	13	36	0.33	161	1
$Cu_{47}Zr_{45}Al_8$	123	37	108	0.30	-	2
$(Cu_{47}Zr_{45}Al_8)_{96}Lu_4$	119	37	89	0.35	-	2
$(Cu_{47}Zr_{45}Al_8)_{96}Y_4$	97	32	88	0.36	-	2
$(Cu_{47}Zr_{45}Al_8)_{96}Du_4$	119	36	85	0.36	-	2

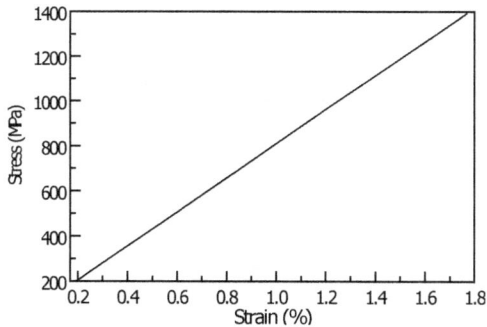

Fig. 5.1. A stress-strain loading curve for $Zr_{61}Cu_{27}Fe_2Al_{10}$ bulk sample up to 1400 MPa.

These elastic constants are found to be sensitive probes for evaluation of the structure changes and analyzing phase transformations in glassy alloys. The elastic moduli of the $Zr_{65}Ni_{10}Cu_5Al_{7.5}Pd_{12.5}$ alloy were studied in the glassy, quasicrystalline and crystalline states [3]. A relation between the structure and elasticity of the glassy alloy was established. The high Poisson's ratio, large values of K, λ (Lame' parameter) and K/G ratio were found in the as-cast state (Fig. 5.2). The high Poisson's ratio of this alloy (partly possibly owing to low: 110 mass ppm (0.06 at.%) oxygen content) correlates with large compressive plasticity of Zr-based BMGs containing Pd [4]. The reduction of K and λ on heating is associated with the formation of an icosahedral phase.

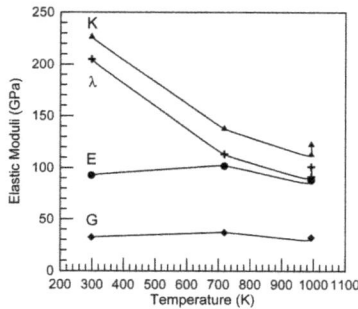

Fig. 5.2. Elastic parameters as a function of the annealing temperature (and time at 1000 K -the last points). Reproduced from [3] with permission of Elsevier.

5.1.2 Mechanical strength, hardness and plasticity

Owing to the absence of crystalline lattice and dislocations, a unique deformation mechanism is realized in bulk metallic glassy alloys [5], which thus, exhibit high strength, high hardness, good wear resistance [6] and large elastic deformation [7]. For example, the $(Al_{0.84}Y_{0.09}Ni_{0.05}Co_{0.02})_{95}Sc_5$ glassy alloy has an ultra-high tensile fracture strength exceeding 1500 MPa which surpasses those for all other Al-based fully crystalline and amorphous alloys reported up to date [8]. 1.5-1.8 GPa strength values are typical for Zr- and Pd-based alloys. Cu-Zr-based alloys become stronger with increase in Cu content which is related to the smaller excess volume in Cu-rich alloys [9]. For example, compressive yield strength values for Cu-Zr-Ti glassy alloys are higher than those for Zr-based bulk glassy alloys [10]. The $Cu_{57.5}Zr_{30}Ti_{10}Ni_{2.5}$ bulk glassy alloy showed an ultimate compressive strength of 2.2 MPa [11]. Similar values were found for the Cu–Zr–Ti–(Pd, Ag, Pt and Au) bulk glass-formers [12]. The highest strength value among Cu-based alloys (2.3 GPa) was found for a Cu-Zr-Ti-Co bulk metallic glass [13]. Ferrous-metals based alloys have much higher strength. For example, the Fe–(Co,Ni)–(Zr,Nb,Ta)–(Mo,W)–B system bulk glassy alloys exhibit a high compressive strength of 3.8 GPa and good corrosion resistance [14]. The (Fe, Co)-Cr-Mo-C-B-Tm system glassy alloys prepared in a cylindrical form with a diameter up to 18 mm demonstrate an excellent GFA and high strength exceeding 4 GPa [15]. The $Co_{56}Ta_9B_{35}$ bulk glassy alloy exhibited ultrahigh fracture strength exceeding 5 GPa, high Young's modulus of 275 GPa and high specific strength of 643 N/m·g [16]. Hardness values of bulk metallic glasses range from about 2 GPa for La- and Mg-based BMG to about 5 GPa for Zr-based to 10-12 GPa for Fe-based BMGs and up to 16 GPa for Co-based BMGs.

Mechanical properties of bulk metallic glasses in terms of the yield strength (σ_y), Vickers microhardness (HV) as well as the Young's modulus (E) are summarized in Table 5.2 (compressive) and Table 5.3 (tensile). A correlation is found between HV and σ_y as expected for isotropic solids like BMGs. One can note that the Zr-Cu-Fe-Al alloys with good plasticity showed HV/σ_y values [17] lower than 3 (Fig. 5.3(a)). Such alloys are marked with diamonds. It is interesting to note that this ratio calculated using the tensile strength of Zr-Cu-Ni-Al alloys (Table 5.3) tested in tension is also lower than 3. It can be connected with high purity of Zr used for the samples preparation in that work [18] which might have caused reduction in hardness. Also, the majority of these alloys were reported to exhibit high Charpy impact fracture toughness exceeding 100 kJ/m². Formation of a more open structure with a larger excess volume can also be a reason for higher plasticity of these glasses [19]. Also there is a correlation between strength and the Young's modulus (Fig. 5.3(b)).

Table 5.2. Composition and compressive mechanical properties of bulk glassy alloys. The values of σ_y are rounded to the nearest tens.

r	Content, at. %	σ_y (MPa)	E (GPa)	HV (kgf/mm²)	HV (MPa)	HV/ σ_y	Ref.
1	$Co_{56}Ta_9B_{35}$	5590	247		15900	2.8	16
2	$Cu_{60}Hf_{25}Ti_{15}$	2010	124				20
4	$Cu_{47.5}Zr_{47.5}Al_5$	1550	87				21
5	$Cu_{46}Zr_{46}Al_8$	1900		580	5690	3.0	22
7	$Cu_{47}Zr_{47}Al_6$	1730		580	5690	3.3	23
9	$Fe_{63}Cr_3Mo_{12}P_{10}C_7B_5$	2900		975	9555	3.3	24
10	$La_{45}Al_{45}Ni_{10}$	1080	52	330	3240	3.0	25
11	$La_{45}Al_{35}Ni_{20}$	1000	46	305	2990	3.0	25
12	$La_{50}Al_{35}Ni_{15}$	950	41	290	2850	3.0	25
13	$La_{50}Al_{30}Ni_{20}$	930	41	285	2800	3.0	25
14	$La_{55}Al_{25}Ni_{20}$	740	34	225	2210	3.0	23
15	$Mg_{80}Cu_{10}Y_{10}$	630		220	2160	3.4	26
16	$Zr_{70}Al_{10}Ni_{20}$	1410	61	432	4240	3.0	27
17	$Zr_{65}Al_{10}Ni_{25}$	1580	65	484	4750	3.0	27
19	$Zr_{65}Al_{15}Ni_{20}$	1610	71	494	4850	3.0	27
20	$Zr_{60}Al_{15}Ni_{25}$	1640	73	502	4930	3.0	27
21	$Zr_{60}Ni_{20}Al_{20}$	1790	78	549	5390	3.0	28
22	$Zr_{70}Ni_{20}Al_{10}$	1410	61	432	4240	3.0	29
23	$Zr_{65}Ni_{25}Al_{10}$	1520	65	484	4750	3.1	29
24	$Zr_{65}Ni_{20}Al_{15}$	1610		494	4850	3.0	29
25	$Zr_{60}Ni_{25}Al_{15}$	1640		502	4930	3.0	29
27	$Cu_{58.8}Zr_{29.4}Ti_{9.8}Y_2$	1780	115				30
29	$Zr_{55}Al_{20}Co_{20}Cu_5$	2000	92				31

30	$Zr_{60}Cu_{30}Al_{10}$	1710	93	465	4560	2.7	32
31	$Zr_{60}Cu_{25}Fe_5Al_{10}$	1650	92	479	4700	2.9	32, 33
32	$Zr_{60}Cu_{22.5}Fe_{7.5}Al_{10}$	1720	100	485	4760	2.8	32
33	$Zr_{60}Cu_{20}Fe_{10}Al_{10}$	1710	104	492	4830	2.8	32
34	$Zr_{60}Cu_{17.5}Fe_{12.5}Al_{10}$	1730	95	498	4890	2.8	32
35	$Zr_{62.5}Cu_{22.5}Fe_5Al_{10}$	1580	88	460	4520	2.8	33
36	$Zr_{65}Cu_{20}Fe_5Al_{10}$	1590	91	451	4430	2.8	33
37	$Zr_{67.5}Cu_{17.5}Fe_5Al_{10}$	1500	91	434	4260	2.8	33

Table 5.3. Tensile mechanical properties of the cast Zr-Cu-Ni-Al bulk glassy alloys: σ_t is tensile strength, HV is Vickers microhardness, and E is the Young's modulus [18]. The values σ_y are rounded to the nearest tens.

Nr	Alloy composition	σ_t (MPa)	E (GPa)	HV (kgf/mm^2)	HV (MPa)	HV/σ_t
1	$Zr_{46}Cu_{34}Ni_8Al_{12}$	1780	111	562	5513	3.1
2	$Zr_{48}Cu_{28}Ni_{12}Al_{12}$	1910	102	530	5199	2.7
3	$Zr_{48}Cu_{30}Ni_{10}Al_{12}$	1980	92	528	5180	2.6
4	$Zr_{48}Cu_{32}Ni_8Al_{12}$	2100	102	527	5170	2.5
5	$Zr_{48}Cu_{34}Ni_6Al_{12}$	1900	94	529	5189	2.7
6	$Zr_{50}Cu_{26}Ni_{12}Al_{12}$	1880	88	498	4885	2.6
7	$Zr_{50}Cu_{28}Ni_{10}Al_{12}$	1990	92	517	5072	2.5
8	$Zr_{50}Cu_{30}Ni_8Al_{12}$	1820	92	526	5160	2.8
9	$Zr_{50}Cu_{32}Ni_6Al_{12}$	1880	92	521	5111	2.7
10	$Zr_{50}Cu_{34}Ni_4Al_{12}$	1910	91	517	5072	2.7
11	$Zr_{52}Cu_{26}Ni_{10}Al_{12}$	1960	89	509	4993	2.5
12	$Zr_{52}Cu_{28}Ni_8Al_{12}$	1798	94	512	5023	2.8
13	$Zr_{52}Cu_{30}Ni_6Al_{12}$	1820	93	506	4964	2.7

14	$Zr_{52}Cu_{32}Ni_4Al_{12}$	1780	88	501	4915	2.8
15	$Zr_{48}Cu_{32}Ni_{10}Al_{10}$	1890	94	513	5033	2.7
16	$Zr_{50}Cu_{28}Ni_{12}Al_{10}$	1810	89	518	5082	2.8
17	$Zr_{50}Cu_{32}Ni_8Al_{10}$	1800	87	508	4983	2.8
18	$Zr_{50}Cu_{30}Ni_{10}Al_{10}$	1960	92	509	4993	2.5
19	$Zr_{52}Cu_{32}Ni_6Al_{10}$	1890	86	490	4807	2.5
20	$Zr_{52}Cu_{28}Ni_{10}Al_{10}$	1890	88	498	4885	2.6
21	$Zr_{52}Cu_{30}Ni_8Al_{10}$	1860	89	498	4885	2.6
22	$Zr_{54}Cu_{28}Ni_6Al_{10}$	1780	83	477	4679	2.6
23	$Zr_{54}Cu_{28}Ni_8Al_{10}$	1570	89	485	4758	3.0
24	$Zr_{50}Cu_{28}Ni_{14}Al_8$	1800	86	509	4993	2.8
25	$Zr_{50}Cu_{32}Ni_{10}Al_8$	1960	96	504	4944	2.5
26	$Zr_{50}Cu_{34}Ni_8Al_8$	1890	98	503	4934	2.6
27	$Zr_{52}Cu_{28}Ni_{12}Al_8$	1900	92	490	4807	2.5
28	$Zr_{52}Cu_{30}Ni_{10}Al_8$	1830	84	490	4807	2.6
29	$Zr_{52}Cu_{32}Ni_8Al_8$	1800	95	492	4827	2.7
30	$Zr_{52}Cu_{34}Ni_6Al_8$	1860	92	488	4787	2.6
31	$Zr_{54}Cu_{30}Ni_8Al_8$	1820	97	469	4601	2.5
32	$Zr_{54}Cu_{32}Ni_6Al_8$	1810	97	475	4660	2.6

Deformation behavior of bulk metallic glassy alloys has been under intense investigations since the last century [34]. Although, the majority of bulk metallic glasses tested at ambient temperature fracture shortly after yielding, considerable plasticity was observed in several specific bulk metallic glassy alloys [35,36], particularly, in the Zr-Cu-Fe-Al system alloys [33] (Fig. 5.4(a)), Zr-rich alloys of the Zr-Cu-Ni-Al system [18], Zr-Al-Ni-Pd [37] and Pt-Cu-Ni-P [38] system alloys. Moreover, room temperature tensile ductility of a few percents is found in small samples of Zr-rich metallic glasses at high enough strain rate [39,40]. One should note that mechanical behavior of a BMG at room temperature also significantly depends on the sample size and testing machine stiffness. By using smaller samples and stiffer machines one can obtain higher plasticity [41,42]. A well known indentation size effect is also found in the case of bulk metallic glasses [43,44].

Micro-hardness of the samples measured at different load cannot be directly compared as it gradually decreases with load typically up to about 5 N and then follows a plateau (Fig. 5.4(b)). Thus, one should always report the indentation load used. Moreover strain softening and strain rate dependence can also change the behavior of BMGs.

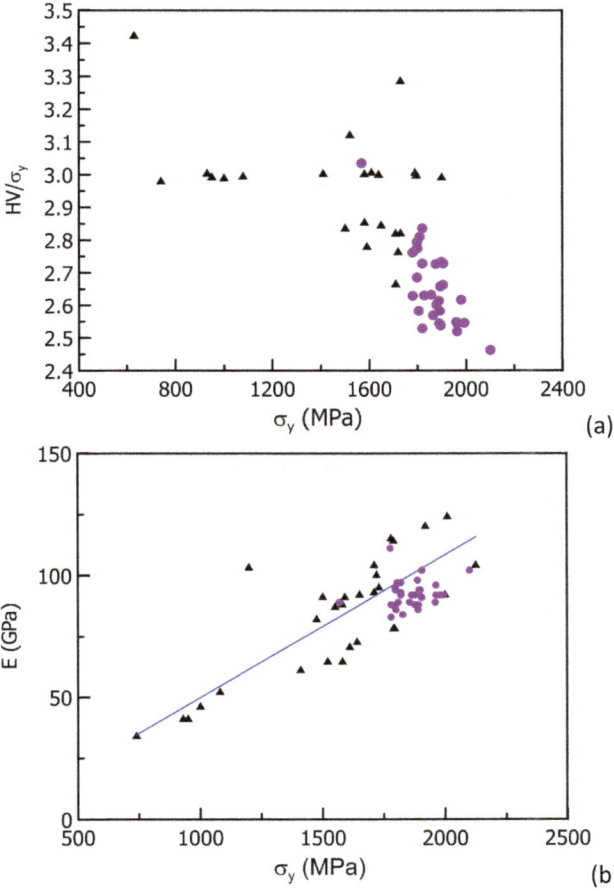

Fig. 5.3. Correlation between the mechanical properties of bulk metallic glasses. (a) HV/σ_y versus σ_y. Here σ_y denotes the yield strength. (b) E as a function of σ_y. Compressive σ_y (triangles) and σ_t tensile (circles).

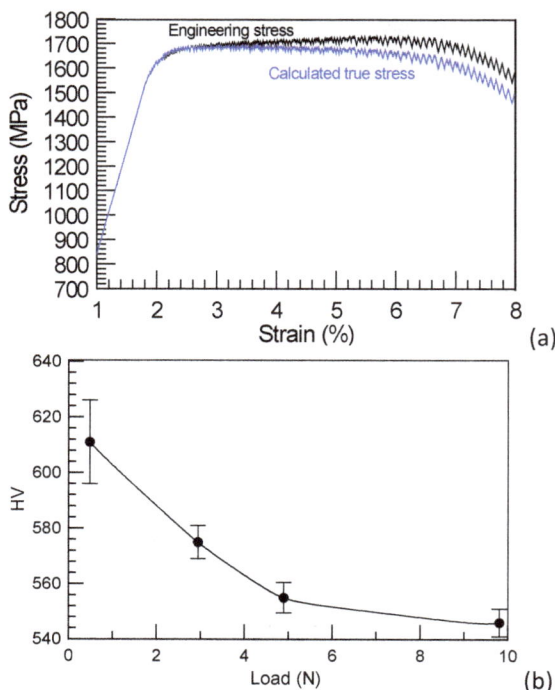

Fig. 5.4. (a) Stress-strain curves of a Zr-Cu-Fe-Al BMG alloy. Note that if the sample is not bended upon compression the calculated true stress is nearly constant before a dominant shear band is formed. Starting at about 7% strain there is an apparent decrease owing to the reduction of the cross-section and formation of cracks in stick-slip mode. (b) Microhardness (HV) of a Cu-Zr-Al BMG measured at different load.

Ductility of the $Pd_{40}Ni_{40}Si_xP_{20-x}$ [45,46] BMGs is drastically improved by application of the fluxing technique with B_2O_3 [46]. It is likely because fluxing removes possible inclusions acting as stress concentrations. Zr-Cu-Fe-Al bulk glassy alloys exhibit phase separation on heating prior to crystallization [47] which also may be responsible for their higher room-temperature ductility compared to the Zr-Cu-Al alloys. In other alloys the improved plasticity can be related to the Poisson's ratio or the elastic moduli ratio [48], in-situ nano-crystallization [49] or glassy phase separation [50]. An empirical rule of the brittle-ductile transition in BMGs was established based on the Poisson's ratio [51]. There is an abrupt brittle–ductile threshold around the Poisson's ratio $v_0 \approx 0.31$. BMGs with a higher

Poisson's ratio have a higher plastic deformation. However, there are some exceptions. For example, the Poisson's ratio of the $Au_{49}Cu_{26.9}Ag_{5.5}Pd_{2.3}Si_{16.3}$ BMG $v = 0.406$ is larger then v_0 but this alloy does not exhibit any plastic deformation at room temperature [52].

Compressive plasticity of the Fe–Cr–Mo–P–C–B BMGs (up to 3.6 %) with fracture strengths of about 3.5 GPa was improved by decreasing the shear modulus [24]. The improvements in ductility were found to be related to the changes in the electronic structure. Flux-treated $Fe_{70-60}Ni_{10-20}P_{13}C_7$ BMG samples of 1 mm diameter (with an aspect ratio of 2:1) also showed high compressive plasticity of more than 50% as well as good bending ductility [53]. Nanocrystals of 1–1.5 nm size were observed in the structure of $Fe_{50}Ni_{30}P_{13}C_7$ BMG by using aberration-corrected high-resolution transmission electron microscopy. This alloy also exhibits a yield strength of $\sigma_y = 2250$ MPa, good plasticity (up to 22 %), a fracture toughness K_C of 50 MPa/m$^{1/2}$. Homogenously dispersed nanocrystals play a key role in the enhancement of mechanical properties [54]. A $(Co_{0.7}Fe_{0.2}Ni_{0.1})_{67.7}B_{21.9}Si_{5.1}Nb_5Cu_{0.3}$ BMG exhibited a high fracture strength of 4770 MPa and a plastic strain of 5.5% [55]. The addition of Cu induced formation of locally ordered regions which enhanced plasticity.

Further investigations on this subject showed that embrittlement of the Mg- and La-based alloys is caused by the inclusions of crystalline oxide particles acting as stress concentrators [56]. This is because oxygen is insoluble in solid Mg and La, while it dissolves very well in Zr and Ti (both up to ~30 at.%) [57]. It explains reduced plasticity of Mg- and La-based metallic glasses.

Like the Hadfield steel in which locking of dislocations takes place [58] BMGs exhibit serrated flow (Fig. 5.4a) with regular stress drops typical for ductile bulk metallic glassy alloys [59,60] and negative strain rate sensitivity. Contrary to crystalline alloys one can see general absence of strain hardening except for the initial transition period from the elastic to plastic deformation mode. The average stress drop magnitude increases with strain, reaching a critical value at fracture.

One should mention that the fracture strength of BMGs varies significantly depending on the existence of defects (likely pores and surface imperfections) which cause premature fracture, for example, as found for the $Zr_{63-x}Cu_{24}Al_xNi_{10}Co_3$ alloys at 15 at. % Al (Table 5.4) [61]. It explains large scattering in the values of the percentage strain from sample to sample [62].

Unlike the prediction by the weighted rule of mixing for the elastic properties of the $Zr_{63-x}Cu_{24}Al_xNi_{10}Co_3$ BMGs, the measured moduli G and E show stiffening while increasing the aluminum content likely owing to partly covalent bonding. The Poisson's ratio exhibits a clear maximum $v = 0.32$ at the aluminum concentration $c_{Al} = 13$ at. %. The enhanced shear softness at $c_{Al} = 13$ at. % was also accompanied by an enhancement of the ductility in terms of the maximal plastic strain ε_p, during a quasi-static compression test, and in terms of creep deformation at constant load and room temperature. According to the density functional theory calculations the presence of an Al atom in the Cu–Zr clusters introduces new well localized states at low energies that are characterized by covalent-like bond promoting a higher mechanical strength [63]. Al atoms alter significantly the electronic structure and significant charge transfer was found to take place upon mechanical deformation.

Table 5.4. Mechanical properties of the $Zr_{63-x}Cu_{24}Al_xNi_{10}Co_3$ alloys obtained by compression testing. $\sigma_{0.2}$ is the proof strength. ε_p is the engineering plastic strain and its maximum value is shown instead of the average one owing to large data scattering. The data is taken from [61] with permission of Elsevier.

Al content	Sample Nr	$\sigma_{0.2}$ (MPa)	ε_p (%)
10 at. %	1	1775	10
	2	1780	2.6
	3	1790	3.6
	4	1780	4.4*
	Average	1781±10	Max: 10
13 at. %	1	1850	12.6
	2	1900	2.5
	3	1900	1.9
	4	1920	3*
	Average	1893±48	Max: 12.6
15 at. %	1	1900	0.2
	2	1815	5.5
	3	1920	0.4
	Average	1878±138	Max: 5.5

* The measurement was stopped to check the shape of the deformed sample.

5.1.3 Plastic deformation mechanism

Room-temperature plastic deformation of ductile crystalline solids generally proceeds via dislocation activity within the slip bands or by the formation of twins producing elongation. Due to work hardening the plastic flow in ductile crystalline alloys is delocalized because the strained regions become more resistant to further deformation. On the other hand the inhomogeneous plastic flow of glassy alloys at a relatively low homologous temperature (at least several tens degrees Kelvin below T_g) occurs by propagation of shear bands [64], which are 10-20 nm thick [65] (shear bands observed in crystalline alloys are much thicker) and make steps on the surface up to several micrometers in height (Fig. 5.5) [66]. Nevertheless, even in crystalline materials deformation by shear band starts to dominate over dislocation slip and twinning after certain strain or at high strain rate [67].

A strongly localized shear deformation at room temperature [68,69] without strain hardening limits practical application of BMGs. However, in some BMGs formation of multiple shear bands solves this problem [33,45]. A certain band once formed propagates within the sample and can gradually reach the other surface of the sample, leaving either a relatively flat or curved interface between two pieces of the sample (see in Fig. 5.5 marked with arrows) further exhibiting the slip-stick process [70] and concentration of the subsequent deformation activity in this band. In some alloys the onset of the shear band activity was observed well below the yield strength in the strain-stress curves [71].

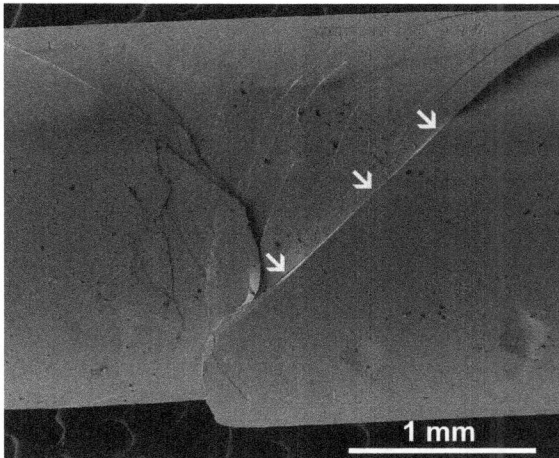

Fig. 5.5. Shear bands in the $Zr_{61.88}Cu_{18}Ni_{10.12}Al_{10}$ BMG with large compressive plasticity, SEM. The mechanical test was stopped after the formation of a dominant shear band (marked with the arrows) and stick-slip behavior leading to large shear offsets on the surface.

As the localized shear deformation is a dominant plastic-deformation mode at room temperature, tensile ductility of metallic glasses is not found except for few special cases in thin sections of hypoeutectic alloys [71] tested at a relatively high strain rate [44,72]. Tensile deformation behavior of Zr-based glassy thin foils has been also studied recently in-situ in TEM and the foils were found to be more ductile than larger samples [73,74]. Hypoeutectic Zr-based alloys were found to be much more ductile compared to hypereutectic ones owing to a larger excess (free) volume [75]. Large plasticity at room temperature was achieved in the Zr-Cu-Ni-Al system alloys with appropriate elastic moduli combination [76]. Deformation of BMGs obeys the Mohr-Coulomb yield criterion [77]. Metallic glasses as well as granular materials like powders, ball, sands exhibit dilatation on shear localization.

The free volume model [78] explains deformation in BMGs as a series of diffusion-like local atomic jumps into vacant sites in the regions of large excess volume. The shear transformation zones model [79] and the atomic-jump free volume model [78] represent the alternative approach. If the atoms are used as the particles in a granular model of a metallic glass, a shear-band thicknesses of 2.5–3.1 nm would be expected according to the Goldschmidt atomic diameters. However as the structure of a metallic glasses is an efficient packing of clusters the outer diameter of the clusters (approximately three times the diameter of the atoms of the base element in the glasses) suggest the shear-band thicknesses shall be close to the observed thickness of 10-20 nm [80].

Formation of a shear band by shear deformation localization as modeled by molecular dynamics (MD) [81] is shown in Fig. 5.6 as the von-Mises shear strain invariant η_s which is defined in terms of the relative displacement of the particle configuration as a function of strain (e) from the initial state. Despite on the rather small sample size compared to the experimental one, one can observe a heterogeneous distribution of the higher η_s regions even in the early stage of 0.2 strain. There is also a clear tendency for strain localization in the shear bands after 0.5 strain as marked with arrows.

Figure 5.7 shows the results of finite element modeling of plastic strain of the non-deformable sample matrix separated by the shear bands. Strain is localized at the intersections of the shear bands while local stress maxima are found in the non-deformed glassy matrix, especially close to the loading punches.

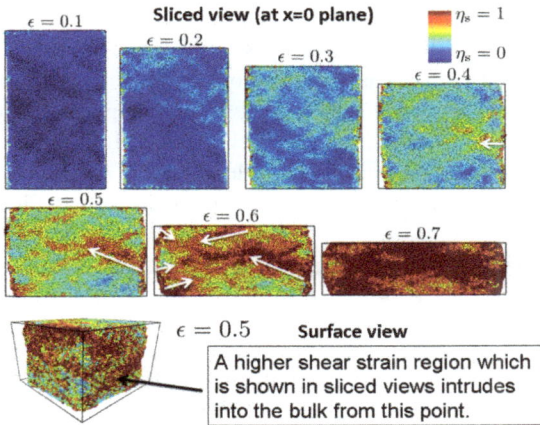

Fig. 5.6. Sliced view (at x=0 plane) of a set of local shear strain maps and a surface view at strain (ε)=0.5 obtained in molecular dynamics simulation of the sample deformed to 0.7 strain at an initial strain rate of $1 \cdot 10^{-3}$. Reproduced from [81] with permission of Elsevier.

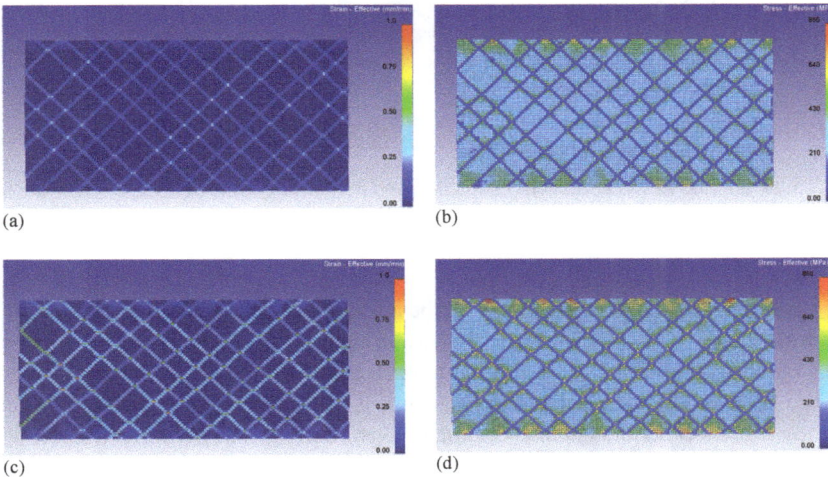

Fig. 5.7. Finite element modeling of plastic deformation of the non-deformable sample matrix separated by the shear bands. (a,c) strain (dimensionless) and (b,d) stress distribution (the scale in MPa) up to 0.03 and 0.07 strain, respectively. Reproduced from [81] with permission of Elsevier.

A dislocation is clearly distinct from the crystal lattice until it disappears on the surface of the sample or at a grain boundary while a shear transformation zone, if it exists, is difficult to identify and may be not distinct after shear [82], though it possesses a larger volume. However, when the spatial displacement distribution was measured around the shear band tip terminated in a deformed bulk metallic glass using a digital image correlation technique good agreement was found between the experimentally observed and theoretically predicted displacement fields for mixed dislocations [83]. Here the shear step on the surface of a BMG sample is considered as a result of propagation of a mixed dislocation (Fig. 5.8).

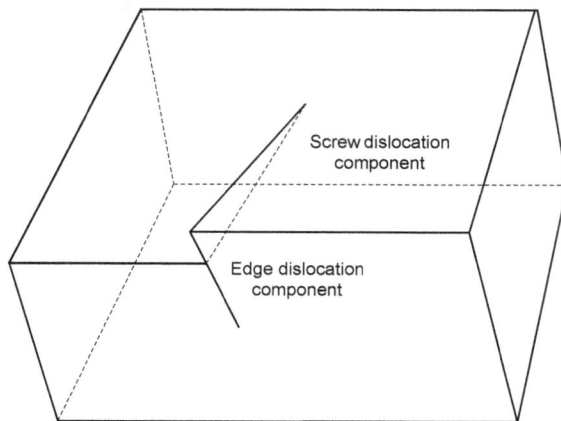

Fig. 5.8. Schematic representation of a shear band in metallic glass having a screw and edge component.

Shear transformation zones are more likely to be found in the areas with a higher structural disorder [9], larger excess volume [84], or areas of high atomic-level stresses [85]. These regions can therefore be treated as fertile sites for shear transformations in metallic glasses [86]. The formation of a shear transformation zone is likely in the areas of a higher local stress near a stress concentrator (flaw or surface notch) leading to its structural disordering [87].

Recently, Greer et al. [88] published a seminal work on shear bands in metallic glasses. A shear band could be initiated from the percolation of the shear transformation zones. Nucleation of the shear band includes: its creation by structural rejuvenation and sliding along the rejuvenated plane. Heterogeneous nucleation of shear bands indicates that the propagating shear band could be divided into the zones that include shear-rejuvenated glass near the shear band tip, glue zone, and liquid or near-liquid tail [86,89]. Activation of the shear transformation zones along the band makes the structure more disordered [90]. High-pressure torsion treatment also caused structural rejuvenation and softening at room temperature [91] while, high pressure torsion at 323 K slightly increased hardness [92]. Variable resolution fluctuation electron microscopy studies indicated that room-temperature plastic deformation of $Zr_{52.5}Cu_{17.9}Ni_{14.6}Al_{10}Ti_5$ caused significant changes in it's the medium-range order [93].

Depending on annealing temperature the glasses can be either structurally relaxed or rejuvenated. Structural rejuvenation within the supercooled region and subsequent water cooling improve plasticity of BMGs [94]. Relaxation enthalpies of a $Cu_{44}Zr_{44}Al_8Hf_2Co_2$ BMG were reported to be varied in the range of 150-1000 J/mol [95]. Also, after annealing at $1.12T_g$ the $Cu_{44}Zr_{44}Al_8Hf_2Co_2$ glass is characterized by a heterogeneous strain distribution. This non-affine strain/stress fields are reported to be responsible for the observed increase in plastic deformability.

Nucleation kinetics of shear bands in metallic glasses was studied by nanoindentation [96]. The strain rate dependence suggested that the nucleation of shear bands exhibits thermally activated behavior. The activation energy for nucleation of shear bands is found to be close to that of the slow-β relaxation in metallic glasses.

The experimental and computational methods indicate that the internal structure of metallic glasses plays important role in their mechanical behavior. Processing history of a bulk metallic glassy alloy, i.e. annealing and cooling rate, leads to a change in the chemical and topological local order, and thus to alteration of the mechanical behavior [97,98].

Computer simulations [99] were performed in order to investigate both the structural effect, in terms of the configurational potential energy and the fraction of full icosahedral clusters and fragmented ones, and the compositional effect in Cu-Zr metallic glasses on their elastic and plastic properties. While the bulk modulus K remains independent on structural change, the shear modulus G and thus the Poisson's ratio v (or G/K-ratio) widely change depending upon the cooling rate during sample preparation: G increases at slower cooling, whereas v decreases. These changes were accompanied by an increase of the potential energy and an increase of the fraction of fully icosahedral clusters which favor the delocalization of shear transformation.

Molecular dynamics (MD) simulation also indicated structural disordering of the icosahedra into less-shear-resistant, fragmented clusters that are prone to structural changes [100]. In the $Cu_{64}Zr_{36}$ glass, the total number of icosahedral units inside the shear band decreased, the number of icosahedra in the $Cu_{36}Zr_{64}$ glass did not change [101]. Both topological and chemical short range order in Cu-Zr metallic glasses are affected inside the shear bands, while the surrounding matrix is slightly affected [101]. The tensile behavior of the $Ni_{60}Nb_{40}$ metallic glass has been investigated by using *ab-initio* simulation with a large cell containing 1024 atoms. With increasing tensile strain, the local atomic strain was found to become more heterogeneous [102].

The excess (free) volume is created inside the shear bands. Nanocrystallization is suggested to take place within the shear bands if nanoscale voids are not formed [103]. The formation of nanovoids reduces the excess volume. Independent of the alloy composition, plastic deformation inside the shear bands was found to increase the fraction of Cu-Cu and Zr-Zr bonds and decrease the fraction of unlike Cu-Zr bonds. On the other hand the chemical composition of the shear band of a Cu-Zr metallic glass is found to be the same as of the non-deformed sample [104]. The authors also showed that the deformation can vary from homogeneous flow to localized flow depending on the excess volume content.

Formation of nanovoids was observed in the shear bands [105,106] and described with a revised classical nucleation theory [107]. Molecular dynamics simulations illustrated the initial stages of cavitation in the $Pd_{82}Si_{18}$ and $Cu_{46}Zr_{54}$ metallic glasses on deformation [108].

The atomic structure of the $Pd_{53}Ni_{34}Cu_{13}$ metallic glass was studied by molecular dynamics simulation at different temperatures and no anisotropy was found in the initial state. After deformation the anisotropy was revealed by significant differences in the partial pair distribution functions. The anisotropy was found to be constrained into the shear plane. Residual anisotropy was found at a low temperature, while close to the glass transition the anisotropy induced by the initial shear process was removed by the subsequent recovery process [109].

The shear band activity was monitored by a high-speed camera [110] and by thermography [111]. In-situ monitoring of the deformation behavior of bulk metallic glasses performed using an infrared camera [112] showed a weak rise of temperature during a shear band operation. The instant melting of BMG specimens at the moment of the failure was captured by thermography while temperature rise in the shear band was much lower.

Shear softening can be a result of local heating or driven by stresses [113,114]. The temperature in the central area of a shear band could be as high as several thousand Kelvin depending on the shear band propagation speed [115]. At the same time much lower temperature rise inside a shear band was also suggested [116,117]. It was shown that upon four-point bending of the Zr-based BMG coated by sputtered tin creation and propagation of shear bands caused the local temperature rise in the BMG and the appropriate calculations of the temperature rise were provided [113].

Nevertheless, regardless of the source of strain localization at high enough strain rate within the shear band the central shear zone heats up due to friction forces [118] and its viscosity can drop several orders in magnitude further increasing the shear speed in the band. The band propagation can be stopped likely owing to the decrease in axial stress waiting for the next round of the increase in stress. At the critical shear offset sudden fracture takes place owing to the low viscosity reduced by the acting friction force heat release. Also, no coating (with sputtered tin, Rose metal and indium) melting was observed later near individual shear bands. Melting occurred only in some places near the fracture surface, near microcracks and near places of shear band concentrations where total heat release was large enough to initiate melting (Fig 5.9). It was suggested that the shear localization and banding is a stress driven process and the temperature rise that was observed in the shear bands is due to the friction forces in the thin liquid layer within the shear bands.

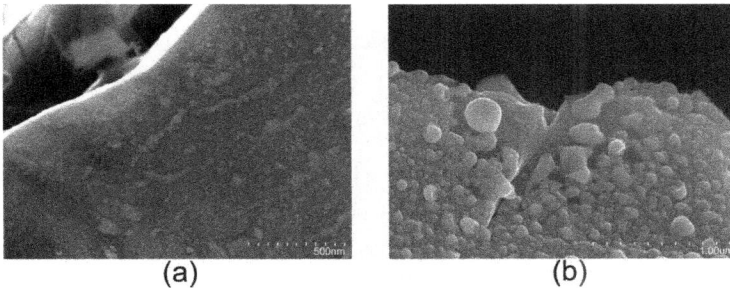

(a)   (b)

Fig. 5.9. SEM images the areas near the fracture surface of the bulk sample coated by indium (a)- and Rose metal (b) after the compression test. Reproduced from [118] with permission of Nature Publishing Group.

Room-temperature deformation process of a Zr-based BMG sample was studied using the Gleeble 3800 thermomechanical simulator [119]. Two types of stress drops were observed: single-stage and double-stage one. The temperature rise during shear band propagation was measured with a high data acquisition frequency using a thermocouple attached to the sample and also calculated based on different approaches according to overall temperature rise of the sample. There is a correlation between the calculated temperature inside the shear band and the double stage character of the stress drop. The increase of the stress drop velocity at the second stage may be associated with the temperature rise in the shear band as a result of friction force acting in the sliding medium. Modeling of heating and cooling processes of the sample upon shear band propagation by finite elements method showed good agreement with calculations based on the experimental results. The samples with a diameter of 3 mm and a length of 6 mm exhibited the stress drop velocities of single-stage and double-stage events of approximately 1150 MPa/s and 2950 MPa/s, respectively. Taking into account the Young's modulus of 76 GPa and the relationship $E=\sigma/e$ (where σ is stress and e is strain) one can obtain 2.5 and 3.8 m/s for single-stage and double-stage events, respectively. If one suggests that the shear bands propagate at 45 degrees to the uniaxial load shear band propagation speed is 6.5 and 10 m/s for single-stage and double-stage shear events, respectively. It is in good correlation with the values for similar not-well developed shear bands in another work [120]. The velocity of shear bands in the $Pd_{40}Ni_{40}P_{20}$ bulk metallic glass was also tested in compression. The data indicate that in case of simultaneous sliding of the entire sample the displacement rate is approximately 2 mm/s while if the displacement occurs as a localized propagating front, the velocity of the front is approximately 2.8 m/s [121]. The velocity of shear bands measured with high-speed video cameras and the operation time was found to be close to their temporal resolution [122]. Digital image correlation procedure was applied to the images to determine the displacement adjacent to the shear band as a function of time and position across the width of the sample [123].

The connection between macroscopic inhomogeneous deformation at room temperature and microscopic deformation was also tested by thermal imaging of heat evolution. The serrated flow is correlated with shear band propagation [124].

By using the free volume model it was shown that a sharp increase of the excess/free volume at yielding is mainly responsible for the low viscosity of a propagating shear band [125]. MD simulation showed the excess volume production on the deformation process in metallic glasses [126] as a result of the reformation of the atomic bonds during the deformation process. The volume change is an important factor in the mechanical response of metallic glasses subjected to mechanical deformation. It is suggested that the adiabatic heating may not be necessary as a mechanism for shear band formation and at large strains, the excess volume change could reach or even exceed that of the supercooled liquid at the glass transition.

Exceptionally high diffusion coefficient in the shear band also indicates high excess (free) volume concentration [127]. Generation of a large fraction of the excess volume during deformation of a metallic glassy sample leading to enormous softening was observed by nanoindentation in terms of hardness [128], by using MD simulation [126] and by an atomic force microscope in a tapping mode in terms of the reduction of the Young's modulus [129].

The idea of generation and redistribution of the excess volume in a shear band is supported by the work where a large decrease in the hardness within a zone up to 0.1 mm and even more was observed near the dominant shear band [128]. Thickness of the softened zone (centered on the shear band) increases with increasing the shear offset and total strain. Measurements of dimension change perpendicular to the shear band suggested dilatation within the band of about 1 %. DSC exhibited high enthalpy of relaxation in that area which is reduced by annealing.

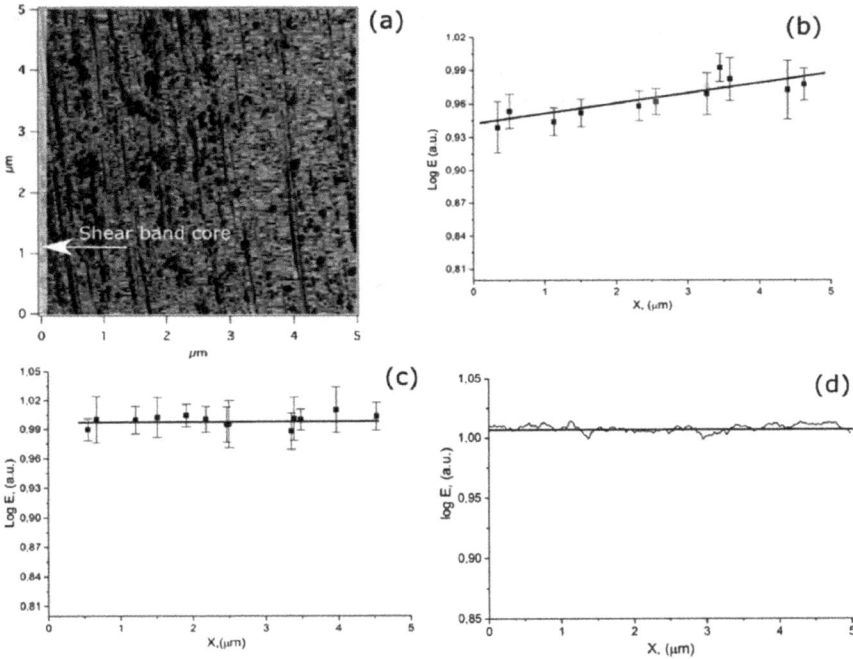

Fig. 5.10. The Young's modulus distribution image (a) and the normalized logE values picked from chosen areas of the polished sample near the shear band core (initial offset of the shear band was about 2 μm) as a function of the coordinate X (b). The normalized logE of the region far from the shear band edge (more than 40 μm away) (c). The normalized logE measured near the shear band core after the annealing (d). Reproduced from [129] with permission of Elsevier.

The elastic modulus variation in the surface regions of the deformed $Zr_{60}Cu_{20}Co_{10}Al_{10}$ BMG near and far from a shear band with a shear step of about 2 μm was measured using the atomic force microscope quantitative nanomechanical mapping method (Fig. 5.10). It was found that the Young's modulus decreases gradually towards the shear band's central region on the micrometer scale. Maximum decrease in the Young's modulus near the shear band's central region is about 70%. Again this effect is attributed to the generation and migration of the excess volume during the shear band operation [129].

It was suggested that the macroscopic yield point and completion of the so-called strain hardening in the beginning of plastic deformation corresponds to the transecting of the shear bands [130]. Also, formation of a dominant shear band must be accompanied by the reduction of the apparent flow stress owing to reduction of the effective cross-section area (X·Y in Fig. 5.11) while many BMGs flow at nearly constant stress after the macroscopic yield point (Fig. 5.4) owing to relatively homogeneous deformation induced by several shear bands.

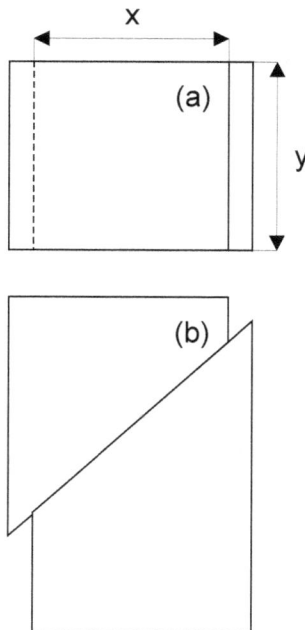

Fig. 5.11. Reduction of the effective cross-section area of the rectangular sample with a dominant shear band upon uniaxial compression (top view (a) and side view (b)) propagating by the stick-slip process.

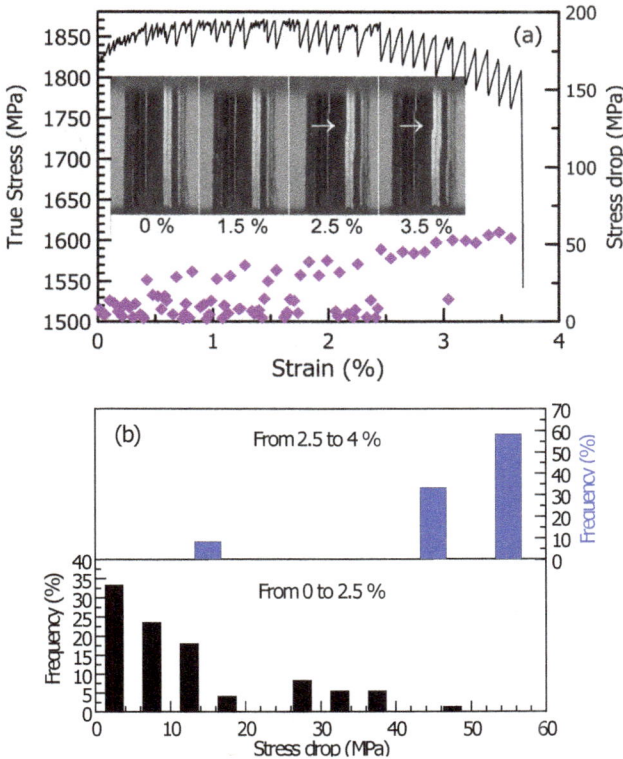

Fig. 5.12. (a) Stress and stress drops (diamonds) as a function of plastic strain for the cylindrical $Zr_{62.5}Cu_{22.5}Fe_5Al_{10}$ sample of 2:1 aspect ratio upon the compression test. The inserts in (a) illustrate a sequence of the video file snapshots corresponding to (0 %) the initial sample before deformation, (1.5 %) the sample deformed plastically to 1.5 %, (2.5 %) the sample deformed plastically to 2.5 % and (3.5 %) the sample before fracture (deformed plastically to 3.5 %). (b) The frequency distribution of the magnitude of stress drops calculated from the continuous strain-stress curve in (a) in two ranges, namely, from 0 to 2.5 % and from 2.5 to about 4 % in plastic deformation. Reproduced from [132] with permission of Elsevier.

The plasticity of BMGs was found to depend on the competition between the formation of new shear bands and the reactivation of the existing shear bands during plastic deformation [131]. Clear transition to a dominant shear band was observed in the Zr-Cu-Fe-Al bulk metallic glassy samples [132]. Fig. 5.12 (a) represents the continuous loading stress-plastic strain curve of the cylindrical $Zr_{62.5}Cu_{22.5}Fe_5Al_{10}$ glassy alloy in its plastic deformation range. The stress drop spectrum changes its shape and the flow stress starts to decrease after 2.5 % of deformation. The frequency distribution of the magnitude of stress drops (Fig. 5.12 (b)) calculated from the continuous strain-stress curve in Fig. 5.12 (a) in two ranges of plastic deformation clearly indicates different deformation behaviors in these two ranges. Formation of new shear bands and propagation of the existing shear bands takes place at the first stage while the formation of a major shear band throughout the sample and its propagation takes place at the second stage. Such a behavior is also illustrated by the changes in the stress drop and strain per stress drop plots (Fig. 5.12 (b)). The snapshots from a video file recorded during its deformation shown in the insert in Fig. 5.12 (a) clearly demonstrate the formation and propagation of a major/dominant shear band in the second part of the deformation process. Formation of the dominant shear band is a stochastic process and in various samples it takes place at different strain values. A similar behavior was observed in the case of the $Pd_{40}Ni_{40}Si_4P_{16}$ BMG and $Ti_{43.6}Zr_{7.9}Cu_{40.4}Ni_{7.1}Co_1$ BMG samples containing 1000 ppm B.

The two-stage deformation process is schematically illustrated in Fig. 5.13. These BMG samples initially show multiple shear bands and microscopically relatively homogeneous deformation at nearly constant flow stress and the formation of a dominant shear band at the late deformation stage which leads to a significant decrease in the flow stress owing to the decrease in the efficient sample cross-section (Fig. 5.11). These two types of the deformation regimes may be responsible for the observed bimodal distribution of the shear offsets measured by nanoindentation [133].

The $Zr_{65}Cu_{15}Ni_{10}Al_{10}$ alloy demonstrated the chaotic nature of the stress serrations by showing the existence of a finite correlation dimension and a positive Lyapunov exponent [134]. In $Cu_{47.5}Zr_{47.5}Al_5$ alloy the distributions of stress drop magnitudes and their time durations appear to follow a power-law scaling form reminiscent of a self-organized critical state. The chaotic dynamics of the less ductile BMG was consistent with a single shear band sliding. In the $Cu_{47.5}Zr_{47.5}Al_5$ alloy , the high ductility appears to be related to concurrent nucleation of a large number of shear bands throughout the sample, which in turn, can give rise to a hierarchy of length scales over which shear bands propagate. In the $Ni_{50}Pd_{30}P_{20}$ bulk metallic glass the serrated flow behavior also appeared to be chaotic and a stochastic model suggested [135] that the underlying serrated slow dynamics initially relies on the appearance of new shear bands on the material, but as the experiment proceeds the nature of these dynamics change, and strain takes place by the extension of the shear bands already present on the BMG sample.

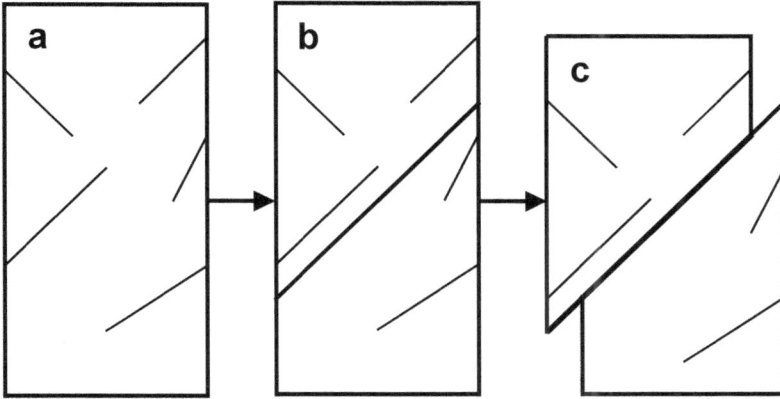

Fig. 5.13. Two-stage deformation mode: a) multiple not well developed shear bands mostly not crossing the entire sample, b) formation of a dominant shear band from one side to another leading to c) a stick-slip flow behavior within this band. Partly reproduced from [132] with permission of Elsevier.

However, the two stage deformation observed in the typical sample having 2:1 aspect ratio (sample height to width or to diameter ratio) may not be the case in the samples of significantly larger or smaller size or different geometry because the engineering plasticity of bulk metallic glasses depends upon their geometry and aspect ratio [136]. At low aspect ratio, shear bands are terminated at the interface with the loading platen and become confined. Also, the friction between the sample and the loading platen becomes more efficient. Different specimen geometries influence the shear-band formation and deformation processes, resulting in variations of the mechanical properties.

The samples with 1:2 aspect ratio can be deformed to large strain values without formation of a dominant shear band [81]. Fig. 5.14 and the insert (a) represents the continuous loading stress-plastic strain curve of the cylindrical $Zr_{61}Cu_{27}Fe_2Al_{10}$ glassy alloy. The true stress values were calculated from nominal stress assuming homogeneous deformation and constant volume of the deforming sample throughout the test. As one can see in Fig. 5.14 (b) the samples deformed for 0.24 and 0.68 strain were intact while the sample deformed to 0.80 was destroyed to pieces.

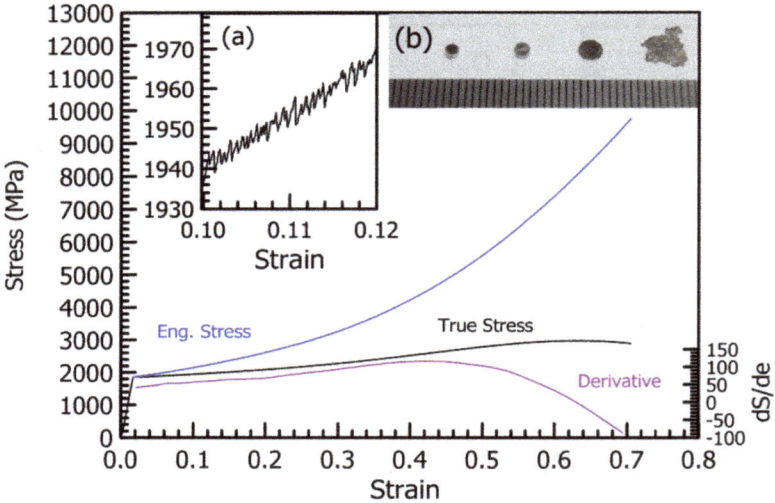

Fig. 5.14. The engineering and calculated compressive true stress-strain curves of the sample having 1:2 aspect ratio tested at room temperature at 5 x 10⁻⁴ s⁻¹ (left axis) and the derivative dσ/de (right axis). The inset (a) a close-up of the curve between 0.10 and 0.12 strain. (b) An optical image of the samples before compression and deformed for 0.24, 0.68 and 0.80 strain, respectively, from left to right. Reproduced from [81] with permission of Elsevier.

The XRD patterns obtained from the deformed samples show no crystalline peaks. As can be seen from shear offsets on the lateral surface in Fig. 5.15 the sample shows formation of multiple shear bands.

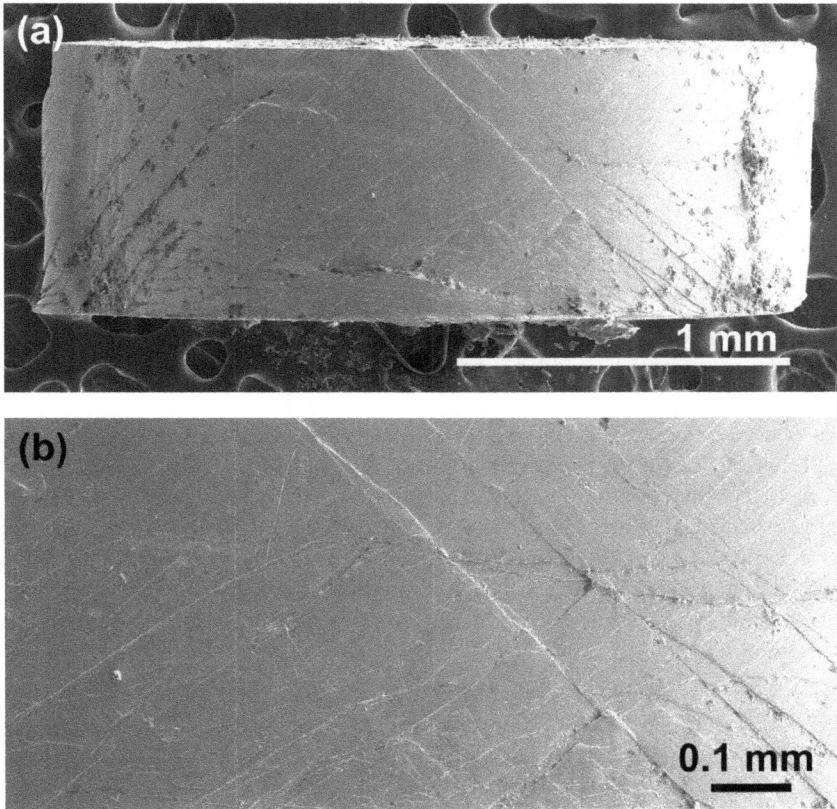

Fig. 5.15. Lateral surface of the sample deformed at 5 x 10⁻⁴ s⁻¹ for 0.24 strain at different magnification (a,b). Reproduced from [81] with permission of Elsevier.

The length of the shear steps (L) on the lateral surface per its area (A) at first approximation is supposed to correlate with the fraction of shear bands (shear surfaces) per sample volume. The thermal properties of the samples were tested by DSC and the resulted values (glass-transition temperature (T_g), crystallization temperature (T_x), exothermic peak temperature (T_p), as well as enthalpies of relaxation (H_r), overshoot (H_o) and crystallization (H_c) are also shown in Table 5.5.

Table 5.5. Engineering (ε_e) and true strain (ε_t), Vickers Hardness (HV), T_g, T_x, T_p, ΔH_r, ΔH_o and ΔH_c values of the studied alloy. Reproduced from Ref. [81] with permission of Elsevier.

ε_e	ε_t	L/A nm^{-1}	HV	T_g (K)	T_x (K)	T_p (K)	ΔH_r (J/mol)	ΔH_o (J/mol)	ΔH_c (J/mol)
0	0	0	457±5	670	763	767	204	14	3236
0.24	0.27	90±7	466±4	666	758	765	637	34	3290
0.68	1.14	138±6	497±3	667	759	765	756	36	3313
0.80	1.61	-	-	672	758	765	262	24	3244

The volume fraction of the shear bands is only $9 \cdot 10^{-10}$. This is negligibly small value to cause significant changes in the thermal properties of materials if the rest of the sample is not affected. According to the XRD data obtained with a relatively high signal to noise ratio the samples remain glassy after severe plastic deformation and even failure of the sample. Even after fracture of the sample deformed to 0.8 engineering strain (1.61 true strain) the enthalpy of crystallization is not affected. The reduced values of T_x and T_p after deformation may be connected with ease of heterogeneous nucleation at the shear steps on the surface or inside the shear bands.

The Vickers microhardness measured in the direction of applied stress increases after deformation. This may be connected with large fraction of the elastic stresses in the glassy matrix [137] induced by so-called not well developed shear bands, the shear bands which shear front did not cross the entire sample.

Although it is well known that structural relaxation below T_g causes embrittlement of the BMG samples $Zr_{63}Cu_{22}Fe_5Al_{10}$ metallic glass even in the relaxed state exhibits nearly as high plasticity as in the as-cast state (Fig. 5.16 (a)) though the structure remains clearly glassy (Fig. 5.16 (b)) [138]. Also, no hardening but even mechanical softening is observed after annealing even though the samples showed a reduction in size and volume typical for fully relaxed Zr-based alloys. The reason for such behavior is connected with nanoscale phase separation at about 10 nm length-scale within the glassy phase on to Cu- and Zr-rich areas on annealing at low temperature which is comparable in size to the width of shear bands (Fig. 5.17). Fe is also placed in antiphase with Cu and likely further promotes phase separation.

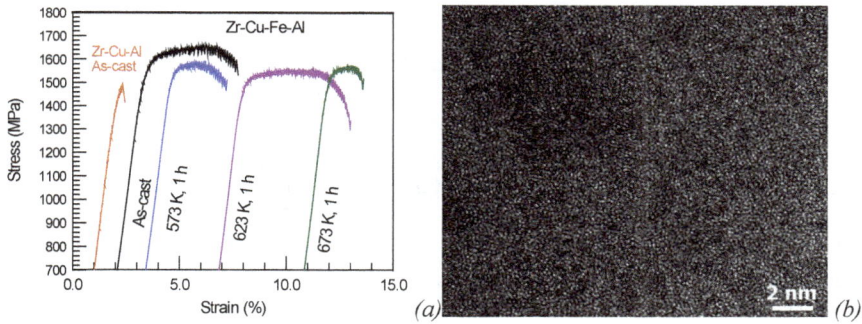

Fig. 5.16. (a) Engineering compressive stress-strain curves of the $Zr_{63}Cu_{27}Al_{10}$ (as-cast) and $Zr_{63}Cu_{22}Fe_5Al_{10}$ (as-cast and heat-treated, as indicated) glassy alloys. (b) HRTEM image of the $Zr_{63}Cu_{22}Fe_5Al_{10}$ sample annealed for 1 h at 623 K. Reproduced from [138] with permission of Elsevier.

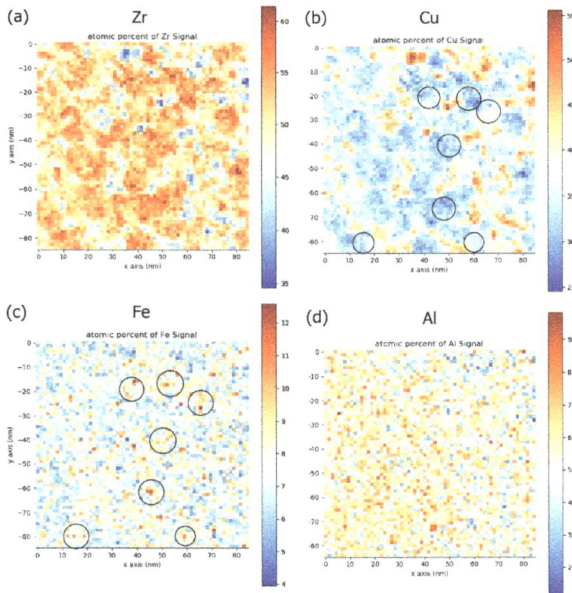

Fig. 5.17. The elemental maps for Zr, Cu, Fe and Al (a-d), respectively, plotted for the Zr-Cu-Fe-Al sample annealed for 1 h at 623 K. Reproduced from [138] with permission of Elsevier.

5.1.4 Structural rejuvenation by deformation and thermal treatment

Structural relaxation (usually observed on heating) decreases the sample volume and causes embrittlement. As has been shown above for the $Zr_{61}Cu_{27}Fe_2Al_{10}$ glassy alloy [81] and also described in other works [139,140] an opposite effect called rejuvenation brings back the glass to a higher energy state. The heat of relaxation increased significantly after deformation (Table 5.5) and the excess heat of relaxation gain as a result of loading to $\varepsilon=0.24$ (of 433 J/mol) is about 9% of the energy used to deform the sample. The stored plastic strain energy at large deformation was also found in a Pd-based alloy [141]. This may indicate that intensive plastic deformation with a large number of the shear bands brought the bulk metallic glass to a higher energy state which is released only after the failure of the sample. On the other hand the heat of the overshoot (ΔH_o) after glass-transition, ordinary indicative of the degree of structural relaxation [142] also increased. Intensive plastic deformation of a bulk metallic glass with a large number of the shear bands before intensive crack initiation has a twofold effect on its properties. In local areas a higher energy state is reached (similar to the glassy phase rejuvenation) which is released after the failure of the sample. The plastically (within the shear bands) and elastically (at the edges of not well developed shear bands) deformed areas release a significant amount of extra heat upon structural relaxation while the decreased T_x may indicate ease of crystallization connected with the preferential states for heterogeneous nucleation in the locally deformed areas. On the other hand another part of the glassy matrix may suffer from the opposite effect of structural relaxation leading to an overall lower T_g and larger heat overshoot of the DSC signal after glass-transition indicative of a more relaxed glass.

Ion irradiation was used to improve the tensile ductility of a metallic glass [143]. Mechanical loading is another method of rejuvenation upon elastostatic loading at stresses well below the yield stress [144] as well as cold rolling [145] and severe plastic deformation [146]. It was also possible to induce residual stresses in a bulk metallic glassy sample by shot peening method which improved the mechanical performance, in particular plasticity [147,148].

Uniaxial compression of a BMG close to the yield strength induced creep of the samples, accompanied by an increase in the stored heat of relaxation [149]. The sample exhibited an increase in compressive plasticity on the subsequent test [150]. This creep is homogeneous and leads to decreases in density and in the elastic moduli. Also, the increase in heat of relaxation and in plasticity (of BMGs) resulting from static loading are reported to depend on the atomic packing density [151].

Plastic deformation under triaxial compression at room temperature can rejuvenate a BMG sample and enable strain-hardening. It suppresses shear-banding in bulk samples in uniaxial (tensile or compressive) tests and prevents catastrophic failure [152]. The energy of the BMG is raised by deformation under the constrained conditions rejuvenating it and creating STZs. When the BMG is deformed under tensile, less-constrained conditions, structural relaxation takes place. The atomic packing density increases and the excess volume introduced by the earlier deformation sinks, causing the flow stress to increase and leading to work hardening.

The energy stored in deformed conventional metals and alloys in the form of increased defect densities of about 10% of the work done is also typical for BMGs. In the glasses, the increased enthalpy after deformation is a sign of rejuvenation. Creep of metallic glasses at room temperature by compressive loading at stresses well below the macroscopic yield stress also leads to rejuvenation and the increases in the enthalpy is found to be more than the work done on the samples. One can suggest that the energy comes from the surroundings because the processing induces an endothermic disordering process in BMGs [153].

As discussed in Chapter 4 cryogenic thermal cycling is another non-destructive method to rejuvenate the glassy structures [154,155]. Contrary to the absence of any effect expected, thermal cycling of metallic glasses induces rejuvenation, reaching less relaxed states of higher energy because of intrinsic non-uniformity of the glassy structure, implying a non-uniform coefficient of thermal expansion. Such a thermal cycling of BMGs also produces desirable improvements in plasticity. A Liquid-state rejuvenated metallic glass was found to be softer and more plastic with a lower shear plane formation energy and a larger STZ volume and size [156].

5.1.5 Strain-rate sensitivity

While the strain-rate sensitivity of polycrystalline alloys is generally positive [157] (except for special cases like Hadfield steel described above), for bulk metallic glasses different results were reported.

The value of strain rate sensitivity parameter (m) can be calculated from the true stress-strain curves according to the relationship:

$$m = \left(\frac{\partial \ln \sigma}{\partial \ln \dot{\varepsilon}} \right)_{\varepsilon, \mathrm{T}}$$

(5.2)

where σ is the flow stress and $\dot{\varepsilon}$ is the strain rate.

A negative m value was reported for the $Zr_{57}Ti_5Cu_{20}Ni_8Al_{10}$ alloy at strain rates ranging from 10^{-4} to $3 \cdot 10^3$ s^{-1} [158]. m value of about -0.002 at room temperature was reported for $Zr_{52.5}Ti_5Cu_{17.9}Ni_{14.6}Al_{10}$ [159] and $Cu_{50}Zr_{50}$ BMG [160] but not in oxide glasses [161]. The $Zr_{61}Cu_{27}Fe_2Al_{10}$ BMG sample also exhibited negative strain rate sensitivity (Fig. 5.18) [81]. At about 0.06 plastic strain the strain rate was changed from $5 \cdot 10^{-3}$ s^{-1} to $5 \cdot 10^{-2}$ s^{-1} while the flow stress decreased from ~1765 MPa to ~1757 MPa leading to m=-0.002. Opposite changes occur when the strain rate was decreased afterwards. There is a little stress overshoot behavior right after strain rate decrease. Also a negative value was obtained [162] for the $Zr_{52.5}Ti_5Cu_{17.9}Ni_{14.6}Al_{10}$ and for a composite, at strain rates of $3.3 \cdot 10^{-4}$ and $3.7 \cdot 10^{-3}$ s^{-1}. The $Zr_{41.2}Ti_{13.8}Cu_{12.5}Ni_{10}Be_{22.5}$ alloy was reported [163] to be insensitive to the strain rate below 473 K.

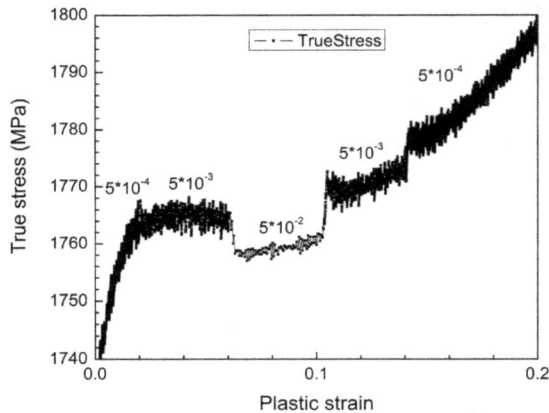

Fig. 5.18. The compressive calculated true stress - plastic strain curves of the $Zr_{61}Cu_{27}Fe_2Al_{10}$ BMG sample of 2 mm in diameter having 1:2 aspect ratio tested at different strain rate as indicated in s^{-1}. Reproduced from [81] with permission of Elsevier.

In Zr-based metallic-glass coatings with micrometer-scale thickness the increase of the penetration rate also leads to the decrease of the hardness [164]. At the same time most of nanoindentation studies indicate positive strain-rate sensitivity parameter values [161,165,166]. Such a difference could be mainly attributed to the stress state which is closer to three-axial compression [167, 168] when tested by indentation compared to uniaxial compression. Possible influence of the mechanical friction between the deforming medium should be considered as a possible source of positive strain-rate sensitivity values. Recently it was found that the strain rate sensitivity on indentation is significantly load dependent and decreases if a high load is used [169].

The $Zr_{65}Cu_{20}Fe_5Al_{10}$, one of relatively ductile BMG glassy alloys samples, was tested in compression at room temperature and almost no strain rate sensitivity was found at low strain rate while it was negative at higher rates (Fig. 5.19) [170]. The $Zr_{65}Cu_{20}Fe_5Al_{10}$ BMG alloys exhibited negative strain rate sensitivity at high strain rates and the parameter $m=-0.0026$ was calculated from the true stress-strain curves.

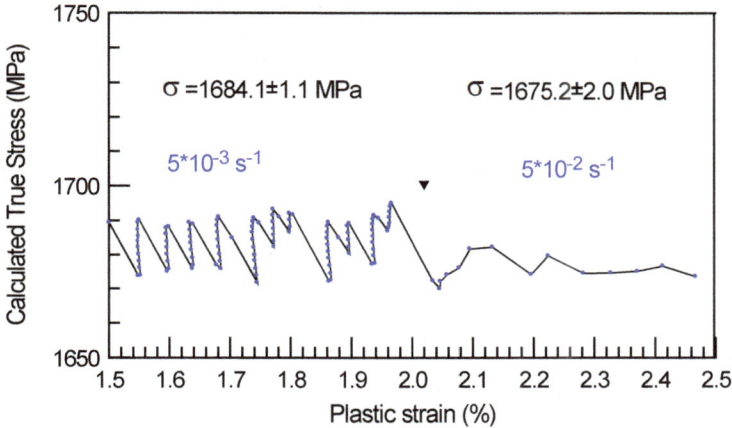

Fig. 5.19. True stress-plastic strain curves (elastic strain is subtracted) performed at strain rates of $5 \cdot 10^{-3}$ and $5 \cdot 10^{-2}$ s^{-1}. The data is taken from Ref. [170] with permission of Elsevier.

The mean and maximum flow stresses and the confidence interval (probability $P = 0.95$) have been calculated in 0.25 % plastic deformation interval using true stress values at both sides of each point of the strain rate change as shown in Table 5.6. Another parameter, which is an average maximum stress before each stress drop (Table 5.6) satisfied normality tests and produced similar results. Not only the flow stress but also the average maximum stress before each stress drop corresponding to a serration for each test is practically the same at each loading rate in the vicinity of the strain rate change point and the differences are practically within the confidence interval which indicates that the $Zr_{65}Cu_{20}Fe_5Al_{10}$ BMG is strain rate insensitive ($m \sim 0$) in the range of strain rates from $5x10^{-6}$ to $5x10^{-4}$ s^{-1}. However, the strain-rate sensitivity becomes meaningfully negative when the strain rate increases from

$5 \cdot 10^{-3}$ s^{-1} to $5 \cdot 10^{-2}$ s^{-1} because the strain rate is so fast that the relaxation time is not enough to build up the stress which explains the decrease in the flow stress.

Table 5.6. Mean (σ_m) and mean maximum (σ_m^{max}) stress values for test 1 and test 2 calculated considering a 0.25 % plastic deformation interval at both sides of the strain rate (ε') change. The mean maximum stress indicates average value of maximum stress before each stress drop. The confidence intervals were calculated for 0.95 % probability. Reproduced from [170] with permission of Elsevier.

Test 1				Test 2			
ε': $5 \cdot 10^{-4}$ - $5 \cdot 10^{-5}$ (s^{-1})		ε': $5 \cdot 10^{-5}$ - $5 \cdot 10^{-6}$ (s^{-1})		ε': $5 \cdot 10^{-6}$ - $5 \cdot 10^{-5}$ (s^{-1})		ε': $5 \cdot 10^{-5}$ - $5 \cdot 10^{-4}$ (s^{-1})	
ε' (s^{-1})	σ_m (MPa)	ε' (s^{-1})	σ_m (MPa)	ε' (s^{-1})	σ_m (MPa)	ε' (s^{-1})	σ_m (MPa)
$5 \cdot 10^{-4}$	1609.4 ± 1.2	$5 \cdot 10^{-5}$	1611.9 ± 0.5	$5 \cdot 10^{-6}$	1543.2 ± 0.2	$5 \cdot 10^{-5}$	1543.8 ± 0.3
$5 \cdot 10^{-5}$	1611.9 ± 0.5	$5 \cdot 10^{-6}$	1612.0 ± 0.2	$5 \cdot 10^{-5}$	1543.5 ± 0.4	$5 \cdot 10^{-4}$	1542.4 ± 0.9
ε' (s^{-1})	σ_m^{max} (MPa)	ε' (s^{-1})	σ_m^{max} (MPa)	ε' (s^{-1})	σ_m^{max} (MPa)	ε' (s^{-1})	σ_m^{max} (MPa)
$5 \cdot 10^{-4}$	1614.6 ± 1.4	$5 \cdot 10^{-5}$	1618.1 ± 1.2	$5 \cdot 10^{-6}$	1548.2 ± 3.2	$5 \cdot 10^{-5}$	1549.7 ± 1.3
$5 \cdot 10^{-5}$	1619.7 ± 3.0	$5 \cdot 10^{-6}$	1618.1 ± 3.1	$5 \cdot 10^{-5}$	1549.1 ± 2.3	$5 \cdot 10^{-4}$	1547.4 ± 1.1

5.1.6 Size effect and in-situ deformation studies

There is a strong size effect on deformation behavior upon indentation in which the deformed volume is significantly smaller while local plastic deformation is larger than in compression and metallic glasses are proven to deform homogeneously on nanoscale [73,74,171]. With decreasing specimen size smaller than the equivalent critical shear offset, the shear deformation of metallic glass is changed from unstable to stable, which leads to a transition from global brittleness on the macroscale to large plasticity on the microscale [172].

Plastic deformation in metallic glasses was performed using the in-situ TEM technique [173,174,175]. After initial observation the beam current was reduced and the sample deformed in a straining holder in tension [175]. This was an important precaution to prevent electron-beam heating of the sample upon deformation. This procedure (straining and subsequent observation) has been repeated several times until some cracks appeared. High-resolution in-situ TEM investigation showed that the $Zr_{65}Ni_{10}Cu_5Al_{7.5}Pd_{12.5}$ ribbon sample retains its glassy structure after in-situ deformation and crack propagation (Fig. 5.20) [175]. No nanocrystallization was observed even near the crack tip.

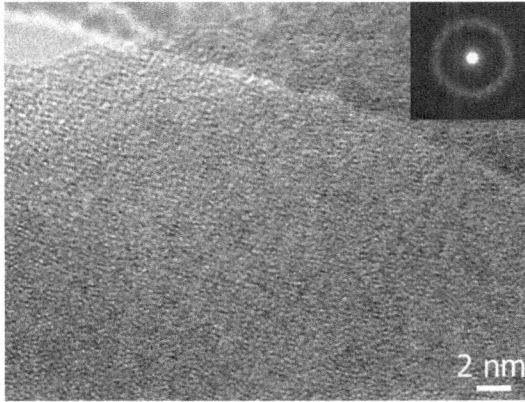

Fig. 5.20. High-resolution TEM image obtained upon in-situ deformation in a TEM near the crack in the $Zr_{65}Ni_{10}Cu_5Al_{7.5}Pd_{12.5}$ alloy. The insert shows a nanobeam diffraction (NBD) pattern produced from the deformed region close to the crack tip.

The homogeneous deformation of the thinned $Zr_{65}Ni_{10}Cu_5Al_{7.5}Pd_{12.5}$ alloy is clearly seen in front of the crack tip upon loading. The crack opening speed was quite low – about tens of nanometers per second or even less. A narrow featureless deformation zone formed ahead of the crack tip is much wider than the typical (10-20 nm) width of a shear band. Similar results were obtained for submicron (about 200 nm) thick $Zr_{52.5}Cu_{17.9}Al_{10}Ni_{14.6}Ti_5$ alloy sample [73].

The in-situ experiments demonstrate a fundamental possibility of a homogeneous-like like plastic flow to occur in a localized area of glassy metal thin foil at room temperature [176]. According to calculations the thermal energy release and rise in temperature depends on the deformation rate and shear offset, in addition to the thermal conductivity and the strength of the glassy phase [177,178]. However, the deformation areas observed in the sample tested in situ in TEM are much wider compared to the highly localized deformation bands observed in bulk glassy sample. Also no significant heating could have taken place as the heat generated upon the deformation is dissipated over a larger volume. Homogeneous deformation of the $Zr_{62.5}Cu_{22.5}Fe_5Al_{10}$ metallic glass was also observed upon in situ nanoindentation in TEM [179].

5.1.7 Fatigue, cyclic and long-term creep deformation in the elastic region.

BMGs were originally considered to have relatively poor fatigue resistance, possibly also connected to the lack of strain hardening, but in fact they show a wide range of fatigue behavior influenced by the intrinsic and extrinsic factors [180,181]. The fatigue-endurance limits of some Zr-based BMGs are comparable with those of the high-strength crystalline alloys [182,183]. Under compressive-compressive fatigue of a Cu-Zr-based BMG fracture occurs in shear at roughly the same angle as seen in continuous monotonic compression [184]. It is at 41° to the loading axis (i.e. the normal vector to the fracture surface is at 49° to the loading axis). The fatigue limits in the Co- and Fe-based BMGs exceeded 2 GPa, and those in the Ni- and Ti-based BMG surpassed about 1.5 GPa [185]. A $Ti_{32.8}Zr_{30.2}Ni_{5.3}Cu_9Be_{22.7}$ BMG is found to possess high fatigue resistance with the fatigue endurance limit of ~530 MPa [186]. The proliferation of shear bands prevents the fatigue crack propagation. Low scattering of fatigue data points indicate a lower sensitivity to defects in fatigue behavior of this BMG.

In-situ observations of the mechanical-damage behavior of BMGs during both fatigue and tensile testing with a video camera thermography showed the relationship with temperature evolutions. The stress–strain diagrams during high-cycle fatigue indicate three effects: the thermoelastic effect, the inelastic effect, and the heat-transfer effect [111]. The specimen temperature was observed to oscillate regularly corresponding to the stress oscillation at each fatigue cycle. A thermoelastic-degradation behavior was observed at the center of the specimen. This behavior was attributed to the free-volume accumulation inside the specimen during fatigue. However, no shear bands were observed before failure during the fatigue experiments, which indicated that the crack-initiation mechanisms of BMGs during fatigue testing could be surface-voids and defects based, and different from the shear-band mechanisms during tensile testing.

The mechanical properties of a BMG also could be altered by prior cyclic nanoindentation in the nominally elastic range [187] inducing a hardening effect. MD simulations were used to study the mechanisms of the hardening [188], and successfully showed the threshold and saturation effects. It was concluded that during the cycling there are microplastic events. The loading and unloading is apparently elastic, though the atomic rearrangements are not individually reversed; there is an accumulation of permanent structural changes [188].

The beginning of nanocrystallization was observed in the $Zr_{62.5}Fe_5Cu_{22.5}Al_{10}$ bulk metallic glassy cubic samples (edge size equal to 4 mm) under a low stress of 586 ± 242 MPa in the elastic region while no such an effect was seen in the rod shaped samples of 2 mm in diameter [189]. This alloy also showed phase separation on heating prior to crystallization [190]. Later the $Zr_{62.5}Fe_5Cu_{22.5}Al_{10}$ BMG samples of 3 mm diameter were cyclically loaded at higher stresses of 800±200 MPa and 1000±200 MPa which are still significantly lower than the yield strength. It was found that kinetically frozen anelastic deformation accumulates at room temperature and causes crystallization of metallic glassy phase. An increase in the crystallization enthalpy after cyclic loading for 1000 cycles at 1000±200 MPa from 44 J/g to 48 J/g is likely connected with the formation of the nuclei of a non-equilibrium phase which together with the residual glassy phase may represent a state with higher energy (Fig. 5.21) [191]. This also led to a better plasticity of the sample. An anelastic contribution can arise from localized viscous flow of structural defects [5,192,193]. The volume changes were locally accommodated by cumulative anelastic deformation and led to subsequent partial nanocrystallization of the glass.

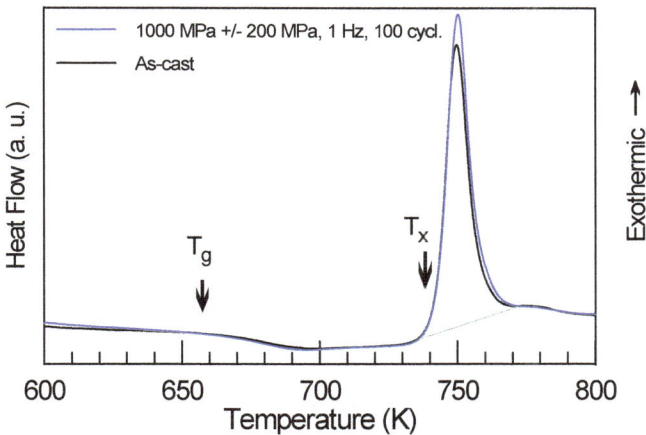

Fig. 5.21. DSC traces of the as-cast sample and after cyclic loading at 1000±200 MPa at 1 Hz for 100 cycles. Reproduced from [191] with permission of Elsevier.

The storage modulus (E') and the internal friction modulation upon cyclic loading of the $Zr_{61}Cu_{27}Fe_2Al_{10}$ bulk metallic glassy samples was studied using dynamical mechanical loading (DMA) within quasi-reversible deformation regime [194]. An increase in the storage modulus indicating stiffening of the sample is observed upon cyclic loading. Apparent cyclic softening (decrease in E' in the amplitude dependent bending DMA test) with the number of loading cycles observed at low stress was found to be connected with the instability of the DMA instrument when testing a ceramic sample. Local atomic changes leading to the detectable increase in storage modulus of the BMG sample take place at the critical elastic bending strain value of 0.3 % (Fig. 5.22 (a)). The observed increase in E' with the number of cycles after about 0.3 % deformation most probably indicates structural changes within so-called weakly bonded regions in this BMG which may trigger activation of the local atomic rearrangements leading to the increased stiffness and even local nanocrystallization. After a certain number of cycles the activated zones of larger volume become inactive owing to structural changes and the E' modulus saturates at a certain value.

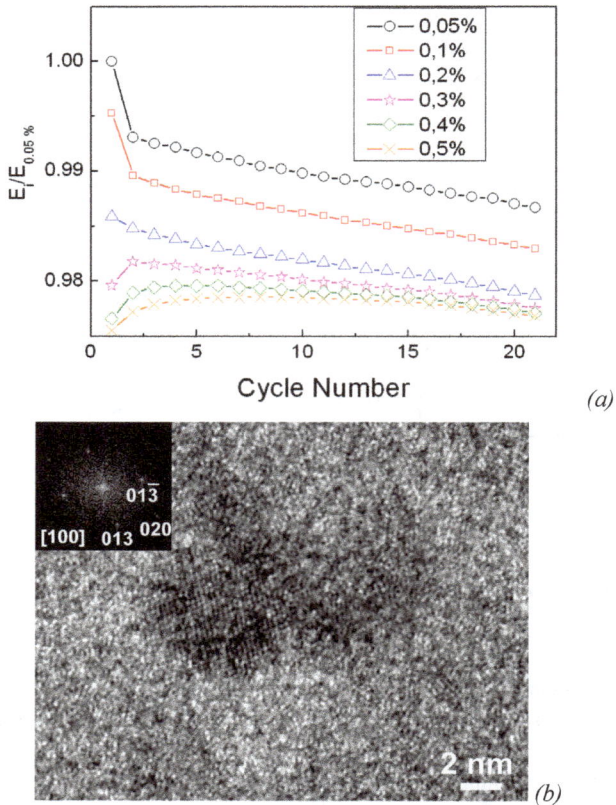

(a)

(b)

Fig. 5.22. (a) Evolution of the relative Storage modulus of the $Zr_{61}Cu_{27}Fe_2Al_{10}$ alloy as a function of the deformation cycle at different strain level. High-resolution TEM image of the BMG sample after mechanical cycling (b). The inset shows a Fourier Transform of large central crystalline particle. Reproduced from [194] with permission of Springer.

The structure of as-cast sample and the sample after a series of cyclic loading tests with maximal deflection 500 μm was studied by TEM. A small fraction of nanocrystals of about 5 nm in size is present in the elastically deformed sample (Fig. 5.22 (b)).

The $Zr_{61}Cu_{27}Fe_2Al_{10}$ bulk metallic glass also indicated complex changes in the properties after 10 loading cycles of $50\pm37.5\%$ of the yield stress in uniaxial compression at room temperature [195]. The surface microhardness measured on the sample faces normal to the loading axis increases by ~8%. This effect progressively decreased on ageing at room temperature and after 45 days the samples have recovered their initial hardness. The resulting stored anelastic strains lead to non-uniformity of the effective hardness measured on the end-faces. The cyclic loading appears to accelerate the development of anelastic strain relative to that on static loading because the overall Young's modulus increases by ~0.8% in 10 cycles. The relaxation and crystallization enthalpies of the sample increased as a result of structural rejuvenation. The excess specific heat capacity, $\Delta C_p = 1.85$ J/mol·K (Fig. 5.23), of an elastically cycled glass over an as-cast glass also suggests that cycling takes the metallic glass into a more excited state.

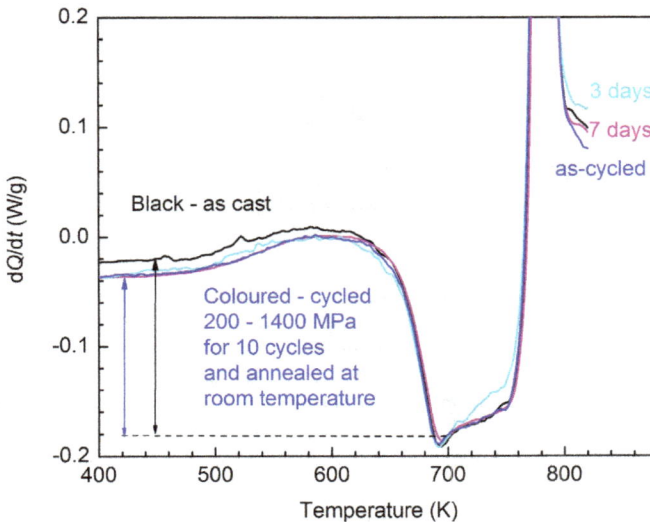

Fig. 5.23. DSC traces indicating the difference in C_p values between the supercooled liquid and glass in as-cast state, immediately after elastic cycling as well as 3 and 7 days later. Each time the sample was cut from the centre of the elastically cycled cylindrical sample. Reproduced from [195] with permission of Elsevier.

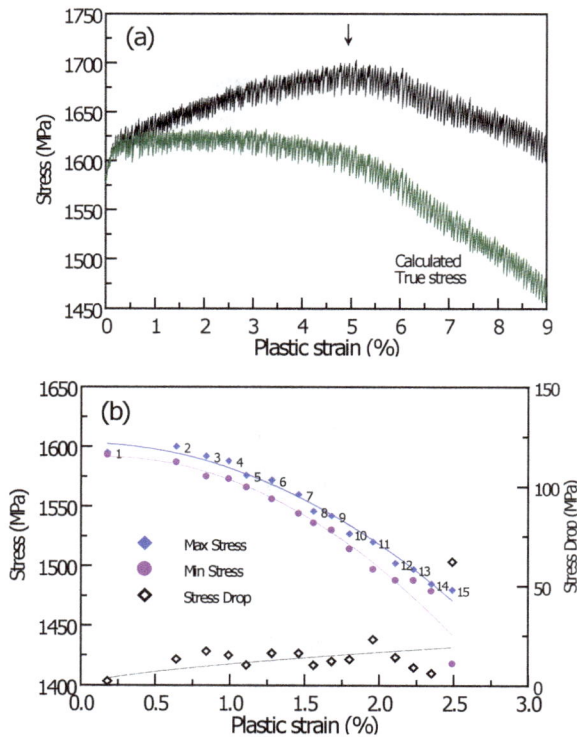

Fig. 5.24. Compression testing of the $Zr_{62.5}Cu_{22.5}Fe_5Al_{10}$ BMG. The samples are cylinders 2 mm in diameter and 4 mm in height (i.e. aspect ratio 2:1). (a) The stress-strain curves (nominal and calculated true stress) for monotonic loading to failure. The arrow indicates the strain at which there appears to be a change in deformation mode (stage 1 to stage 2). (b) The results of cyclic loading (the stress ratio r = 0) are shown by the data points labeled with the cycle number. For each cycle, the maximum stress at the onset of yielding, the minimum stress at the end of the stress drop under load, and the value of the stress drop are given. All the results are shown as a function of plastic strain. Reproduced from [198] with permission of Springer.

There have been very few studies of low-cycle fatigue (LCF) of metallic glasses and BMGs [196,197]. Samples of the $Zr_{62.5}Fe_5Cu_{22.5}Al_{10}$ bulk metallic glass were tested for LCF. The samples were unloaded, either completely or partially according to the stress ratio $r = \sigma_0/\sigma_{max}$. Mostly $r = 0$ (i.e. complete unloading), but in some cases the applied stress was reduced not to zero but to 200 MPa, giving $r \approx 0.12$. Comparison of the tests in monotonic loading and cyclic loading (repeated loading to the onset of plasticity, then unloading) shows that the compressive plasticity of the glass is drastically reduced under cyclic loading (Fig. 5.24) [198]. This effect arises from stress reversal accelerating concentration of shear in the dominant shear band.

5.1.8 Fracture toughness

The relatively large dimensions of bulk metallic glasses with good glass-forming ability permitted performing fracture toughness tests. The fracture toughness, (K_{Ic}) of BMGs varies significantly from about 1 MPa·m$^{1/2}$ for RE-based (rare-earth) and some of Fe-based BMGs, close to that of oxide glasses, up to 100 MPa·m$^{1/2}$ and even more for Zr- and Pd-based BMGs [199]. One of the reasons for low K_{Ic} of RE, Mg and Fe-based BMGs is low solubility of oxygen in the base elements and these glassy alloys which leads to the formation of the oxide particles acting as the stress concentrators [200] whereas Zr- and Pd-based BMGs dissolve a large amount of oxygen interstitially. The $Zr_{61}Ti_2Cu_{25}Al_{12}$ BMG alloy surprisingly showed fatigue pre-cracked fracture toughness in excess of 100 MPa·m$^{1/2}$ [201]. The correlations of the toughness were made with the Poisson's ratio, the product of shear modulus and molar volume (μV_m), and the glass transition temperature. The $Pd_{79}Ag_{3.5}P_6Si_{9.5}Ge_2$ BMG also showed an exceptionally high fracture toughness of 200 MPa·m$^{1/2}$ [202]. A significant reduction in the fracture toughness is found at cryogenic temperature [203]. Toughness of the $Zr_{52.5}Cu_{17.9}Ni_{14.6}Al_{10}Ti_5$ BMG decreases from 101 ± 7 MPa·m$^{1/2}$ at room temperature to 58 ± 10 MPa·m$^{1/2}$ at 123 K. At the same time, it is known that as an engineering property fracture toughness is strongly dependent on the sample size and geometry. Thus, for reproducible values samples of the same, preferably larger than a certain critical size should be studied [204].

5.1.9 Fracture mechanisms

The fracture surface of a Zr-based BMG (Fig. 5.25) exhibits alternating areas of vein and flat fracture surfaces formed by penetration of air into a liquidlike layer [205,206] after initial cleavage-like behavior starting from the lateral surface of the rod sample [207]. The cleavage-like and vein patterns also alternate in the body of the sample in the direction normal to shear deformation. The vein fracture patterns, which are specific to metallic glasses, may signify local melting of the material at the surface upon fracture [208,209].

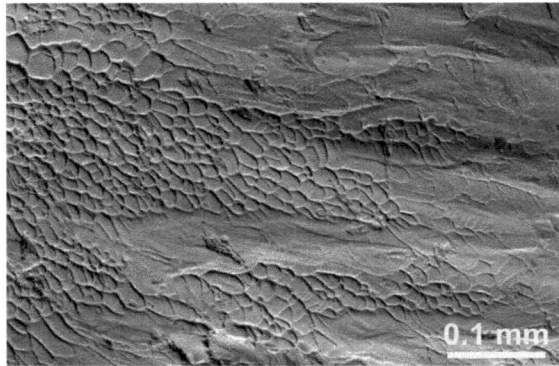

Fig. 5.25. Fracture surface of the $Zr_{61.88}Cu_{18}Ni_{10.12}Al_{10}$ BMG after failure of the sample containing vein pattern and smooth areas.

Brittle BMGs such as Mg-, Fe- or RE-based often destroy to several pieces on fracture while ductile BMGs exhibit a smaller number of fragments. Nanoscale wavy steps generated by crack front waves were observed on the fracture surface of the $Pd_{40}Ni_{40}Si_4P_{16}$ bulk metallic glass [210]. The microbranching, also found in other alloys [211,212], results from dissipating energy accumulated at the crack front by creating the additional fracture surface area when the crack speed exceeds the critical value.

5.1.10 Dynamic mechanical properties

As it is schematically shown in Fig. 5.26 [213,214] the fracture strength of the $Zr_{57}Ti_5Cu_{20}Ni_8Al_{10}$ alloy tested at strain rates from 10^{-4} to 10^3 s^{-1} is lowered under dynamic compression while shear-band propagation remains the dominant deformation mechanism. In dynamic loading experiment at high strain rate macroscopic plasticity also dropped to zero due to the lack of time for stress relaxation. High-speed cinematography indicated that at such conditions fracture occurs on a single major plane, accompanied by the emission of visible light which was also observed during an ordinary nearly quasistatic loading testing conditions.

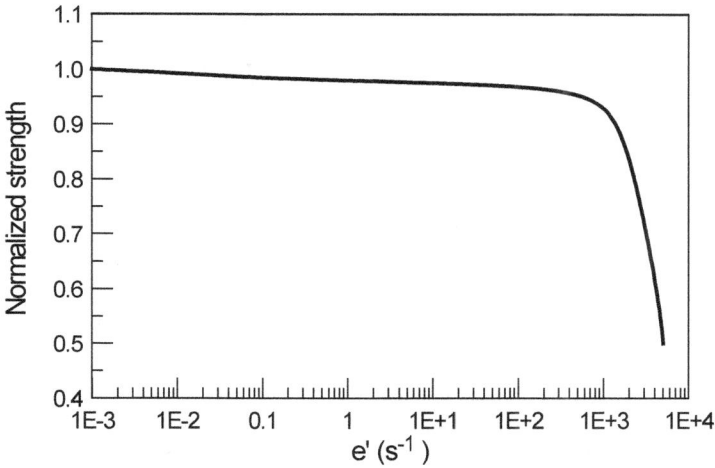

Fig. 5.26. Fracture stress of typical BMGs as a function of the strain rate (e'), schematic representation.

The impact fracture energy of 63 kJ/m^2 and stress of 1615 MPa for the bulk amorphous $Zr_{50}Al_{10}Cu_{30}Ni_5$ alloy were measured by using the Charpy test specimen with U-shape notch [215]. The impact fracture stress was nearly the same as the static tensile fracture strength of 1570 MPa obtained in a wide strain rate range of 10^{-4} to 10^{-1} s^{-1}. The fracture surface appearance changes from the vein pattern in the region near the U-shape notch to an equiaxed dimple-like pattern in the inner region.

5.1.11 Wear resistance

5.1.11.1 Macroscopic wear properties

Macroscopic wear resistance of metallic glasses has been studied extensively [216,217]. Fe-, Ni- and Co-based metallic glasses were found to have high wear resistance [218,219]. La-, Mg-, Pd- and Zr- bulk metallic glasses exhibited classical behavior for abrasive wear and the wear resistance increased linearly with hardness [220]. It was shown that some metallic glasses exhibit better wear resistance compared to the crystalline state [221]. On the other hand, macroscopic wear resistance of other metallic glasses was found to be lower than in the corresponding crystalline states [222 , 223]. However, usually BMGs embrittle on crystallization owing to the formation of primary intermetallics. The improvement of wear resistance of a Zr-based BMG composite containing β-Zr crystalline phase was attributed to its plastic deformation and work hardening, which decrease strain accumulation and release strain energy in the glassy matrix [224]. It was found that tribological behavior of BMGs is governed by both mechanical and mechanochemical phenomena [225]. The optimization of the tribological behavior of BMGs significantly depends on the counterparts as a function of contact conditions (contact pressure and environment).

5.1.11.2 Nano-scale and atomic-scale wear resistance

Successful applications of metallic glasses in micro-electro-mechanical devices was demonstrated [226,227]. This implies that any mechanical contact between the component surfaces leading to wear is a significant factor limiting their durability. When the size of the mechanical components is reduced to the micrometer and sub-micrometer level, the native surface oxide layer begins to play an important role in the contact mechanical behavior of metallic glasses. The nanoscale tribological properties of the $Ni_{62}Nb_{38}$ and Pt-Cu-Ni-P metallic glasses were studied [228,229,230]. $W_{33}N_{32}B_{35}$ glassy films deposited by sputtering having a high hardness (24 GPa) show wear resistance comparable to industrially used coatings of TiN [231].

The nanoscale and atomic-scale scratch wear resistance of typical engineering material type metallic glasses: $Ti_{43}Zr_{10}Cu_{36}Ni_9Sn_2$, $Zr_{62.5}Cu_{22.5}Fe_5Al_{10}$ and $Mg_{65}Cu_{25}Gd_{10}$ having native surface oxides were studied with the diamond single crystal cantilevers [232]. The $Ti_{43}Zr_{10}Cu_{36}Ni_9Sn_2$ and $Zr_{62.5}Cu_{22.5}Fe_5Al_{10}$ metallic glasses showed nano plowing type wear mechanism with high quality of the wear trace as shown in Fig. 5.27. Isometric projection of a typical scratch is shown in Fig. 5.28.

The $Zr_{62.5}Cu_{22.5}Fe_5Al_{10}$ is the most scratch resistant alloy with average wear resistance $Rw=1/k_w=0.093\pm0.016$ $\mu N/nm^2$. It is followed by the $Ti_{43}Zr_{10}Cu_{36}Ni_9Sn_2$ (average $1/k_w=0.051\pm0.020$ $\mu N/nm^2$) and $Mg_{65}Cu_{25}Gd_{10}$ (average $1/k_w=0.033\pm0.007$ $\mu N/nm^2$) metallic glasses. The inverse wear coefficient $(Rw=1/k_w)$ is an indicator of wear resistance. At low load the wear resistance mostly correlates with HV_c of the surface oxides (crystalline state values) on the $Ti_{43}Zr_{10}Cu_{36}Ni_9Sn_2$, $Zr_{62.5}Cu_{22.5}Fe_5Al_{10}$ and $Mg_{65}Cu_{25}Gd_{10}$ metallic glasses: TiO_2 $(HV_c=14)$, ZrO_2 $(HV_c=18)$ and MgO $(HV_c=6)$. At high loads, when the oxide film is supposed to be partly destroyed the wear resistance is also connected with a lower hardness of the $Mg_{65}Cu_{25}Gd_{10}$ metallic glass compared to the $Ti_{43}Zr_{10}Cu_{36}Ni_9Sn_2$ and $Zr_{62.5}Cu_{22.5}Fe_5Al_{10}$ metallic glasses. As it was shown, the change in the wear rate as a function of load for the $Ti_{43}Zr_{10}Cu_{36}Ni_9Sn_2$, $Zr_{62.5}Cu_{22.5}Fe_5Al_{10}$ and $Mg_{65}Cu_{25}Gd_{10}$ metallic glasses takes place at a load of about 1 μN when the oxide film break likely takes place (Fig. 5.29).

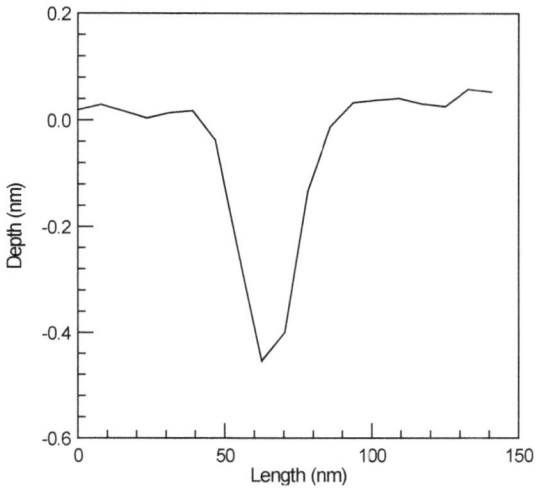

Fig. 5.27. The integrated profile on the surface of the $Ti_{43}Zr_{10}Cu_{36}Ni_9Sn_2$ metallic glass. Reproduced from [232] with permission of Elsevier.

Fig. 5.28. Isometric projection of a nanoscale scratch on the surface of the $Ti_{43}Zr_{10}Cu_{36}Ni_9Sn_2$ metallic glass.

The wear resistance was found to generally decrease with load in accordance with the idea of surface oxides as a protective medium versus wear (Fig. 5.30) which is destroyed at a higher load.

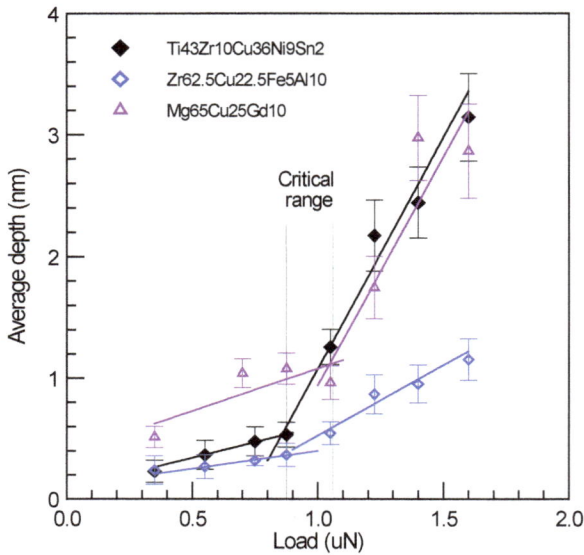

Fig. 5.29. The scratch depth as a function of load. Reproduced from [232] with permission of Elsevier.

It is known that the surface oxide can have a beneficial effect on reducing wear during the sliding of metals and alloys, particularly by preventing the metal to metal contact. An oxide thickness of a few nanometers is typical for metallic glassy alloys. It was found that the $Ni_{62}Nb_{38}$ metallic glass forms an amorphous niobium oxide of about 3-4 nm thick at ambient conditions [228,233]. Multicomponent metallic glasses, on the other hand, mostly form complex amorphous oxides on their surface. For example, in the case of the $Cu_{47}Zr_{45}Al_8$ one, it is a mixture of amorphous ZrO_2 and Al_2O_3 with nanoscale inclusions of the Cu_2O crystalline phase [234]. X-ray photoemission spectroscopy the $Ti_{45}Zr_{10}Pd_{10}Cu_{31}Sn_4$ alloy surface showed that the oxides contains all elements except for Pd [235] followed by the Pd and Cu enriched metallic glassy layer underneath. A similar Cu-enriched layer was also observed in the case of the $Cu_{47}Zr_{45}Al_8$ BMG [218]. This is also consistent with a very low Ni content in the surface niobium oxide of the Ni-Nb metallic glass [233].

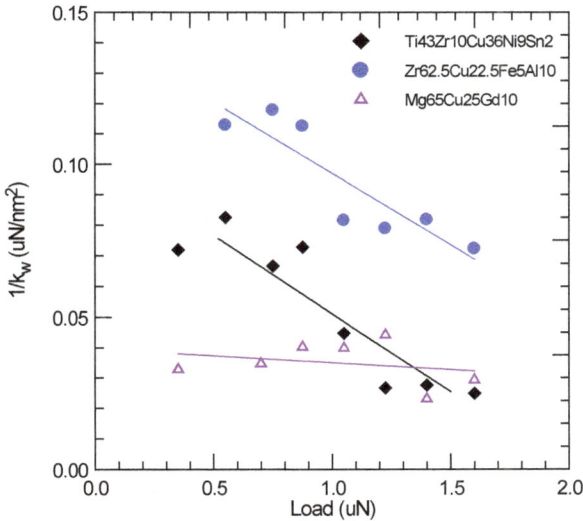

Fig. 5.30. The inverse wear coefficient as wear resistance ($R_w = 1/k_w$) as a function of load.

5.1.12 Bauschinger-type effect

Under constant torque in the $Cu_{46}Zr_{42}Al_7Y_5$ metallic glassy sample [236], the angle of twist shows a time dependence that is interpreted as due to recoverable anelastic deformation. Torque reversal shows a Bauschinger-type effect in the anelastic deformation; in particular, the relative deformation rate increases by a factor of 4.5 on the first reversal. This result considered in terms of the operation of shear transformation zones suggested that stress reversal in the elastic regime is particularly effective in inducing local structural changes.

5.2 Plastic flow of bulk metallic glasses on heating

Structural relaxation on annealing below T_g significantly influences plasticity/toughness of BMGs, their yield strength, elastic moduli and hardness values. Annealing below and above the glass transition temperature leading to structural relaxation showed that the shear modulus was found to increase with the annealing time, while the strain at fracture decreased [237].

When tested in-situ on heating BMGs first show non-Newtonian flow near T_g and slightly above it, and then Newtonian flow in the supercooled liquid region [238]. However, the boundaries of all these temperature regions (inhomogeneous deformation, non-Newtonian flow and Newtonian flow) are strain rate dependent, i.e. transition from one region to another can take place with decrease in the strain rate near T_g. Anelastic atomic rearrangements at room temperature induced by applied stress below the yield stress are limited to immediate neighborhood of atoms [239]. At a temperature close to T_g atomic rearrangements are spatially more extended and creep deformation occurs as macroscopically anelastic behavior.

Some of the glassy alloys exhibit "superplasticity" (actually good fluidity) on heating to a supercooled liquid region [240]. Because of this bulk glassy alloys can be thermo-mechanically shaped or welded in the supercooled liquid region before they crystallize. They can be used for creation of micro-gears with diameters ranging from 0.1 to 1 mm by the Newtonian flow. Various micro- and nano-scale patterns [241,242] can be produced on bulk metallic glasses in the supercooled liquid region (Fig. 5.31) [243]. Bulk metallic glasses can be deformed in a manner similar to plastics by blow molding. This allows to net shape complex geometries including shapes, which can not be produced by other ways.

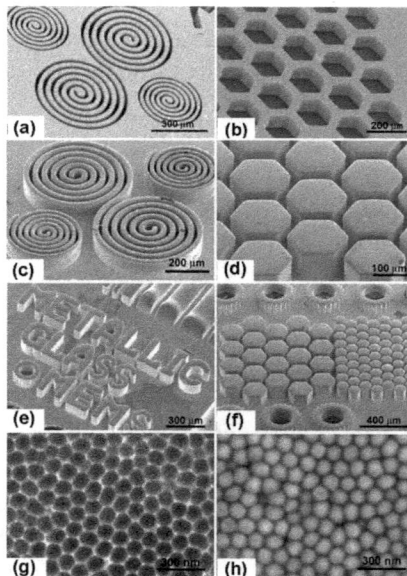

Fig. 5.31. (a) and (b) images of Si molds, (c)-(f) the patterns fabricated by using Si molds on the $Pd_{40}Ni_{40}Si_4P_{16}$ metallic glassy surface, (g) is porous alumina and (h) is the corresponding nano rods fabricated on the $Pd_{40}Ni_{40}Si_4P_{16}$ metallic glassy surface with the porous alumina (g). Reproduced from [243] with permission of Springer.

Shaping of BMGs can also be done by the electromechanical shaping technology at low applied stresses due to the high electrical resistivity of glassy alloys [244]. Bonding of glassy alloys can be achieved by Joule heating [245] as well as by laser [246,247], electron-beam [248,249] and friction welding [250]. When the temperature increases to T_g the BMG samples, in general, become more ductile and the yield stress decreases. Above T_g, metallic glasses transform into the supercooled liquid state, and as summarized in Fig. 5.32 (a) [251], only at a high strain rate they exhibit an inhomogeneous deformation.

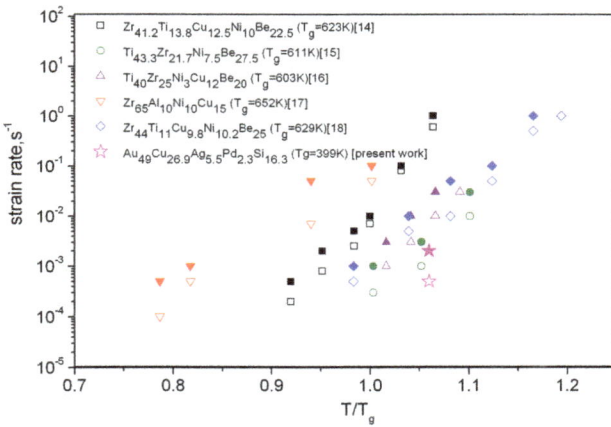

(a)

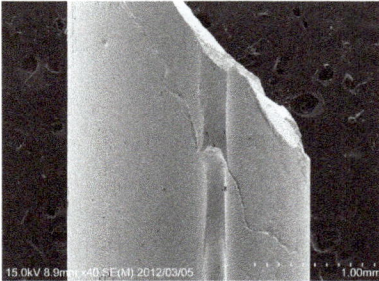

(b)

(c)

Fig. 5.32. Transition from inhomogeneous (filled symbols) to homogeneous (open symbols) deformation mode at the elevated temperature for several BMGs in compression as a function of the strain rate and temperature: literature and current research data. Reproduced from [251] with permission of the American Institute of Physics. The lateral (b) and fracture surface (c) appearance for the $Au_{49}Cu_{26.9}Ag_{5.5}Pd_{2.3}Si_{16.3}$ BMG rod tested at T_g.

The $Au_{49}Cu_{26.9}Ag_{5.5}Pd_{2.3}Si_{16.3}$ BMG exhibits a mechanical behavior, which is distinct from the other metallic glasses [251]. Even at a low strain rate of $1.25 \cdot 10^{-4} s^{-1}$, which is generally considered as quasi-static deformation, it shows brittle fracture at temperatures near T_g (Fig. 5.32 (b,c)) and only at the temperature of 1.08 T_g the samples began to show plastic behavior (Fig. 5.32 (a)). A reason for such a behavior may be connected with the structure of the Au-based amorphous phase with large number of the icosahedral clusters [252] and possibly strong covalent bonds with Si. Thus, the internal structure of Au-based BMG can make a strong influence on the temperature of brittle–ductile transition.

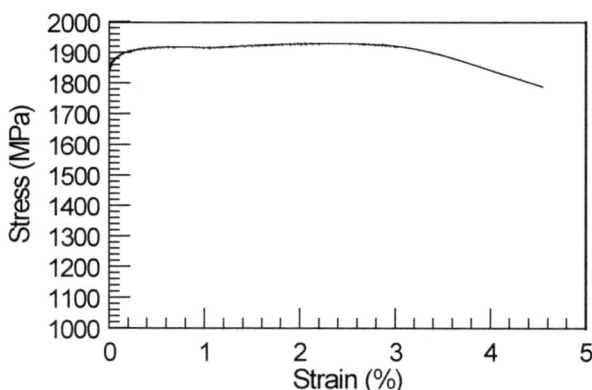

Fig. 5.33. Stress-strain diagrams of the $Zr_{64.13}Ni_{10.12}Cu_{15.75}Al_{10}$ glassy alloy tested at boiling liquid nitrogen temperature.

5.3   Deformation of bulk metallic glasses at cryogenic temperature

The deformation behavior of the Zr-based BMG alloys was also tested at the boiling liquid nitrogen temperature [253,254,255]. The sample tested in liquid nitrogen exhibits higher compression strength compared to the sample tested at room temperature. Also no clear serrated flow is observed [255]. The typical stress-strain diagram of the $Zr_{64.13}Ni_{10.12}Cu_{15.75}Al_{10}$ glassy alloy tested at boiling liquid nitrogen temperature is shown in Fig. 5.33. The mechanical behavior and the kinetics of shear deformation in bulk metallic glasses were also investigated at room and liquid nitrogen temperature using the acoustic emission technique. It was demonstrated that the intensive acoustic emission reflecting the activity of strongly localized shear bands at room temperature vanishes at the transition from serrated to non-serrated plastic flow at low temperature. The disappearance of the acoustic emission signals clearly indicates that the shear band propagation velocity significantly decreases at low temperature, and sliding along the dominant shear band takes place at the machine-driven rate [256].

The samples tested in liquid nitrogen still exhibited formation of the localized shear deformation bands (Fig. 5.34). The fracture surfaces of the samples tested in liquid nitrogen also show the mirror looking cleavage-like areas mixed up with typical vein patterns in nearly equal proportions like it is shown in Fig. 5.25 for the sample tested at room temperature.

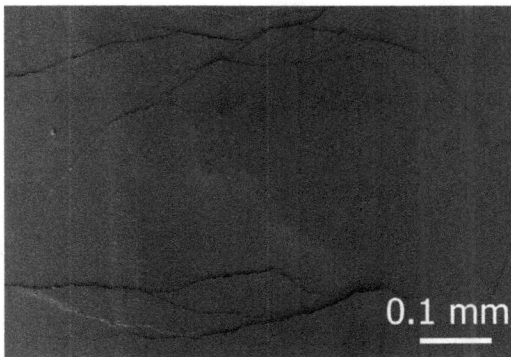

Fig. 5.34. SEM image in secondary electrons of the lateral surface of the $Zr_{64.13}Ni_{10.12}Cu_{15.75}Al_{10}$ alloy sample after mechanical testing at boiling liquid nitrogen temperature.

5.4   Non-uniform materials with glassy phase

5.4.1   Porous glasses

Porous glassy materials are very promising for structural applications owing to their low density and good mechanical properties. Open-pore type Zr-based BMGs [257] were produced by the salt-replication technique while closed-pore type porous Pd-based BMGs [258,259] were produced by holding the melt under pressurized hydrogen which is released when the pressure is reduced. Porous $Zr_{55}Cu_{30}Al_{10}Ni_5$ alloy samples were also produced by spark plasma sintering (SPS) [260]. The porosity was controlled by the sintering temperature and loading pressure in the SPS process. The thermal stability of the sintered porous glassy alloy specimens was similar to that of the original powder. The sintered porous glassy alloy specimens exhibited a larger plastic ductility, a lower Young's modulus and a lower fracture strength than the as-cast alloy specimen.

Closed-pore type glassy alloys were obtained for the $Pd_{42.5}Cu_{30}Ni_{7.5}P_{20}$ glassy alloy has an extremely high glass-forming ability among metallic alloys. Pd absorbs large quantities of hydrogen. Once the hydrogen pressure is reduced the dissolved hydrogen starts to precipitate from the melt causing its bubbling. The melt was subjected to different hydrogenation treatments following water or oil cooling. In order to produce relatively small finely dispersed pores the foaming pressure before cooling was just 10 % less than the hydrogenation pressure (Fig. 5.35 (a)) while the sample remained glassy (Fig. 5.35 (b)). When the pressure is reduced by more than about 10 % the samples with much larger pores and higher porosity can be produced. As the initial hydrogen pressure at which hydrogen is dissolved increases porosity decreases, pore number density increases and pore size becomes smaller.

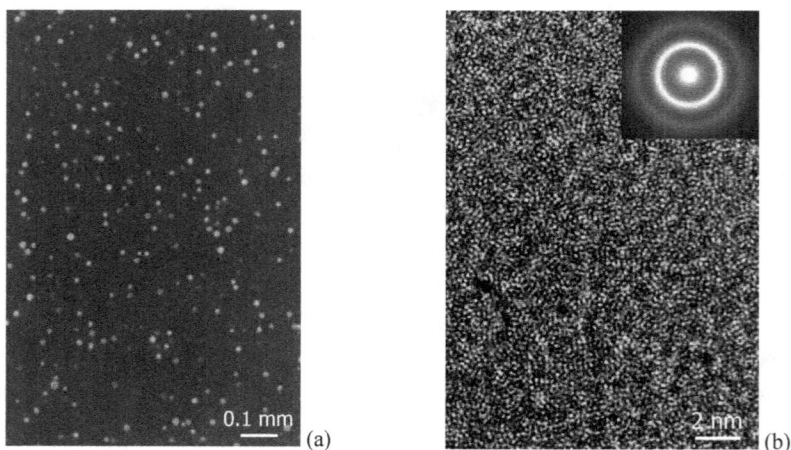

Fig. 5.35. SEM (a) image of the porous structure obtained in the $Pd_{42.5}Cu_{30}Ni_{7.5}P_{20}$ BMG with initial pressure of 15 MPa and high-resolution TEM (b) image of the glassy phase with the SAED pattern inserted.

The mechanical properties are summarized in Table 5.7. As one can see the yield strength at 3.7 % porosity is 0.93 of the monolithic glass while its plasticity reached 18%.

Table 5.7. Porosity, 0.2 % proof stress ($\sigma_{0.2}$), the Young's modulus (E) and plastic strain (ε) of the porous Pd$_{42.5}$Cu$_{30}$Ni$_{7.5}$P$_{20}$ glassy rods tested in compression. The data from Ref. [261].

Porosity (%)	$\sigma_{0.2}$ (MPa)	E (GPa)	ε (%)
0	1630	102	0
1.7	1580	94	9.5
2.2	1550	98	4.5
3.0	1520	98	7.5
3.7	1520	88	18

Pore size and distribution can be controlled by the preparation condition depending on temperature and pressure of the melt. Another method to control the pore size and distribution is subsequent annealing of the porous alloy within the supercooled liquid region which causes further expansion of the pores leading to the materials with porosity over 10 %. The expansion of pores takes place owing to significantly reduced viscosity of a liquid compared to a glass.

5.4.2 Glassy-crystalline dual phase alloys

The formation of heterogeneous microstructures combining a glassy matrix with crystalline particles enhances plasticity of BMGs [262,263]. The mechanisms of ductilization of bulk metallic glassy alloys by crystalline and quasicrystalline inclusions are associated with blocking and branching of the propagating localized shear bands by the structural inhomogeneities for example, crystalline particles [263]. This enables multiplication, branching and termination of the shear bands similar to the composites where cracks are either blocked or blunted by the secondary phases.

Various other bulk metallic glass composites with inclusions of a crystalline phase(s) were produced starting from the glassy sample [264,265,266,267]. Ductile BMGs samples containing dendritic crystals were produced in the La-Al-(Cu,Ni) [268] (HCP α-La phase as the dendritic phase) and Zr-Ti-Nb-Cu-Be (in which the dendritic phase is a body-centered cubic (Zr,Ti,Nb) solid solution) with large Be content [269 , 270] as well as Ti$_{46}$Zr$_{20}$V$_{12}$Cu$_5$Be$_{17}$ [271] alloys. These materials were created using the strategy of microstructural toughening and ductility enhancement in metallic glasses. Although the Zr-Ti-Nb-Cu-Be β-Zr/BMG dual phase alloys show necking owing to stress softening of the glassy phase, it was reported that the Cu-Zr-Al-Co [272,273] and Cu-Zr-Al composites [274] show strain hardening owing to transformation-induced plasticity upon martensitic transformation. 1 mm diameter rods of Ti-Cu-Ni glassy/crystal alloys also exhibited tensile plasticity [275]. Thus, BMG alloys and the composites containing glassy phase are promising structural and functional materials of the present century. Some BMGs undergo crystallization during deformation, which also leads to their ductilization [276]. Deformation-induced nanocrystallization [277,278] is also helpful in ductilization of BMGs.

The $Ni_{40}Cu_{10}Ti_{33}Zr_{17}$ alloy [279] produced by copper mold casting shows a superior combination of strength and ductility not only due to a composite effect resulting from a multiphase structure, but also due to a strain-induced martensitic transformation. The microstructure consisted of the austenitic cP2 cubic (Ni,Cu)(Ti,Zr) phase embedded in the glassy matrix in the as-cast alloy, while mP4 martensitic (Ni,Cu)(Ti,Zr) phase is formed during plastic deformation. The crystals further act as strong barriers for shear band propagation, promoting increase in stress and subsequent shear band formation and branching, thus extending the ductility and preventing a premature brittle fracture. The sample showed the ultimate strength of 2000 MPa, the yield stress of 650 MPa and the compressive ductility of 15 % [279]. Unlike common observations of serrated plastic flow associated with the large scale shear bands in monolithic metallic glasses, propagation of the shear bands in the studied alloy does not exhibit the stress drops, although the shear bands were readily seen on the lateral surface of the samples tested to fracture in addition to multiple slip bands in cP2 crystalline grains.

Amplitude dependent internal friction measurement was preformed for the $Ni_{40}Cu_{10}Ti_{33}Zr_{17}$ alloy samples. The internal friction values at initial loading were always higher than the subsequent ones. These facts are in line with the irreversible martensitic cP2→mP4 transformation mechanism, which is also responsible for the formation of a hysteresis loop. Owing to relatively high sensitivity of the DMA machine to tiny structural changes the occurrence of martensitic transformation is detected even in the linear quasi-elastic deformation mode [280].

Various Ni-Cu-Ti-Zr alloys were further studied [281] and more complex systems alloys were produced [282] (Table 5.8). The formation of the mixed glassy-crystal structure is illustrated in Fig. 5.36. The samples demonstrate effects of transformation-induced plasticity (TRIP) and superelasticity. It was found that the addition of Y does not lead to increase in the glass forming ability of Ni-Ti-Cu-Zr system alloys while a large amount of Y (more than 0.5 at.%) induces $NiTi_2$ phase precipitates. The additions of Co and Nb increase yield strength of these alloys by 350 and 600 MPa respectively. At the same time the alloys exhibit large plasticity. The additions of Al and B lead to double increase of the yield strength, but fully suppress the martensitic transformation. These alloys have a lower plasticity compared to the alloys with Nb.

Table 5.8. The values of maximum stress σ_{max}, stress of the beginning of plastic deformation σ_p, the stress initiating the martensitic transformation σ_m and the relative strain ε. Phase composition cP2 + glassy phase, $Ni_{50}Zr_{17}Ti_{33}$ also contains the $Ni_{10}Zr_7$ one. Ref. indicates literature source.

Alloy	σ_{max} (MPa)	σ_p (MPa)	σ_m (MPa)	ε (%)	Ref.
$Ni_{25}Cu_{25}Ti_{33}Zr_{17}$	1605	1455	-	1.1	281
$Ni_{35}Cu_{15}Ti_{33}Zr_{17}$	1790	1475	635	9.4	281
$Ni_{40}Cu_{10}Ti_{35}Zr_{15}$	1870	1350	270	18.2	281
$Ni_{40}Cu_{10}Ti_{40}Zr_{10}$	1810	1325	420	16.4	281
$Ni_{45}Cu_5Ti_{42}Zr_8$	2700	1600	165	22	281
$Ni_{50}Zr_{17}Ti_{33}$	2450	2200	900	2	281
$Ti_{42.0}Ni_{39.0}Cu_9Zr_{10}$	2150	1970	320	18.3	282
$Ti_{40.0}Ni_{39.5}Cu_8Zr_{10}Co_2Y_{0.5}$	2620	1955	780	25	282
$Ti_{39.5}Ni_{39.0}Cu_8Zr_8Co_2Y_{0.5}Nb_3$	2520	1730	1210	25	282

The $Ti_{42.3}Zr_{7.7}Cu_{41.7}Ni_{7.3}Co_1$ sample with 1000 ppm of B containing a crystalline phase embedded in the glassy matrix (Fig. 5.37) [283] also exhibited formation of multiple shear offsets on the lateral surface of the sample (Fig. 5.37, insert). Strain hardening and no detectable serrated flow were observed at the initial stage of deformation while weak stress drops of 1-3.5 MPa are seen after 4.5 % of plastic deformation. Fracture occurs when strain hardening ($d\sigma/d\varepsilon$) (stress σ and strain ε) mechanisms in the crystalline phase are exhausted. Deformation of such a dual-phase material likely takes place by a series of smaller sample displacements in the shear bands though the shear banding mechanism is still active in this case (Fig. 5.37 insert).

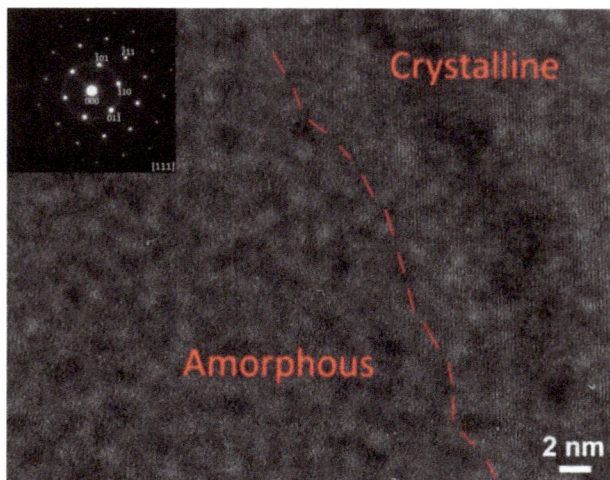

Fig. 5.36. HRTEM of the as-cast $Ti_{40}Ni_{39.5}Cu_8Zr_{10}Co_2Y_{0.5}$ with a typical electron diffraction pattern containing both sharp rings from the crystalline particle and a halo pattern from the surrounding glassy phase (inset). Reproduced from [282] with permission of Elsevier.

A strong influence of particles distributed in the amorphous matrix on shear band propagation was confirmed using the numerical methods [284,285,286] and experimental studies [287,288]. For example, Ta-rich particles with bimodal distribution act as discrete obstacles, separating and restricting the highly localized shear-banding, avoiding catastrophic shear-through of the whole sample and enhancing plasticity of the monolithic Zr-Cu-Ni-Al BMG [287].

The Zr-based bulk metallic glassy samples containing the rare earth elements with a maximum diameter of 5 mm were successfully prepared by copper mold casting method at low vacuum conditions [289]. The structure of the alloys consisted of the amorphous matrix with a small fraction of RE oxides. The maximum compression strength was achieved in the $Zr_{62}Cu_{22.5}Fe_5Al_{10}Dy_{0.5}$ alloy. At the same time, the optimal combination of strength and plasticity (fracture strength of 1890 MPa and plastic deformation of 3.3 %) was observed in the $Zr_{62}Cu_{22}Fe_5Al_{10}Gd_1$ alloy.

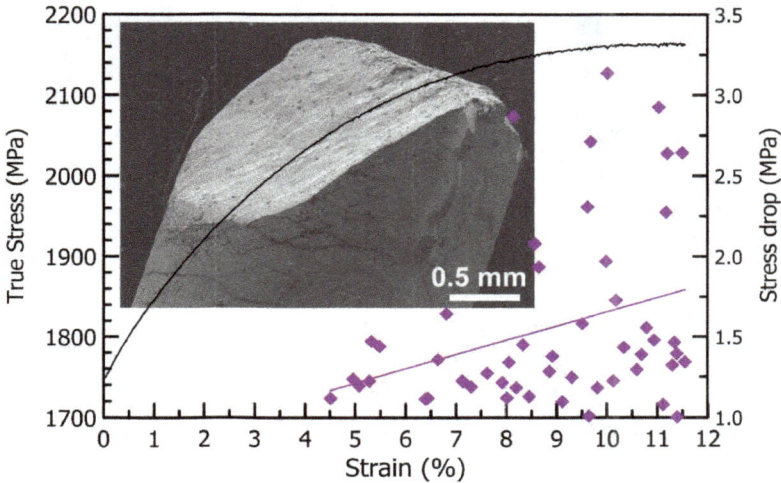

Fig. 5.37. Stress and stress drop values as a function of plastic strain for the cylindrical Ti$_{42.3}$Zr$_{7.7}$Cu$_{41.7}$Ni$_{7.3}$Co$_1$ crystal-glassy sample with 1000 ppm of B of 2:1 aspect ratio upon the compression test. The insert represents a lateral surface of the sample tested to fracture. The diamond symbols denote the stress drop values exceeding 1 MPa. Reproduced from [132] with permission of Elsevier.

The finite element model FEM analysis [81,290,291] indicated that the oxide particles with equiaxed shape can block propagation of a shear band and promote generation of new shear bands facilitating plasticity. The results of finite element deformation modeling of two microstructure types: (1) with a shear band and (2) with both a shear band and a non-deformable oxide particle indicated that, strain is localized in shear band in both cases, while the local stress maxima are, as expected, on the surface of the sample in the region close to shear band in the first model and on the surface of the particle in the second model (Fig. 5.38). The deformation behavior of the first model sample was similar to experimental behavior of the BMG: after elastic deformation the plastic flow takes place without significant strain hardening, which usually leads to catastrophic shear band propagation and failure. The presence of a non-deformable particle of a favorable (round) shape hampers the shear band propagation. The fracture strength of RE oxides under soft deformation condition is more than 2500 MPa [292] which is enough for this purpose. At the same time in the other regions of the sample the level of stress will be increased to generate new shear bands which will improve overall plasticity.

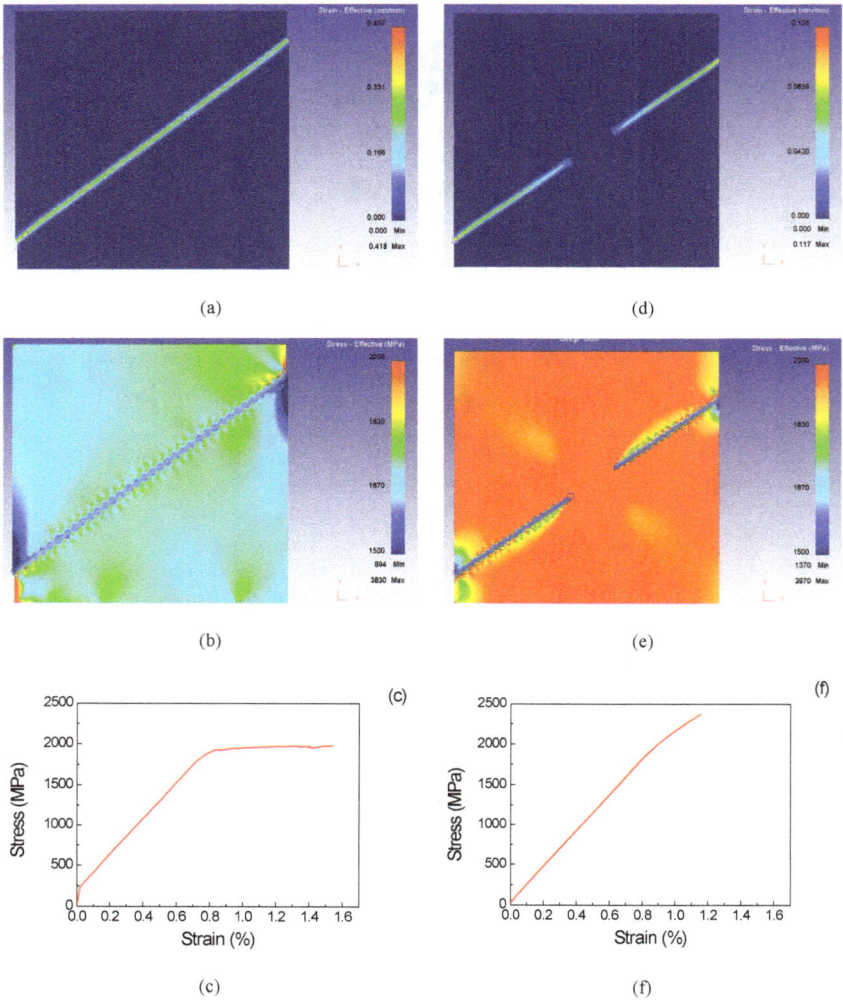

Fig. 5.38. Finite element modeling of plastic deformation of the non-deformable sample matrix with a shear band (a,b,c) and of the non-deformable sample matrix with both a shear band and a non-deformable particle (d,e,f): strain distribution at 1.5 % strain (a,d), stress distribution at 1.5 % strain (b,e) and stress-strain curves (c,f). Reproduced from [289] with permission of Elsevier.

The sample with 1 at. % of Gd containing in the RE metal oxide particles showed the formation of multiple shear bands in contradistinction to the alloy without oxide particles. One should also mention that glassy-polymer composite materials developed recently combine attractive properties of both [293].

5.4.3 Mechanical properties of nanocomposite alloys

Mechanical strength and ductility of glassy alloys can be improved by the precipitates of nanocrystalline or nanoquasicrystalline phase. The blockage of the crack-tip area by nanoparticles stabilizing plastic deformation was observed in the $Cu_{50}Zr_{50}$ and the $Zr_{65}Al_{7.5}Ni_{10}Pd_{17.5}$ glassy alloys [294]. The $Zr_{65}Al_{7.5}Cu_{7.5}Ni_{10}Pd_{10}$ alloy having nanoscale icosahedral phase particles embedded in the glassy matrix showed a better combination of the mechanical properties compared to the as-cast glassy sample [295] without precipitates. Because the single-phase icosahedral phase alloys are extremely brittle it can be considered that the inclusions of a precipitating phase in this alloy cause deviation and blockage of the operating shear bands which improve the plasticity of the alloy. Thus, good mechanical properties are attributed to the existence of the residual intergranular glassy phase while the icosahedral particles can act as a resisting medium against the shear deformation. Nanocrystalline precipitates increase the room-temperature mechanical strength of the Zr-Al-Cu-Pd [296], Zr-Al-Cu-Pd-Fe and (Zr/Ti)-Cu-Al-Ni [297] bulk glassy alloys. The compressive room-temperature stress-strain curves of the $Zr_{65}Al_{7.5}Ni_{10}Cu_{17.5}$ and $Zr_{65}Al_{7.5}Ni_{10}Pd_{17.5}$ bulk metallic glassy alloys revealed significantly different plasticity of these alloys, though both of the alloys possess a glassy structure [298].

An alloy, which crystallizes by the nucleation and growth mechanism, is supposed to be less prone to nanocrystallization during deformation while the alloy having pre-existing nuclei is predisposed to show such a behavior [299]. Precipitation of the nanoscale α-Al particles at room temperature was observed in the Al-TM-RE glass-formers (like $Al_{90}Fe_5Gd_5$ or $Al_{90}Fe_5Ce_5$) within the shear bands on deformation [300,301], while a very limited fraction or complete absence of nanocrystals (depending on RE metal) was observed after significant plastic deformation in the $Al_{85}RE_8Ni_5Co_2$ amorphous and glassy samples (see Chapter 4) which undergo a clear nucleation and growth behavior.

The nanocrystallization within the shear bands was also observed in the $Cu_{47.5}Zr_{47.5}Al_5$ glassy alloy which was reported to be strain-hardenable [277], in a Ti-based alloy [302] and many others [303]. The effect of strain rate on the flow stress [304] and ductility was studied [305] and the results showed that the higher strain rate used facilitates plasticity. The $Cu_{60}Zr_{20}Ti_{20}$ bulk metallic glass was rolled at room temperature and at cryogenic temperature [306]. Nanocrystallization and the free volume decrease were found at room temperature. Cu-enriched glassy regions as precursors to nanocrystallization were observed at the cryogenic temperature. A strong composition dependence of plasticity of $(Zr,Cu)_{100-x}Al_x$ BMGs was observed. The compositions with primary nanoscale phase exhibit a large plastic strain facilitated by deformation-induced nanocrystallization in the shear bands [307]. Yavari et al. [308] using a synchrotron X-ray beam of 2 µm in diameter showed formation of deformation-induced nanocrystals on the compressed side of the $Pd_{40}Cu_{30}Ni_{10}P_{20}$ alloy tested in bending. As crystallization is accompanied by a decrease in the molar volume the applied pressure increases the driving force for crystallization.

Al-RE-TM amorphous alloys posses a high tensile strength exceeding 1200 MPa and good bend ductility [309,310]. The homogeneous dispersion of the nanoscale FCC α-Al particles in the amorphous matrix causes a drastic increase in the tensile fracture strength to 1560 MPa [311]. It is considered that the nanoscale α-Al particles can act as an effective barrier against shear deformation of the amorphous matrix [312]. At the same time, it was also suggested [313] that the hardening could be attributed mainly to solute enrichment of the residual glassy matrix due to lowering of the Al content.

However, in some cases, formation of the primary α-Al particles was found to deteriorate the mechanical properties: decrease the tensile strength and hardness. Partial substitution of Ni for Cu in the $Al_{85}Y_8Ni_5Co_2$ metallic glass caused formation of the nanoscale α-Al particles and drastically decreased the tensile strength and hardness values of the alloy [314]. Cu has much lower absolute value of heat of mixing with Al, Y and Co than Ni may be responsible for such a decrease. The α-Al inter-particular distances in the $Al_{85}Y_8Ni_3Co_2Cu_2$ metallic glass significantly exceed the particle size. Thus, here, α-Al particles cannot act as an effective barriers against the shear deformation in the amorphous matrix.

Also contrary to thermally produced α-Al particles on annealing which strengthen the Al-Y-Ni-Co metallic glasses α-Al particles precipitated upon mechanical deformation within the shear bands do not have such an effect [315]. Softening of the metallic glass after cold rolling (Table 5.9) can be understood taking into account that the shear bands generate extra excess volume which makes the material softer (HV=333) and enhances subsequent sliding within these shear bands. Similar softening behavior was observed on rolling the Zr-Cu-Ni-Al alloys [316] while shot-peening [317] induced both softening by introduction of shear bands and hardening by the residual stress. The $Al_{85}Y_8Ni_5Co_2$ glassy alloy sample rolled at 473 K exhibited even lower hardness value which is only 70 % of its original hardness. Such a drastic decrease in hardness after rolling at 473 K in contrast to annealing at the same temperature or hardening upon phase transformation on annealing at 573 K can be explained by the excess volume generated within the shear bands.

Table. 5.9. Vickers microhardness (HV) of the samples in as-prepared state, after annealing for 15 min at different temperature and after rolling at different temperature.

	Room temperature	At 373 K	At 473 K
Not rolled	HV 411±20	HV 414±10	HV 427±3
After rolling	HV 333±20	HV 417±22	HV 286±14

As a result of mechanical rolling performed at 473 K the diffusion process leading to redistribution of the excess volume is accelerated and a larger volume in the vicinity of the shear bands can be softened. Although α-Al nanocrystals formed upon rolling at 473 K can be the barriers for shear bands their efficiency in doing so depends on the size and volume fraction [318]. One can suggest that the nanocrystals impede the formation of new shear bands but do not affect future propagation of the already existing bands. In turn, a decrease in volume during the formation of α-Al nanocrystals with densely packed atomic structure may further increase the excess (or free) volume within the shear bands and facilitate further sliding.

It is suggested that not only α-Al particles but also the amorphous matrix has some role in the hardening and embrittlement of the alloy. The $(Al_{0.84}Y_{0.09}Ni_{0.05}Co_{0.02})_{95}Sc_5$ amorphous alloy has an ultra-high tensile fracture strength slightly exceeding 1500 MPa which surpasses those for all other Al-based fully crystalline and fully amorphous alloys reported up to date [8] and can be likely further strengthened by nanoscale crystallization.

References

[1] W. H. Wang, The elastic properties, elastic models and elastic perspectives of metallic glasses, Progress in Materials Science, 57 (2012) 487–656. https://doi.org/10.1016/j.pmatsci.2011.07.001

[2] L. Deng, B. Zhou, H. Yang, X. Jiang, B. Jiang, X. Zhang, Roles of minor rare-earth elements addition in formation and properties of Cu–Zr–Al bulk metallic glasses, Journal of Alloys and Compounds, 632 (2015) 429-434. https://doi.org/10.1016/j.jallcom.2015.01.036

[3] D. V. Louzguine-Luzgin, M. Fukuhara, A. Inoue, Specific volume and elastic properties of glassy, icosahedral quasicrystalline and crystalline phases in Zr-Ni-Cu-Al-Pd alloy, Acta Materialia, 55 (2007) 1009-1015. https://doi.org/10.1016/j.actamat.2006.09.013

[4] H. Kato, J. Saida, A. Inoue, Influence of hydrostatic pressure during casting on as cast structure and mechanical properties in $Zr_{65}Al_{7.5}Ni_{10}Cu_{17.5-x}Pd_x$ (x = 0, 17.5) alloys, Scripta Mater., 51 (2004) 1063-1068. https://doi.org/10.1016/j.scriptamat.2004.08.004

[5] A. S. Argon, Mechanisms of inelastic deformation in metallic lasses, Journal of Physics and Chemistry of Solids, 43 (1982) 945-949.
https://doi.org/10.1016/0022-3697(82)90111-1

[6] N. Togashi, M. Ishida, N. Nishiyama, A. Inoue, Wear resistance of metallic glass bearings, Rev. Adv. Mater. Sci., 18 (2008) 93-97.

[7] A. R. Yavari, J. J. Lewandowski, J. Eckert, Mechanical properties of bulk metallic, MRS Bull., 32 (2007) 635-638. https://doi.org/10.1557/mrs2007.125

[8] A. Inoue, S. Sobu, D. V. Louzguine, H. Kimura, K. Sasamori, Ultrahigh strength Al-based amorphous alloys containing, J. Mater. Res., 19 (2004) 1539-1543.
https://doi.org/10.1557/JMR.2004.0206

[9] M. Wakeda, Y. Shibutani, S. Ogata, J.Y. Park, Relationship between local geometrical factors and mechanical properties for Cu–Zr amorphous alloys, Intermetallics, 15 (2007) 139–144. https://doi.org/10.1016/j.intermet.2006.04.002

[10] A. Inoue, Stabilization of metallic supercooled liquid and bulk amorphous alloys, Acta Mater. 48 (2000) 279-306. https://doi.org/10.1016/S1359-6454(99)00300-6

[11] D.V. Louzguine, A. Inoue, Influence of Ni and Co additions on supercooled liquid region, devitrification behaviour and mechanical properties of Cu-Zr-Ti bulk metallic glass, Mater. Sci. Forum/J. Metastable Nanocryst. Mater., 15–16 (2003) 31–36.
https://doi.org/10.4028/www.scientific.net/JMNM.15-16.31

[12] D.V. Louzguine, H. Kato, A. Inoue, Investigation of mechanical properties and devitrification of Cu-based bulk glass formers alloyed with noble metals, Science and Technology of Advanced Materials, 4 (2003) 327–331.
https://doi.org/10.1016/j.stam.2003.09.003

[13] D. V Louzguine, A. Inoue, Structural and thermal investigations of a high-strength Cu-Zr-Ti-Co bulk metallic glass, Philosophical Magazine Letters, 83 (2003) 191-196.
https://doi.org/10.1080/0950083031000066126

[14] A. Inoue, T. Zhang, A. Takeuchi, Bulk amorphous alloys with high mechanical strength and good soft magnetic properties in Fe–TM–B (TM= IV–VIII group transition metal) system, Appl. Phys. Lett., 71 (1997) 464-466. https://doi.org/10.1063/1.119580

[15] K. Amiya and A. Inoue, Fe-(Cr, Mo)-(C, B)-Tm bulk metallic glasses with high, strength and high glass-forming ability, Materials Transactions, 47 (2006) 1615-1618. https://doi.org/10.2320/matertrans.47.1615

[16] J.F. Wang, R. Li, H. Nengbin, T. Zhang, Co-based ternary bulk metallic glasses with ultrahigh strength and plasticity, Journal of Materials Research, 26 (2011) 2072-2079. https://doi.org/10.1557/jmr.2011.187

[17] L.A. Davis, Hardness/strength ratio of metallic glasses, Scripta Metallurgica, 9 (1975) 431-435. https://doi.org/10.1016/0036-9748(75)90177-5

[18] Y. Yokoyama, A. Inoue, Compositional dependence of thermal and mechanical properties of quaternary Zr-Cu-Ni-Al bulk glassy alloys, Materials Transactions, 48 (2007) 1282 – 1287. https://doi.org/10.2320/matertrans.MF200622

[19] E. Ma, J. Ding, Tailoring structural inhomogeneities in metallic glasses to enable tensile ductility at room temperature, Mater. Today, 19 (2016) 568–579. https://doi.org/10.1016/j.mattod.2016.04.001

[20] A. Inoue, W. Zhang, T. Zhang, K. Kurosaka, Formation and mechanical properties of Cu–Hf–Ti bulk glassy alloys, J. Mater. Res., 16 (2001) 2836-2844. https://doi.org/10.1557/JMR.2001.0391

[21] K. B. Kim, J. Das, M.H. Lee, S. Yi, E. Fleury, Z.F. Zhang, W.H. Wang, J. Eckert, Propagation of shear bands in a $Cu_{47.5}Zr_{47.5}Al_5$ bulk metallic glass, J. Mater. Res., 23 (2008) 6-12. https://doi.org/10.1557/JMR.2008.0025

[22] Q. Zhang, W. Zhang, G. Xie, A. Inoue, Glass-forming ability and mechanical properties of the ternary Cu-Zr-Al and quaternary Cu-Zr-Al-Ag bulk metallic glasses, Materials Transactions, 48 (2007) 1626-1630. https://doi.org/10.2320/matertrans.MJ200704

[23] J. Bhatt, W. Jiang, X. Junhai, W. Qing, C. Dong, B.S. Murty, Optimization of bulk metallic glass forming compositions in Zr-Cu-Al system by thermodynamic modeling, Intermetallics, 15 (2007) 716-721. https://doi.org/10.1016/j.intermet.2006.10.018

[24] X. J. Gu, S. J. Poon, G. J. Shiflet and M. Widom, Ductility improvement of amorphous steels: roles of shear modulus and electronic structure, Acta Mater., 56 (2008) 88–94. https://doi.org/10.1016/j.actamat.2007.09.011

[25] A. Inoue, T. Zhang, T. Masumoto, Al-La-Ni amorphous alloys with a wide supercooled liquid region, Materials Transactions JIM, 30 (1989) 965-972. https://doi.org/10.2320/matertrans1989.30.965

[26] A. Inoue, A. Kato, T. Zhang, S. Kim, T. Masumoto, Mg-Cu-Y amorphous alloys with high mechanical strengths produced by metallic mold casting method, Materials Transactions JIM, 32 (1991) 609-616. https://doi.org/10.2320/matertrans1989.32.609

[27] A. Inoue, T. Zhang, T. Masumoto, Zr-Al-Ni amorphous alloys with high class transition temperature and significant supercooled liquid region, Materials Transactions JIM, 31 (1990) 177-183. https://doi.org/10.2320/matertrans1989.31.177

[28] Q. Jing, Y. Zhang, D. Wang, Y. Li, A study of the glass forming ability in Zr-Ni-Al alloys, Materials Science and Engineering A, 441 (2006) 106–111. https://doi.org/10.1016/j.msea.2006.08.109

[29] A. Inoue, T. Zhang, T. Masumoto, Zr-Al-Ni amorphous alloys with high glass transition temperature and significant supercooled liquid region, Materials Transactions JIM, 31 (1990) 177-183. https://doi.org/10.2320/matertrans1989.31.177

[30] T. Zhang, K. Kurosaka, A. Inoue, Thermal and mechanical properties of Cu-based Cu-Zr- Ti-Y bulk glassy alloys, Materials Transactions, 42 (2001) 2042-2045. https://doi.org/10.2320/matertrans.42.2042

[31] T. Wada, T. Zhang, A. Inoue, Formation and high mechanical strength of bulk glassy alloys in Zr–Al–Co–Cu system, Materials Transactions, 44 (2003) 1839-1844. https://doi.org/10.2320/matertrans.44.1839

[32] Q.S. Zhang, W. Zhang, D.V. Louzguine-Luzgin, A. Inoue, High glass-forming ability and unusual deformation behavior of new Zr-Cu-Fe-Al bulk metallic glasses, Materials Science Forum, 654-656 (2010) 1042-1045. https://doi.org/10.4028/www.scientific.net/MSF.654-656.1042

[33] Q.S. Zhang, W. Zhang, G.Q. Xie, D.V. Louzguine-Luzgin, A. Inoue, Stable flowing of localized shear bands in soft bulk metallic glasses, Acta Materialia, 58 (2010) 904–909. https://doi.org/10.1016/j.actamat.2009.10.005

[34] A. L. Greer, E. Ma, Bulk metallic glasses: at the cutting edge of metals research, MRS Bull., 32 (2007) 611-619. https://doi.org/10.1557/mrs2007.121

[35] A. R. Yavari, J. J. Lewandowski, J. Eckert, Mechanical properties of bulk metallic glasses, MRS Bul., 32 (2007) 635-638. https://doi.org/10.1557/mrs2007.125

[36] J. Pan, K.C. Chan, Q. Chen, L. Liu, Enhanced plasticity by introducing icosahedral medium-range order in ZrCuNiAl metallic glass, Intermetallics 24 (2012) 79–83. https://doi.org/10.1016/j.intermet.2012.01.006

[37] H. Kato, J. Saida, A. Inoue, Influence of hydrostatic pressure during casting on as cast structure and mechanical properties in $Zr_{65}Al_{7.5}Ni_{10}Cu_{17.5-x}Pd_x$ (x= 0, 17.5) alloys, Scripta Mater., 51 (2004) 1063-1068. https://doi.org/10.1016/j.scriptamat.2004.08.004

[38] J. Schroers, W. L. Johnson, Ductile bulk metallic glass, Phys. Rev. Lett., 93 (2004) 255506-255510. https://doi.org/10.1103/PhysRevLett.93.255506

[39] Y. Yokoyama, K. Fujita, A.R. Yavari, A. Inoue, Malleable hypoeutectic Zr–Ni–Cu–Al bulk glassy alloys with tensile plastic elongation at room temperature, Philos. Mag. Lett., 89 (2009) 322-334. https://doi.org/10.1080/09500830902873575

[40] K.K. Song, S. Y. Pauly, S. Zhang, P. Scudino, K.B. Gargarella, U. Surreddi, K. Kühn, J. Eckert, Significant tensile ductility induced by cold rolling in $Cu_{47.5}Zr_{47.5}Al_5$ bulk metallic glass, Intermetallics, 19 (2011) 1394-1398. https://doi.org/10.1016/j.intermet.2011.05.001

[41] Z. Han, W.F. Wu, Y. Li, Y.J. Wei, H.J. Gao, An instability index of shear band for plasticity in metallic glasses, Acta Materialia, 57 (2009) 1367-1372. https://doi.org/10.1016/j.actamat.2008.11.018

[42] Y. Yang, C.T. Liu, Size effect on stability of shear-band propagation in bulk metallic glasses: an overview. J. Mater. Sci., 47 (2012) 55–67. https://doi.org/10.1007/s10853-011-5915-8

[43] W. J. Wright, R. Saha, W. D. Nix, Deformation Mechanisms of the $Zr_{40}Ti_{14}Ni_{10}Cu_{12}Be_{24}$ Bulk Metallic Glass, Materials Transactions, 42 (2001) 642-649. https://doi.org/10.2320/matertrans.42.642

[44] N. Vansteenberge, J. Sort, A. Concustell, J. Das, S. Scudino, S. Surinach, J. Eckert, M. Baro, Dynamic softening and indentation size effect in a Zr-based bulk glass-forming alloy, Scripta Materialia, 56 (2007) 605. https://doi.org/10.1016/j.scriptamat.2006.12.014

[45] N. Chen, D. V. Louzguine-Luzgin, G.Q. Xie, T. Wada, A. Inoue, Influence of minor Si addition on glass forming ability and mechanical properties of $Pd_{40}Ni_{40}P_{20}$ alloy, Acta Materialia, 57 (2009) 2775-2780. https://doi.org/10.1016/j.actamat.2009.02.028

[46] N. Chen, D. Pan, D. V. Louzguine-Luzgin, G. Q. Xie, M. W. Chen, A. Inoue, Improved thermal stability and ductility of flux-treated $Pd_{40}Ni_{40}Si_4P_{16}$ BMG, Scripta Materialia, 62 (2010) 17-20. https://doi.org/10.1016/j.scriptamat.2009.09.013

[47] D. V. Louzguine-Luzgin, G. Xie, Q. Zhang, A. Inoue, Effect of Fe on the glass-forming ability, structure and devitrification behavior of Zr-Cu-Al bulk glass-forming alloys, Philosophical Magazine, 90 (2010) 1955–1968. https://doi.org/10.1080/14786430903571495

[48] J. J. Lewandowski, W. H. Wang, A. L. Greer, Intrinsic plasticity or brittleness of metallic glasses, Philos. Mag. Lett., 85 (2005) 77-81. https://doi.org/10.1080/09500830500080474

[49] K. Hajlaoui, A. R. Yavari, B. Doisneau, A. LeMoulec, W. J. Botta, G. Vaughan, A. Greer L., A. Inoue, W. Zhang, A. Kvick, Shear delocalization and crack blunting of a metallic glass containing nanoparticles: In situ deformation in TEM analysis, Scripta Mater., 54 (2006) 1829-1834. https://doi.org/10.1016/j.scriptamat.2006.02.030

[50] K. B. Kim, J. Das, F. Baier, M. B. Tang, W. H. Wang, J. Eckert, Heterogeneity of a $Cu_{47.5}Zr_{47.5}Al_5$ bulk metallic glass, Appl. Phys. Lett., 88 (2006) 051911-051914. https://doi.org/10.1063/1.2171472

[51] G. N. Greaves, A. L. Greer, R. S. Lakes, T. Rouxel, Poisson's ratio and modern materials Nature mater., 10 (2011) 823-837. https://doi.org/10.1038/nmat3134

[52] S. V. Ketov, N. Chen, A. Caron, A. Inoue, D. V. Louzguine-Luzgin, Structural features and high quasi-static strain rate sensitivity of $Au_{49}Cu_{26.9}Ag_{5.5}Pd_{2.3}Si_{16.3}$ bulk metallic glass, Applied Physics Letters, 101 (2012) 241905-241908. https://doi.org/10.1063/1.4770072

[53] C. Wan, W.M. Yang, H.S. Liu, M.Q. Zuo, Q. Li, Z.G. Ma, Y.C. Zhao & A. Inoue Ductile Fe-based bulk metallic glasses at room temperature, Materials Science and Technology, 34 (2018) 751-756. https://doi.org/10.1080/02670836.2017.1412037

[54] B. Sarac, Y.P. Ivanov, A. Chuvilin, T. Schöberl, M. Stoica, Z. Zhang and J. Eckert, Origin of large plasticity and multiscale effects in iron-based metallic glasses, Nature Communications, 9 (2018) 1333. https://doi.org/10.1038/s41467-018-06284-0

[55] Q. Wang, J. Zhou, Q. Zeng, L. Sun, B. Shen, Ductile Co-based bulk metallic glass with superhigh strength and excellent soft magnetic properties induced by modulation of structural heterogeneity, Materialia, 9 (2020) 100561. https://doi.org/10.1016/j.mtla.2019.100561

[56] S.V. Madge, D.V. Louzguine-Luzgin, J.J. Lewandowski, A.L. Greer, Toughness, extrinsic effects and Poisson's ratio of bulk metallic glasses, Acta Materialia, 60 (2012) 4800-4809. https://doi.org/10.1016/j.actamat.2012.05.025

[57] W. F. Gale, T. C. Totemeier, editors. Smithells Metals Reference Book 8-th Edition, Elsevier Butterworth-Heinemann Ltd., Oxford UK (2004) 11-1.

[58] Y.N. Dastur, W.C. Leslie, Mechanism of work hardening in Hadfield manganese steel, Metallurgical Transactions A, 12 (1981) 749–759. https://doi.org/10.1007/BF02648339

[59] W. J. Wright, R. B Schwarz, W. D. Nix, Localized heating during serrated plastic flow in bulk metallic glasses, Materials Science and Engineering A, 219 (2001) 319-321. https://doi.org/10.1016/S0921-5093(01)01066-8

[60] F. H. Dalla Torre, A. Dubach, M. E. Siegrist, J.F. Löffler, Negative strain rate sensitivity in bulk metallic glass and its similarities with dynamic strain aging effect during deformation, Appl. Phys. Lett., 89 (2006) 091918-1-3. https://doi.org/10.1063/1.2234309

[61] A. Caron, R. Wunderlich, D.V. Louzguine-Luzgin, G. Xie, A. Inoue, H.-J. Fecht, Influence of minor aluminum concentration changes in zirconium-based bulk metallic glasses on the elastic, anelastic, and plastic properties, Acta Materialia, 58 (2010) 2004-2013. https://doi.org/10.1016/j.actamat.2009.11.043

[62] H. B. Yu, W. H. Wang, J. L. Zhang, C. H. Shek, H. Y. Bai, Statistic analysis of the mechanical behavior of bulk metallic glasses, Adv. Eng. Mater., 11 (2009) 370-375. https://doi.org/10.1002/adem.200800380

[63] Ch.E. Lekka, Cu-Zr and Cu-Zr-Al clusters: Bonding characteristics and mechanical properties, Journal of Alloys and Compounds, 504 (2010) 190-S193. https://doi.org/10.1016/j.jallcom.2010.02.067

[64] R. D. Conner, Y. Li, W. D. Nix, W. L. Johnson, Shear band spacing under bending of Zr-based metallic glass plates, Acta Materialia, 52 (2004) 2429-2434. https://doi.org/10.1016/j.actamat.2004.01.034

[65] P. E. Donovan, W. M. Stobbs, The structure of shear bands in metallic glasses, Acta Metall., 29 (1981) 1419-1436. https://doi.org/10.1016/0001-6160(81)90177-2

[66] D. P.Wang, B. A.Sun, X. R.Niu, Y. Yang, W.H. Wang, C.T. Liu, Mutual interaction of shear bands in metallic glasses, Intermetallics, 85 (2017) 48-53. https://doi.org/10.1016/j.intermet.2017.01.015

[67] N. Jia, P. Eisenlohr, F. Roters, D. Raabe, X. Zhao, Orientation dependence of shear banding in face-centered-cubic single crystals, Acta Materialia, 60 (2012) 3415–3434. https://doi.org/10.1016/j.actamat.2012.03.005

[68] H. S. Chen, Plastic flow in metallic glasses under compression, Scripta Metallurgica, 7 (1973) 931-935. https://doi.org/10.1016/0036-9748(73)90143-9

[69] F. O. Mear, T Wada, D. V Louzguine-Luzgin, A. Inoue, Highly inhomogeneous compressive plasticity in nanocrystal-toughened Zr–Cu–Ni–Al bulk metallic glass, Philosophical Magazine Letters, 89 (2009) 276-281. https://doi.org/10.1080/09500830902817861

[70] D. Klaumünzer, R. Maaß, J. F. Löffler, Stick-slip dynamics and recent insights into shear banding in metallic glasses, Journal of Materials Research, 26 (2011) 1453-1463. https://doi.org/10.1557/jmr.2011.178

[71] S. Gonzalez, N. Chen, Q.S. Zhang, D.V. Louzguine-Luzgin, J.H. Perepezko, A. Inoue, Effect of shear bands initiated in the pre-yield region on the deformation behaviour of Zr-based metallic glasses, Scripta Materialia, 64 (2011) 713–716. https://doi.org/10.1016/j.scriptamat.2010.12.025

[72] J. H. Na, M. D. Demetriou, M. Floyd, A. Hoff, G. R. Garrett, and W. L. Johnson, Compositional landscape for glass formation in metal alloys, 111 (2014) 9031. https://doi.org/10.1073/pnas.1407780111

[73] H. Guo, P. F. Yan, Y. B. Wang, J. Tan, Z. F. Zhang, M. L. Sui, E. Ma, Tensile ductility and necking of metallic glass, Nature Materials, 6 (2007) 735-739. https://doi.org/10.1038/nmat1984

[74] D. V. Louzguine-Luzgin, A. R. Yavari, G. Q. Xie, S. Madge, S. Li, J. Saida, A. L. Greer, A. Inoue, Tensile deformation behaviour of Zr-based glassy alloys, Phil. Mag. Letters, 90 (2010) 139-148. https://doi.org/10.1080/09500830903485544

[75] M. Aljerf, K. Georgarakis, A.R. Yavari, Shaping of metallic glasses by stress-annealing without thermal embrittlement, Acta Materialia, 59 (2011) 3817–3824. https://doi.org/10.1016/j.actamat.2011.02.039

[76] Y. H. Liu, G. Wang, R. J. Wang, D. Q. Zhao, M. X. Pan, W. H. Wang, Super plastic bulk metallic glasses at room temperature, Science, 315 (2007) 1385-1388. https://doi.org/10.1126/science.1136726

[77] P. E. Donovan, A yield criterion for $Pd_{40}Ni_{40}P_{20}$ metallic glass, Acta Metall., 37 (1989) 445-456. https://doi.org/10.1016/0001-6160(89)90228-9

[78] F. Spaepen, A microscopic mechanism for steady state inhomogeneous flow in metallic glasses, Acta Mater., 25 (1977) 407–415. https://doi.org/10.1016/0001-6160(77)90232-2

[79] A. Argon, Plastic deformation in metallic glasses, Acta Metall., 27 (1979) 47-58. https://doi.org/10.1016/0001-6160(79)90055-5

[80] Y. Zhang, A. L. Greer, Thickness of shear bands in metallic glasses, Appl. Phys. Lett., 89 (2006) 071907. https://doi.org/10.1063/1.2336598

[81] D.V. Louzguine-Luzgin, S.V. Ketov, Z. Wang, M.J. Miyama, A.A. Tsarkov, and A.Yu. Churyumov, Plastic deformation studies of Zr-based bulk metallic glassy samples with a low aspect ratio, Materials Science and Engineering A, 616 (2014) 288–296. https://doi.org/10.1016/j.msea.2014.08.006

[82] T. Egami, T. Iwashita, and W. Dmowski, Mechanical properties of metallic glasses, Metals, 3 (2013) 77-113. https://doi.org/10.3390/met3010077

[83] A. Vinogradov, M. Seleznev, and I. S. Yasnikov, Dislocation characteristics of shear bands in metallic glasses, Scripta Materialia, 130 (2017) 138–142. https://doi.org/10.1016/j.scriptamat.2016.11.017

[84] F. Delogu, Identification and characterization of potential shear transformation zones in metallic glasses, Physical Review Letters, 100 (2008) 255901. https://doi.org/10.1103/PhysRevLett.100.255901

[85] D. Srolovitz, V. Vitek, and T. Egami, An atomistic study of deformation of amorphous metals, Acta Metallurgica, 31 (1983) 335–352. https://doi.org/10.1016/0001-6160(83)90110-4

[86] J. H. Perepezko, S. D. Imhoff, M.-W. Chen, J.-Q. Wang, and S. Gonzalez, Nucleation of shear bands in amorphous alloys, PNAS, 111 (2014) 3938-3942. https://doi.org/10.1073/pnas.1321518111

[87] Z. Y. Liu, Y. Yang, and C. T. Liu, Yielding and shear banding of metallic glasses, Acta Materialia, 61 (2013) 5928-5936. https://doi.org/10.1016/j.actamat.2013.06.025

[88] A. L. Greer, Y. Q. Cheng, and E. Ma, Shear bands in metallic glasses, Materials Science and Engineering R, 74 (2013) 71-132. https://doi.org/10.1016/j.mser.2013.04.001

[89] F. Shimizu, S. Ogata, and J. Li, Yield point of metallic glass, Acta Materialia, 54 (2006) 4293–4298. https://doi.org/10.1016/j.actamat.2006.05.024

[90] D. Klaumünzer, R. Maaß and J.F. Löffler, Journal of Materials Research, 26 (2011) 1453–1463. https://doi.org/10.1557/jmr.2011.178

[91] W. Dmowski, Y. Yokoyama, A. Chuang, Y. Ren, M. Umemoto, K. Tsuchiya, A. Inoue, T. Egami, Acta Mater., 58 (2010) 429. https://doi.org/10.1016/j.actamat.2009.09.021

[92] E. V. Boltynjuk, D. V. Gunderov, E. V. Ubyivovk, A. V. Lukianov, A. M. Kshumanev, A. Bednarz, and R. Z. Valiev, AIP Conference Proceedings, 1748, (2016) 030006.

[93] S. Hilke, H. Rösner, D. Geissler, A. Gebert, M. Peterlechner, G.Wilde, The influence of deformation on the medium-range order of a Zr-based bulk metallic glass characterized by variable resolution fluctuation electron microscopy, Acta Materialia, 171 (2019) 275-281. https://doi.org/10.1016/j.actamat.2019.04.023

[94] J. Saida, R. Yamada, M. Wakeda and S. Ogata, Thermal rejuvenation in metallic glasses, Science and Technology of Advanced Materials, 18 (2017) 152-162. https://doi.org/10.1080/14686996.2017.1280369

[95] K. Kosiba, D. Şopu, S. Scudino, J. Bednarcik and S. Pauly, Modulating heterogeneity and plasticity in bulk metallic glasses: Role of interfaces on shear banding, International Journal of Plasticity, 119 (2019) 156-170. https://doi.org/10.1016/j.ijplas.2019.03.007

[96] J.Q. Wang and J.H. Perepezko, Nucleation kinetics of shear bands in metallic glass, Journal of Chemical Physics, 145 (2016) 211803.

[97] J.J. Lewandowski, Effects of annealing and changes in stress state on fracture toughness of bulk metallic glass, Mater. Trans. JIM, 42 (2001) 633-637. https://doi.org/10.2320/matertrans.42.633

[98] U. Ramamurty, M.L. Lee, J. Basu, and Y. Li, Embrittlement of a bulk metallic glass due to low-temperature annealing, Scripta Mater., 47 (2002) 107-111. https://doi.org/10.1016/S1359-6462(02)00102-1

[99] Y.Q. Cheng, A.J. Cao, and E. Ma, Correlation between the elastic modulus and the intrinsic plastic behavior of metallic glasses: The roles of atomic configuration and alloy composition, Acta Mater., 57 (2009) 3253-3267. https://doi.org/10.1016/j.actamat.2009.03.027

[100] A.J.Cao, Y.Q. Cheng, and E. Ma, Structural processes that initiate shear localization in metallic glass, Acta Mater., 57 (2009) 5146–5155. https://doi.org/10.1016/j.actamat.2009.07.016

[101] Y. Ritter, and K. Albe, Chemical and topological order in shear bands of $Cu_{64}Zr_{36}$ and $Cu_{36}Zr_{64}$ glasses, J. Appl. Phys., 111 (2012) 103527. https://doi.org/10.1063/1.4717748

[102] X. D. Wang, S. Aryal, C. Zhong, W. Y. Ching, H. W. Sheng, H. Zhang, D. X. Zhang, Q. P. Cao and J. Z. Jiang, Atomic picture of elastic deformation in a metallic glass, Scientific Reports, 5 (2015) 9184. https://doi.org/10.1038/srep09184

[103] W. H. Jiang, and M. Atzmon, The effect of compression and tension on shear-band structure and nanocrystallization in amorphous $Al_{90}Fe_5Gd_5$: a high-resolution transmission electron microscopy study, Acta Materialia, 51 (2003) 4095–4105. https://doi.org/10.1016/S1359-6454(03)00229-5

[104] C. Zhong, H. Zhang, Q. P. Cao, X. D. Wang, D. X. Zhang, U. Ramamurty, and J. Z. Jiang, Deformation behavior of metallic glasses with shear band like atomic structure: a molecular dynamics study, Scientific Reports, 6 (2016) 30935-30939. https://doi.org/10.1038/srep30935

[105] Y. Shao, G. N. Yang, K. F. Yao, and X. Liu, Direct experimental evidence of nano-voids formation and coalescence within shear bands, Appl. Phys. Lett., 105 (2014) 181909. https://doi.org/10.1063/1.4901281

[106] A.S. Aronin and D.V. Louzguine-Luzgin, On nanovoids formation in shear bands of an amorphous Al-based alloy, Mechanics of Materials, 113 (2017) 19–23. https://doi.org/10.1016/j.mechmat.2017.07.007

[107] Q. An, G. Garrett, K. Samwer, Y. Liu, S. V. Zybin, S.-N. Luo, M. D. Demetriou, W. L. Johnson, and W. A. Goddard III, Atomistic characterization of stochastic cavitation of a binary metallic liquid under negative pressure, J. Phys. Chem. Lett., 2 (2011) 1320–1323. https://doi.org/10.1021/jz200351m

[108] Q. An, K. Samwer, M. D. Demetriou, M. C. Floyd, D. O. Duggins, W. L. Johnson, and W. A. Goddard III, How the toughness in metallic glasses depends on topological and chemical heterogeneity, Proceedings of the National Academy of Sciences of the United States of America, 113 (2016) 7053–7058. https://doi.org/10.1073/pnas.1607506113

[109] J. E. Velasco, A. Concustell, E. Pineda, and D. Crespo, Plastic deformation induced anisotropy in metallic glasses: A molecular dynamics study, Journal of Alloys and Compounds, 707 (2017) 102–107. https://doi.org/10.1016/j.jallcom.2016.12.233

[110] S.X. Song, X.L. Wang, and T.G. Nieh, Capturing shear band propagation in a Zr-based metallic glass using a high-speed camera, Scripta Mater., 62 (2010) 847–850. https://doi.org/10.1016/j.scriptamat.2010.02.017

[111] B. Yang, P.K. Liaw, G. Wang, M. Morrison, C.T. Liu, R.A. Buchanan, and Y. Yokoyama, In-situ thermographic observation of mechanical damage in bulk-metallic glasses during fatigue and tensile experiments, Intermetallics, 12 (2004) 1265–1274. https://doi.org/10.1016/j.intermet.2004.04.006

[112] G. Wang, Q. Feng, B. Yang, W. Jiang, P.K. Liaw, and C.T. Liu, Thermographic studies of temperature evolutions in bulk metallic glasses, Intermetallics, 30 (2012) 1-11. https://doi.org/10.1016/j.intermet.2012.03.022

[113] J. J. Lewandowski, and A. L. Greer, Temperature rise at shear bands in metallic glasses, Nature Materials, 5 (2006) 15 - 18. https://doi.org/10.1038/nmat1536

[114] P. Guan, M. W. Chen, and T. Egami, Stress-temperature scaling for steady-state flow in metallic glasses, Phys. Rev. Lett., 104 (2010) 205701-205704. https://doi.org/10.1103/PhysRevLett.104.205701

[115] Y. Zhang, N.A. Stelmashenko, Z.H. Barber, W.H. Wang, J.J. Lewandowski, and A.L. Greer, Local temperature rises during mechanical testing of metallic glasses, J. Mater. Res, 22 (2007) 419-427. https://doi.org/10.1557/jmr.2007.0068

[116] K. M. Flores, and R. H. Dauskard, Local heating associated with crack tip plasticity in Zr–Ti–Ni–Cu–Be bulk amorphous metals, J. Mater. Res., 14 (1999) 638-643. https://doi.org/10.1557/JMR.1999.0642

[117] M. Zhao, and M. Li, Local heating in shear banding of bulk metallic glasses, Scr. Mater., 65 (2011) 493-496. https://doi.org/10.1016/j.scriptamat.2011.06.007

[118] S. V. Ketov and D.V. Louzguine-Luzgin, Localized shear deformation and softening of bulk metallic glass: stress or temperature driven? Scientific Reports, 3 (2013) 2798-2803. https://doi.org/10.1038/srep02798

[119] A.I. Bazlov, A.Y. Churyumov, M. Buchet, D.V. Louzguine-Luzgin, On temperature rise within the shear bands in bulk metallic glasses, Metals and Materials International, 24 (2018) 481-488. https://doi.org/10.1007/s12540-018-0072-0

[120] M. Seleznev, I.S. Yasnikov, A. Vinogradov, On the shear band velocity in metallic glasses: A high-speed imaging study, Materials Letters, 225 (2018) 105-108. https://doi.org/10.1016/j.matlet.2018.04.116

[121] W.J. Wright, M.W. Samale, T.C. Hufnagel, M.M. LeBlanc, and J.N. Florando, Studies of shear band velocity using spatially and temporally resolved measurements of strain during quasistatic compression of a bulk metallic glass, Acta Mater., 57 (2009) 4639–4648. https://doi.org/10.1016/j.actamat.2009.06.013

[122] B. Hou, M. Zhao, P. Yang, Y. L. Li, Capture of shear crack propagation in metallic glass by high-speed camera and in situ SEM, Key Engineering Materials, 626 (2015) 162-170. https://doi.org/10.4028/www.scientific.net/KEM.626.162

[123] W. J. Wright, R. R. Byer, and X. Gu, High-speed imaging of a bulk metallic glass during uniaxial compression, Appl. Phys. Lett., 102 (2013) 241920. https://doi.org/10.1063/1.4811744

[124] X.Xie, Y.-C. Lo, Y. Tong, Y. Gao, P.K. Liaw, Origin of serrated flow in bulk metallic glasses, Journal of the Mechanics and Physics of Solids, 124 (2019) 634-642. https://doi.org/10.1016/j.jmps.2018.11.015

[125] S.X. Song, and T.G. Nieh, Flow serration and shear-band viscosity during inhomogeneous deformation of a Zr-based bulk metallic glass, Intermetallics, 17 (2009) 762–767. https://doi.org/10.1016/j.intermet.2009.03.005

[126] Q.K. Li, and M. Li, Free volume evolution in metallic glasses subjected to mechanical deformation, Materials Transactions, 48 (2007) 1816 -1821. https://doi.org/10.2320/matertrans.MJ200785

[127] J. Bokeloh, S. V. Divinski, G. Reglitz, and G. Wilde, Tracer measurement of atomic diffusion inside shear bands of a bulk metallic glass, Phys. Rev. Lett. 107 (2011) 235503-235507. https://doi.org/10.1103/PhysRevLett.107.235503

[128] J. Pan, Q. Chen, L. Liu, and Y. Li, Softening and dilatation in a single shear band, Acta Materialia, 59 (2011) 5146–5158. https://doi.org/10.1016/j.actamat.2011.04.047

[129] S.V. Ketov, H.K. Nguyen, A.S. Trifonov, K. Nakajima, and D.V. Louzguine-Luzgin, Huge reduction of Young's modulus near a shear band in metallic glass, Journal of Alloys and Compounds, 687 (2016) 221-226. https://doi.org/10.1016/j.jallcom.2016.06.116

[130] R.T. Qu, Z.Q. Liu, G. Wang, and Z.F. Zhang, Progressive shear band propagation in metallic glasses under compression, Acta Materialia, 91 (2015) 19–33. https://doi.org/10.1016/j.actamat.2015.03.026

[131] Y.H. Liu, C.T. Liu, A. Gali, A. Inoue, and M.W. Chen, Evolution of shear bands and its correlation with mechanical response of a ductile $Zr_{55}Pd_{10}Cu_{20}Ni_5Al_{10}$ bulk metallic glass, Intermetallics, 18 (2010) 1455–1464. https://doi.org/10.1016/j.intermet.2010.03.037

[132] D.V. Louzguine-Luzgin, V.Yu. Zadorozhnyy, N. Chen, and S.V. Ketov, Evidence of the existence of two deformation stages in bulk metallic glasses, Journal of Non-Crystalline Solids, 396–397 (2014) 20–24. https://doi.org/10.1016/j.jnoncrysol.2014.04.014

[133] J.H. Perepezko, S.D. Imhoff, M.W. Chen, S. Gonzalez, and A. Inoue, Nucleation reactions during deformation and crystallization of metallic glass, Journal of Alloys and Compounds, 536 (2012) 55-59. https://doi.org/10.1016/j.jallcom.2011.12.064

[134] R. Sarmah, G. Ananthakrishna, B.A. Sun, and W.H. Wang, Hidden order in serrated flow of metallic glasses, Acta Materialia, 59 (2011) 4482–4493. https://doi.org/10.1016/j.actamat.2011.03.071

[135] D.V. Louzguine-Luzgin, D.M. Packwood, G. Xie, and A.Yu. Churyumov, On deformation behavior of a Ni-based bulk metallic glass produced by flux treatment, Journal of Alloys and Compounds, 561 (2013) 241–246. https://doi.org/10.1016/j.jallcom.2013.01.193

[136] F.X. Liu, P.K. Liaw, G.Y. Wang, C.L. Chiang, D.A Smith, P.D Rack, P.P.Chu, and R.A. Buchanan, Specimen-geometry effects on mechanical behavior of metallic glasses, Intermetallics, 14 (2006) 1014-1018. https://doi.org/10.1016/j.intermet.2006.01.043

[137] C.E. Packard and C.A. Schuh, Initiation of shear bands near a stress concentration in metallic glass, Acta Mater., 55 (2007), 5348-5358. https://doi.org/10.1016/j.actamat.2007.05.054

[138] D.V. Louzguine-Luzgin, J. Jiang, A.I. Bazlov, V.S. Zolotorevzky, H. Mao, Yu P. Ivanov, A.L. Greer, Phase separation process preventing thermal embrittlement of a Zr-Cu-Fe-Al bulk metallic glass, Scripta Materialia, 167 (2019) 31-36. https://doi.org/10.1016/j.scriptamat.2019.03.030

[139] Y. Sun, A. Concustell, and A. L. Greer, Thermomechanical processing of metallic glasses: extending the range of the glassy state, Nature Reviews Materials, 1 (2016) 16039. https://doi.org/10.1038/natrevmats.2016.39

[140] Y. Tong, W. Dmowski, H. Bei, Y. Yokoyama, T. Egami, Mechanical rejuvenation in bulk metallic glass induced by thermo-mechanical creep, Acta Materialia, 148 (2018) 384-390. https://doi.org/10.1016/j.actamat.2018.02.019

[141] F.O. Mear, B. Lenk, Y. Zhang, and A.L. Greer, Structural relaxation in a heavily cold-worked metallic glass, Scripta Materialia, 59 (2008) 1243–1246. https://doi.org/10.1016/j.scriptamat.2008.08.023

[142] D. V. Louzguine-Luzgin, I. Seki, T. Yamamoto, H. Kawaji, C. Suryanarayana, and A. Inoue, Structural relaxation and crystallization processes in $Cu_{55}Hf_{25}Ti_{15}Pd_5$ metallic glassy alloy, Intermetallics, 23 (2012) 177-181. https://doi.org/10.1016/j.intermet.2011.11.019

[143] R. Liontas, X.W. Gu, E. Fu, Y. Wang, N. Li, N. Mara, J.R. Greer, Effects of helium implantation on the tensile properties and microstructure of $Ni_{73}P_{27}$ metallic glass nanostructures, Nano Letters, 14 (2014) 5176-5183. https://doi.org/10.1021/nl502074d

[144] K.W. Park, C.M. Lee, M. Wakeda, Y. Shibutani, M.L. Falk, and J.C. Lee, Elastostatically induced structural disordering in amorphous alloys, Acta Materialia. 56 (2008) 5440-5450. https://doi.org/10.1016/j.actamat.2008.07.033

[145] H.S. Chen, Stored energy in a cold-rolled metallic glass, Appl. Phys. Lett., 29 (1976) 328–330. https://doi.org/10.1063/1.89084

[146] D. Dmowski, Y.Yokoyama, A. Chuang, Y. Ren, M. Umemoto, K. Tuchiya, A. Inoue, and T. Egami, Structural rejuvenation in a bulk metallic glass induced by severe plastic deformation, Acta Mater 58 (2010) 429–438. https://doi.org/10.1016/j.actamat.2009.09.021

[147] A. Concustell, F.O. Méar, S. Suriñach, M.D. Baró, and A.L. Greer, Structural relaxation and rejuvenation in a metallic glass induced by shot-peening, Philos. Mag. Lett., 89 (2009) 831−840. https://doi.org/10.1080/09500830903337919

[148] Y. Zhang, W. H. Wang, and A. L. Greer, Making metallic glasses plastic by control of residual stress, Nature Mater., 5 (2006) 857-860. https://doi.org/10.1038/nmat1758

[149] H.B Ke, P. Wen, H.L. Peng, W.H. Wang, and A.L. Greer, Homogeneous deformation of metallic glass at room temperature reveals large dilatation, Scripta Materialia, 64 (2011) 966–969. https://doi.org/10.1016/j.scriptamat.2011.01.047

[150] S.C Lee, C.M. Lee, J.W. Yang, J.C. Lee, Microstructural evolution of an elastostatically compressed amorphous alloy and its influence on the mechanical properties, Scripta Mater., 58 (2008) 591-595. https://doi.org/10.1016/j.scriptamat.2007.11.036

[151] K.W. Park, C.M. Lee, M. Wakeda, Y. Shibutani, M.L. Falk, and J.C. Lee, Elastostatically induced structural disordering in amorphous alloys, Acta Mater., 56 (2008) 5440-5445. https://doi.org/10.1016/j.actamat.2008.07.033

[152] J. Pan, Yu. P. Ivanov, W. H. Zhou, Y. Li & A. L. Greer, Strain-hardening and suppression of shear-banding in rejuvenated bulk metallic glass, Nature, 578 (2020) 559–562. https://doi.org/10.1038/s41586-020-2016-3

[153] A. L. Greer and Y. H. Sun, Stored energy in metallic glasses due to strains within the elastic limit, Philosophical Magazine, 96 (2016) 1643-1663. https://doi.org/10.1080/14786435.2016.1177231

[154] S. V. Ketov, Y. H. Sun, S. Nachum, Z. Lu, A. Checchi, A. R. Beraldin, H. Y. Bai, W. H. Wang, D. V. Louzguine-Luzgin, M. A. Carpenter, and A. L. Greer, Rejuvenation of metallic glasses by non-affine thermal strain, Nature, 524 (2015) 200–203. https://doi.org/10.1038/nature14674

[155] F. Shimizu, S. Ogata, and J. Li, Theory of shear banding in metallic glasses and molecular dynamics calculations, Materials Transactions, 48 (2007) 2923–2927. https://doi.org/10.2320/matertrans.MJ200769

[156] W. Guo, R. Yamada, J. Saida, Rejuvenation and plasticization of metallic glass by deep cryogenic cycling treatment, Intermetallics, 93 (2018) 141-147. https://doi.org/10.1016/j.intermet.2017.11.015

[157] D. A. Woodford, Strain-rate sensitivity as a measure ductility, Transactions of the American Society for Metals, 62 (1969) 291-293.

[158] T. Hufnagel, C. Jiao, T. Li, Y. Xing, and L.Q. Ramesh, Deformation and failure of $Zr_{57}Ti_5Cu_{20}Ni_8Al_{10}$ bulk metallic glass under quasi-static and dynamic compression, J. Mater. Res., 17 (2002) 1441-1447. https://doi.org/10.1557/JMR.2002.0214

[159] S. Song, H. Bei, J, Wadsworth, and T.G. Nieh, Flow serration in a Zr-based bulk metallic glass in compression at low strain rates, Intermetallics, 16 (2008) 813-819. https://doi.org/10.1016/j.intermet.2008.03.007

[160] D. Torre, F.H. Dubach, A. Nelson, and J.F. Löffler, Temperature, strain and strain rate dependence of serrated flow in bulk metallic glasses, Mater. Trans. 48 (2007) 1774-1779. https://doi.org/10.2320/matertrans.MJ200782

[161] R. Limbach, B. P. Rodrigues, L. Wondraczek, Strain-rate sensitivity of glasses, Journal of Non-Crystalline Solids, 404 (2014) 124–134. http://dx.doi.org/10.1016/j.jnoncrysol.2014.08.023

[162] D. Torre, F.H. Dubach, A. Siegrist, and J.F. Löffler, Shear striations and deformation kinetics in highly deformed Zr-based bulk metallic glasses, Appl. Phys. Lett., 89 (2006) 091918.

[163] J. Lu, G. Ravichandran, and W. L. Johnson, Deformation behavior of the $Zr_{41.2}Ti_{13.8}Cu_{12.5}Ni_{10}Be_{22.5}$ bulk metallic glass over a wide range of strain-rates and temperatures, Acta Mater., 51 (2003) 3429-3443. https://doi.org/10.1016/S1359-6454(03)00164-2

[164] F.X. Liu, Y.F. Gao, and P.K. Liaw, Rate-dependent deformation behavior of Zr-based metallic-glass coatings examined by nanoindentation, Metallurgical and Materials Transactions A, 39 (2008) 1862-1867. https://doi.org/10.1007/s11661-007-9399-8

[165] D. Pan, and M.W. Chen, Rate-change instrumented indentation for measuring strain rate sensitivity, Journal of Materials Research, 24 (2009) 1466-1470. https://doi.org/10.1557/jmr.2009.0168

[166] E.V. Boltynjuk, D.V. Gunderov, E.V. Ubyivovk, M.A. Monclús, L.W. Yang, J.M. Molina-Aldareguia, A.I. Tyurin, A.R. Kilmametov, A.A. Churakova, A.Yu. Churyumov, R.Z. Valiev, Journal of Alloys and Compounds, 747 (2018) 595-602. https://doi.org/10.1016/j.jallcom.2018.03.018

[167] D. S. Sanditov, V. V. Mantatov, and S. Sh. Sangadiev, Microhardness and plastic deformation of glasses upon microindentation, Glass Physics and Chemistry, 30 (2004) 415-419. https://doi.org/10.1023/B:GPAC.0000045921.17106.64

[168] G. Srikant, N. Chollacoop, and U. Ramamurty, Plastic strain distribution underneath a Vickers Indenter: Role of yield strength and work hardening exponent, Acta Mater., 54 (2006) 5171-5178. https://doi.org/10.1016/j.actamat.2006.06.032

[169] Q. Zhou, Y. Du, W. Han, Y. Ren, H. Zhai, H. Wang, Identifying the origin of strain rate sensitivity in a high entropy bulk metallic glass, Scripta Materialia, 164 (2019) 121–125. https://doi.org/10.1016/j.scriptamat.2019.02.002

[170] S. González, G.Q. Xie, D.V. Louzguine-Luzgin, J.H. Perepezko, and A. Inoue, Deformation and strain rate sensitivity of a Zr–Cu–Fe–Al metallic glass, Materials Science and Engineering A, 528 (2011) 3506–3512. https://doi.org/10.1016/j.msea.2011.01.049

[171] Y. Dong, S. Liu, J. Biskupek, Q.P. Cao, X.D. Wang, J.Z. Jiang, R. Wunderlich, and H.J. Fecht, Improved tensile ductility by severe plastic deformation for nano-structured metallic glass, Materials, 12 (2019) 1611. https://dx.doi.org/10.3390%2Fma12101611

[172] F.F. Wu, Z.F. Zhang, and S.X. Mao, Size-dependent shear fracture and global tensile plasticity of metallic glasses, Acta Mater., 57 (2009) 257–266. https://doi.org/10.1016/j.actamat.2008.09.012

[173] K. Hajlaoui, A.R Yavari, B. Doisneau, A. LeMoulec, W.J. F. Botta, G. Vaughan, A.L. Greer, A. Inoue, W. Zhang, and A. Kvick, Shear delocalization and crack blunting of a metallic glass containing nanoparticles: In situ deformation in TEM analysis, Scripta Materialia, 54 (2006) 1829–1834. https://doi.org/10.1016/j.scriptamat.2006.02.030

[174] C. Gammer, C. Ophus, T.C. Pekin, J. Eckert, A.M. Minor, Local nanoscale strain mapping of a metallic glass during in situ testing, Appl. Phys. Lett. 112 (2018) 171905. https://doi.org/10.1063/1.5025686

[175] D.V. Louzguine-Luzgin, A.R.Yavari, G.Q. Xie, S. Li, S. Madge, J. Saida, A. Greer, and A. Inoue, Tensile deformation behaviour of Zr-based glassy alloys, Phil. Mag. Letters., 90 (2010) 139-143. https://doi.org/10.1080/09500830903485544

[176] D. Z. Chen, D. Jang, K. M. Guan, Q. An, W. A. Goddard, III, and J. R. Greer, Nanometallic glasses: size reduction brings ductility, surface state drives its extent, Nano Lett. 13 (2013) 4462–4446. https://doi.org/10.1021/nl402384r

[177] K. Georgarakis, M. Aljerf, Y. Li, A. Lemoulec, F. Charlot, A. R. Yavari, K. Chornokhvostenko, E. Tabachnikova, G. A. Evangelakis, D. B. Miracle, A. L. Greer, and T. Zhang, Shear band melting and serrated flow in metallic glasses, Applied Physics Letters, 93 (2008) 031907. https://doi.org/10.1063/1.2956666

[178] J.J. Lewandowski, M. Shazly, A. S. Nouri, Intrinsic and extrinsic toughening of metallic glasses, Scripta Materialia, 54 (2008) 337-341. https://doi.org/10.1016/j.scriptamat.2005.10.010

[179] L. Gu, L. Xu, Q.S. Zhang, D. Pan, N. Chen, D. V. Louzguine-Luzgin, K.-F. Yao, W.H. Wang and Y. Ikuhara, Direct in situ observation of metallic glass deformation by real-time nano-scale indentation, Scientific Reports, 5 (2015) 9122. https://doi.org/10.1038/srep09122

[180] G.Y. Wang, P.K. Liaw, M.L. Morrison, Progress in studying the fatigue behavior of Zr-based bulk-metallic glasses and their composites, Intermetallics, 17 (2009) 579-590. https://doi.org/10.1016/j.intermet.2009.01.017

[181] J.J. Kruzic, Understanding the problem of fatigue in bulk metallic glasses, Metall. Mater. Trans. A. 42 (2011) 1516-1523. https://doi.org/10.1007/s11661-010-0413-1

[182] Z. F. Zhang, J. Eckert, L. Schultz, Fatigue and fracture behavior of bulk metallic glass, Metallurgical and Materials Transactions A, 35 (2004) 3489-3498. https://doi.org/10.1007/s11661-004-0186-5

[183] G. Y. Wang, P. K. Liaw, W. H. Peter, B. Yang, Y. Yokoyama, M. L. Benson, B. A. Green, M. J. Kirkham, S. A. White, T. A. Saleh, R. L. McDaniels, R. V. Steward, R. A. Buchanan, C. T. Liu, C. R. Brook, Fatigue behavior of bulk metallic glasses, Intermetallics, 12 (2004) 885–892. https://doi.org/10.1016/j.intermet.2004.02.043

[184] M. Freels, G.Y. Wang, W. Zhang, P.K. Liaw, A. Inoue, Cyclic compression behavior of a Cu-Zr-Al-Ag bulk metallic glass, Intermetallics, 19 (2011) 1174-1183. https://doi.org/10.1016/j.intermet.2011.03.023

[185] K. Fujita, W. Zhang, B.L. Shen, K. Amiya, C. L. Ma, N. Nishiyama, Fatigue properties in high strength bulk metallic glasses, Intermetallics, 30 (2012) 12-18. https://doi.org/10.1016/j.intermet.2012.03.021

[186] X.D. Wang, P. Liu, Z.W. Zhu, H.F. Zhang and X.C. Ren, High fatigue endurance limit of a Ti-based metallic glass, Intermetallics, 119 (2020) 106716. https://doi.org/10.1016/j.intermet.2020.106716

[187] C.E. Packard, L.M. Witmer, C. A. Schuh, Hardening of a metallic glass during cyclic loading in the elastic range, Appl. Phys. Lett., 92 (2008) 171911. https://doi.org/10.1063/1.2919722

[188] C. Deng, C. A. Schuh, Atomistic mechanisms of cyclic hardening in metallic glass, Appl. Phys. Lett., 100 (2012) 251909. https://doi.org/10.1063/1.4729941

[189] A. Caron, A. Kawashima, H. J. Fecht, D. V. Louzguine-Luzguin, A. Inoue, On the anelasticity and strain induced structural changes in a Zr-based bulk metallic glass, Appl. Phys. Lett., 99 (2011) 171907. https://doi.org/10.1063/1.3655999

[190] D. V. Louzguine-Luzgin, G. Xie, Q. Zhang, A. Inoue, Effect of Fe on the glass-forming ability, structure and devitrification behavior of Zr-Cu-Al bulk glass-forming alloys, Philosophical Magazine, 90 (2010) 1955–1968. https://doi.org/10.1080/14786430903571495

[191] A.Yu. Churyumov, A.I. Bazlov, V.Yu. Zadorozhnyy, A.N. Solonin, A. Caron, D.V. Louzguine-Luzgin, Phase transformations in Zr-based bulk metallic glass cyclically loaded before plastic yielding, Materials Science and Engineering A, 550 (2012) 358–362. https://doi.org/10.1016/j.msea.2012.04.087

[192] J. Y. Cavaille, L David, J.Perez, Relaxation phenomena in non crystalline solids: case of polymeric materials, Material Science Forum, 366–368 (2001) 499–545. https://doi.org/10.4028/www.scientific.net/MSF.366-368.499

[193] T.C. Hufnagel, R.T. Ott, J. Almer, Structural aspects of elastic deformation of a metallic glass, Physical Review B - Condensed Matter and Materials Physics, 73 (2006) 064204. https://doi.org/10.1103/PhysRevB.73.064204

[194] V. Yu. Zadorozhnyy, M. Yu. Zadorozhnyy, A. Yu. Churyumov, S. V. Ketov, I. S. Golovin, D. V. Louzguine-Luzgin, Room-temperature dynamic quasi-elastic mechanical behavior of a Zr–Cu–Fe–Al bulk metallic glass, Phys. Status Solidi A, 213 (2016) 450–456. https://doi.org/10.1002/pssa.201532638

[195] D.V. Louzguine-Luzgin, V. Yu. Zadorozhnyy, S.V. Ketov, Z. Wang, A.A. Tsarkov, A.L. Greer, On room-temperature quasi-elastic mechanical behaviour of bulk metallic glasses, Acta Materialia, 129 (2017) 343–351. https://doi.org/10.1016/j.actamat.2017.02.049

[196] H. Zhang, K.Q. Qiu, Z.G. Wang, Q.S. Zang, H.F. Zhang, Low-cycle fatigue of a bulk amorphous alloy, Acta Metall. Sinica, 39 (2003) 405-408.

[197] C. K. Huang, J.J. Lewandowski, Effects of changes in chemistry on the flex bending fatigue behavior of Al-based amorphous alloy ribbons, Metall. Mater. Trans. A, 43A. (2012) 2687-2696. https://doi.org/10.1007/s11661-011-0853-2

[198] D. V. Louzguine-Luzgin, L. V. Louzguina-Luzgina, S. V. Ketov, V. Yu. Zadorozhnyy, A. L. Greer, Influence of cyclic loading on the onset of failure in a Zr-based bulk metallic glass, Journal of Materials Science, 49 (2014) 6716–6721. https://doi.org/10.1007/s10853-014-8276-2

[199] S. V. Madge, Toughness of bulk metallic glasses, Metals, 5 (2015) 1279-1305. https://doi.org/10.3390/met5031279

[200] S.V Madge, P.Sharma, D.V Louzguine-Luzgin, A.L Greer, A Inoue, New La-based glass-crystal ex situ composites with enhanced toughness, Scr. Mater. 62 (2010) 210–213. https://doi.org/10.1016/j.scriptamat.2009.10.029

[201] Q. He, Y.Q. Cheng, E. Ma, J. Xu, Locating bulk metallic glasses with high fracture toughness: Chemical effects and composition optimization, Acta Materialia. 59 (2011) 202–215. https://doi.org/10.1016/j.actamat.2010.09.025

[202] M. D. Demetriou, M. E. Launey, G. Garrett, P. J. Schramm, D. C. Hofmann, W. L. Johnson and R. O. Ritchie, A damage-tolerant glass, Nature Materials, 10 (2011) 123-128. https://doi.org/10.1038/nmat2930

[203] Y. Zhou, J. Liu, D. Han, X. Chen, G. Wang, Q. Zhai, Reduced fracture toughness of metallic glass at cryogenic temperature, Metals. 7 (2017) 151-157. https://doi.org/10.3390/met7040151

[204] B. Gludovatz, S. E. Naleway, R. O. Ritchie, J. J. Kruzic, Size-dependent fracture toughness of bulk metallic glasses, Acta Mater., 70 (2014) 198-207. https://doi.org/10.1016/j.actamat.2014.01.062

[205] A. S. Argon, M. Salama, The mechanism of fracture in glassy materials capable of some inelastic deformation, Mater. Sci. Eng., 23 (1976) 219-230. https://doi.org/10.1016/0025-5416(76)90198-1

[206] D. V Louzguine-Luzgin, A.Vinogradov, A. R. Yavari, S. Li, G. Xie, A. Inoue, On the deformation and fracture behaviour of a Zr-based glassy alloy, Philosophical Magazine, 88 (2008) 2979-2987. https://doi.org/10.1080/14786430802446674

[207] C. A. Pampillo, Flow and fracture in amorphous alloys, J. Mater. Sci., 10 (1975) 1194-1227. https://doi.org/10.1007/BF00541403

[208] F. Spaepen, Metallic glasses: Must shear bands be hot? Nature Materials 5, (2006) 7 – 8. https://doi.org/10.1038/nmat1552

[209] B. Yang, M. L. Morrison, P.P. Liaw, K. R. A. Buchanan, G.Wang, C. T. Liu, M. Denda, Dynamic evolution of nanoscale shear bands in a bulk-metallic glass, Appl. Phys. Lett. 86 (2005) 141904-141907. https://doi.org/10.1063/1.1891302

[210] N. Chen, D. V. Louzguine-Luzgin, G. Q. Xie, A. Inoue, Nanoscale wavy fracture surface of a Pd-based bulk metallic glass, Applied Physics Letters, 94 (2009) 131906. https://doi.org/10.1063/1.3109797

[211] D. V. Louzguine-Luzgin, A. Vinogradov, A. R. Yavari, S. Li, G. Xie, A. Inoue, On the deformation and fracture behaviour of a Zr-based glassy alloy, Philosophical Magazine, 23 (2008) 2979-2987. https://doi.org/10.1080/14786430802446674

[212] J. X. Meng, Z. Ling, M. Q. Jiang, H. S. Zhang, L. H. Dai, Dynamic fracture instability of tough bulk metallic glass, Appl. Phys. Lett., 92 (2008) 171909. https://doi.org/10.1063/1.2913206

[213] H. A. Bruck, A. J. Rosakis, W.L. Johnson, The dynamic compressive behavior of beryllium bearing bulk metallic glasses, J. Mater. Res., 11 (1996) 503 -511. https://doi.org/10.1557/JMR.1996.0060

[214] T. Hufnagel, T. Jiao, Y. Li, L. Xing, K. Ramesh, Deformation and failure of $Zr_{57}Ti_5Cu_{20}Ni_8Al_{10}$ bulk metallic glass under quasistatic and dynamic compression, Journal of Materials Research, 17 (2002) 1441–1445. https://doi.org/10.1557/JMR.2002.0214

[215] A. Inoue, T. Zhang, Impact fracture energy of bulk amorphous $Zr_{55}Al_{10}Cu_{30}Ni_5$ alloy, Materials Transactions, JIM, 37 (1996) 1726-1729. https://doi.org/10.2320/matertrans1989.37.1726

[216] E. Fleury, S.M. Lee, H.S. Ahn, W.T. Kim, D.H. Kim, Tribological properties of bulk metallic glasses, Materials Science and Engineering A, 375–377 (2004) 276–279. https://doi.org/10.1016/j.msea.2003.10.065

[217] A. L. Greer, K. L. Rutherford & I. M. Hutchings, Wear resistance of amorphous alloys and related materials, International Materials Reviews 47 (2002) 87-112. https://doi.org/10.1179/095066001225001067

[218] R. Moreton, J.K. Lancaster, The friction and wear behaviour of various metallic glasses, Journal of Materials Science Letters, 4 (1985) 133-137. https://doi.org/10.1007/BF00728057

[219] B. Prakash, Abrasive wear behaviour of Fe, Co and Ni based metallic glasses, Wear, 258 (2005) 217–224. https://doi.org/10.1016/j.wear.2004.09.010

[220] A. L. Greer, W.N. Myung, Abrasive wear resistance of bulk metallic glasses, Mat. Res. Soc. Symp. Proc. 644 (2001) L10.4.1. https://doi.org/10.1557/PROC-644-L10.4

[221] J.Bhatt, S.Kumar, C.Dong, B.S Murty, Tribological behaviour of $Cu_{60}Zr_{30}Ti_{10}$ bulk metallic glass,   Materials Science and Engineering A, 458 (2007) 290-294. https://doi.org/10.1016/j.msea.2006.12.060

[222] M. Anis, W.M. Rainforth, H.A. Davies, Wear behavior of rapidly solidified $Fe_{68}Cr_{18}Mo_2B_{12}$, Wear, 172 (1994) 135. https://doi.org/10.1016/0043-1648(94)90281-X

[223] H.W. Jin, R. Ayer, J.Y. Koo, R. Raghavan, U. Ramamurty, Reciprocating wear mechanisms in a Zr-based bulk metallic glass, J. Mater. Res., 22 (2007) 264-273. https://doi.org/10.1557/jmr.2007.0048

[224] X.F. Wu, G.A. Zhang, F.F Wu, Wear behaviour of Zr-based in situ bulk metallic glass matrix composites, Bull Mater Sci., 39 (2016) 703-707. https://doi.org/10.1557/jmr.2007.0048

[225] P.-H.Cornuault, G. Colas, A. Lenain, R. Daudin, S.Gravier, On the diversity of accommodation mechanisms in the tribology of bulk metallic glasses, Tribology International, 141 (2020) 105957. https://doi.org/10.1016/j.triboint.2019.105957

[226] Y. Saotome, S. Miwa, T. Zhang, A. Inoue, The micro-formability of Zr-based amorphous alloys in the supercooled liquid state and their application to micro-dies, Journal of Materials Processing Technology, 113 (2001) 64-69. https://doi.org/10.1016/S0924-0136(01)00605-7

[227] J.W. Lee, Y.C. Lin, N. Chen, D. V. Louzguine, M. Esashi, T. Gessner, Development of the large scanning mirror using Fe-based metallic glass ribbon, Japanese Journal of Applied Physics. 50 (2011) 087301. https://doi.org/10.7567/JJAP.50.087301

[228] A. Caron, C. L. Qin, L. Gu, S. Gonzalez, A. Shluger, H. J. Fecht, D. V. Louzguine-Luzgin A. Inoue, Structure and nano-mechanical characteristics of surface oxide layers on a metallic glass, Nanotechnology 22 (2011) 095704. https://doi.org/10.1088/0957-4484/22/9/095704

[229] A. Caron, P. Sharma, A. Shluger, H. J. Fecht, D. V. Louzguine-Luzguin, A. Inoue, Effect of surface oxidation on the nm-scale wear behavior of a metallic glass, Journal of Applied Physics, 109 (2011) 083515. https://doi.org/10.1063/1.3573778

[230] A. Caron, D. V. Louzguine-Luzgin, R. Bennewitz, Structure vs chemistry: friction and wear of Pt-based metallic surfaces, ACS Appl. Mater. Interfaces, 5 (2013) 11341–11347. https://doi.org/10.1021/am403564a

[231] S.V. Madge, A. Caron, R. Gralla, G. Wilde, S.K. Mishra, Novel W-based metallic glass with high hardness and wear resistance, Intermetallics, 47 (2014) 6–10. https://doi.org/10.1016/j.intermet.2013.12.003

[232] D.V. Louzguine-Luzgin, H.K. Nguyen, K. Nakajima, S.V. Ketov, A.S. Trifonov, A study of the nanoscale and atomic-scale wear resistance of metallic glasses, Materials Letters, 185 (2016) 54–58. https://doi.org/10.1016/j.matlet.2016.08.035

[233] A. S. Trifonov, A. V. Lubenchenko, V. I. Polkin, A. B. Pavolotsky, S. V. Ketov, D. V. Louzguine-Luzgin, Difference in charge transport properties of Ni-Nb thin films with native and artificial oxide, Journal of Applied Physics, 117 (2015) 125704. https://doi.org/10.1063/1.4915935

[234] D.V. Louzguine-Luzgin, C.L. Chen, L.Y. Lin, Z.C. Wang, S.V. Ketov, M.J. Miyama, A.S. Trifonov, A.V. Lubenchenko, Y. Ikuhara, Bulk metallic glassy surface native oxide: Its atomic structure, growth rate and electrical properties, Acta Materialia, 97 (2015) 282–290. https://doi.org/10.1016/j.actamat.2015.06.039

[235] C.L. Qin, J.J. Oak, N. Ohtsu, K. Asami, A. Inoue, XPS study on the surface films of a newly designed Ni-free Ti-based bulk metallic glass, Acta Materialia, 55 (2007) 2057–2063. https://doi.org/10.1016/j.actamat.2006.10.054

[236] Y. H. Sun, D.V. Louzguine-Luzgin, S. Ketov, A.L. Greer, Pure shear stress reversal on a Cu-based bulk metallic glass reveals a Bauschinger-type effect, Journal of Alloys and Compounds, 615 (2014) 75–78. https://doi.org/10.1016/j.jallcom.2013.11.104

[237] G. Kumar, D. Rector, R.D. Conner, and J. Schroers, Embrittlement of Zr-based bulk metallic glasses, Acta Mater., 57 (2009) 3572-3583. https://doi.org/10.1016/j.actamat.2009.04.016

[238] Y. Kawamura, T. Nakamura, A. Inoue, and T. Masumoto, High-strain-rate superplasticity due to Newtonian viscous flow in $La_{55}Al_{25}Ni_{20}$ metallic glass, Mater Trans JIM, 40 (1999) 794-803. https://doi.org/10.2320/matertrans1989.40.794

[239] T. Egami, Y. Tong, and W. Dmowski, Deformation in metallic glasses studied by synchrotron X-ray diffraction, Metals, 6 (2016) 22-27. https://doi.org/10.3390/met6010022

[240] Y. Kawamura, T. Shibata, A. Inoue, and T. Masumoto, Superplastic deformation of $Zr_{65}Al_{10}Ni_{10}Cu_{15}$ metallic glass, Scripta Mater., 37 (1997) 431-436. https://doi.org/10.1016/S1359-6462(97)00105-X

[241] Y. Saotome, K. Ito, T. Zhang, and A. Inoue, Superplastic nanoforming of Pd-based amorphous alloy, Scripta Mater., 44 (2001) 1541-1545. https://doi.org/10.1016/S1359-6462(01)00837-5

[242] J. Schroers, T. M. Hodges, G. Kumar, H. Raman, A. J. Barnes, Q. Pham, T. A. Waniuk, Thermoplastic blow molding of metals, Materials Today, 14 (2011) 14–19. https://doi.org/10.1016/S1369-7021(11)70018-9

[243] N. Chen, H. A. Yang, A. Caron, P. C. Chen, Y. C. Lin, D. V. Louzguine-Luzgin, K. F. Yao, M. Esashi, A. Inoue, Glass-forming ability and thermoplastic formability of a $Pd_{40}Ni_{40}Si_{4}P_{16}$ glassy alloy, Journal of Materials Science, 46 (2011) 2091-2096. https://doi.org/10.1007/s10853-010-5043-x

[244] A R. Yavari, M F. Oliveira, C. S. Kiminami, A. Inoue, W. J. Botta, Electromechanical shaping assembly and engraving of bulk metallic glasses, Mater. Sci. Eng. A 375 (2004) 227-234. https://doi.org/10.1016/j.msea.2003.10.267

[245] M. Oliveria, W. J. Botta F, A. R. Yavari, Connecting, assemblage and electromechanical shaping of bulk metallic glasses, Mater. Trans. JIM, 41 (2000) 1501-1504. https://doi.org/10.2320/matertrans1989.41.1501

[246] D. V. Louzguine-Luzgin, G. Q. Xie, T. Tsumura, H. Fukuda, K. Nakata, H. M. Kimura, A Inoue, Structural investigation of Ni–Nb–Ti–Zr–Co–Cu glassy samples prepared by different welding techniques, Materials Science and Engineering B, 148 (2008) 88-91. https://doi.org/10.1016/j.mseb.2007.09.034

[247] T. Tsumura, K. Nakata, Laser welding of Ni-based metallic glass foil, Welding International, 25 (2011) 491-496. https://doi.org/10.1080/09507111003655366

[248] Y. Yokoyama, N. Abe, K. Fukaura, A. Inoue, Electron Beam Welding of $Zr_{50}Cu_{30}Ni_{10}Al_{10}$ Bulk Glassy Alloys, Mater. Trans. JIM., 43 (2002) 2509-2515. https://doi.org/10.2320/matertrans.43.2509

[249] D. V. Louzguine-Luzgin, Y. Yokoyama, G. Xie, N. Abe, A. Inoue, Transmission electron microscopy investigation of the structure of a welded $Zr_{50}Cu_{30}Ni_{10}Al_{10}$ glassy alloy sample, Philosophical Magazine Letters, 87 (2007) 549-554. https://doi.org/10.1080/09500830701320299

[250] T. Shoji, Y.Ohno, Y. Kawamura, Joining of $Zr_{41}Be_{23}Ti_{14}Cu_{12}Ni_{10}$ bulk metallic glasses by a friction welding method, Materials Transactions, 44 (2003) 1809-1816. https://doi.org/10.2320/matertrans.44.1809

[251] S. V. Ketov, N. Chen, A. Caron, A. Inoue, D. V. Louzguine-Luzgin, Structural features and high quasi-static strain rate sensitivity of $Au_{49}Cu_{26.9}Ag_{5.5}Pd_{2.3}Si_{16.3}$ bulk metallic glass, Applied Physics Letters, 101 (2012) 241905. https://doi.org/10.1063/1.4770072

[252] S. Mechler, E. Yahel, P. S. Pershan, M. Meron, B. Lin, Crystalline monolayer surface of liquid Au–Cu–Si–Ag–Pd: Metallic glass former, App. Phys. Let., 98 (2011) 251915. https://doi.org/10.1063/1.3599515

[253] A. Kawashima, Y. Zeng, M. Fukuhara, H. Kurishita, N. Nishiyama, H. Miki, A. Inoue, Mechanical properties of a $Ni_{60}Pd_{20}P_{17}B_3$ bulk glassy alloy at cryogenic temperatures, Materials Science and Engineering A, 498 (2008) 475-481. https://doi.org/10.1016/j.msea.2008.08.033

[254] E. D. Tabachnikova, A. V Podol'ski, V. Z Bengus, S. N Smirnov, D. V Luzgin, A. Inoue Low-temperature plasticity anomaly in the bulk metallic glass $Zr_{64.13}Cu_{15.75}Ni_{10.12}Al_{10}$, Low Temp. Phys., 34 (2008) 675-677. https://doi.org/10.1063/1.2967517

[255] D.V. Louzguine-Luzgin, A Vinogradov, S Li, A Kawashima, G Xie, A.R Yavari, A. Inoue, Deformation and fracture behavior of metallic glassy alloys and glassy-crystal composites, Metallurgical and Materials Transactions A, 42 (2011) 1504-1510. https://doi.org/10.1007/s11661-010-0391-3

[256] A. Vinogradov, A. Lazarev, D.V. Louzguine-Luzgin, Y. Yokoyama, S. Li, A.R. Yavari, A. Inoue, Propagation of shear bands in metallic glasses and transition from serrated to non-serrated plastic flow at low temperatures, Acta Materialia, 58 (2010) 6736-6743. https://doi.org/10.1016/j.actamat.2010.08.039

[257] A. H. Brothers, D. C. Dunand, Ductile bulk metallic glass foams, Adv. Mater., 17 (2005). 484-486. https://doi.org/10.1002/adma.200400897

[258] T. Wada, A. Inoue, Formation of porous Pd-based bulk glassy alloys by a high hydrogen pressure melting-water quenching method and their mechanical properties, Mater. Trans., 45 (2004) 2761-2765. https://doi.org/10.2320/matertrans.45.2761

[259] A. Inoue, T. Wada, D. V. Louzguine-Luzgin, Improved mechanical properties of bulk glassy alloys containing spherical pores, Materials Science and Engineering: A, 471 (2007)144-150. https://doi.org/10.1016/j.msea.2006.10.172

[260] G. Xie, W. Zhang, D. V. Louzguine-Luzgin, H. Kimura, A. Inoue, Fabrication of porous Zr–Cu–Al–Ni bulk metallic glass by spark plasma sintering process, Scripta Materialia, 55 (2006) 687–690. https://doi.org/10.1016/j.scriptamat.2006.06.034

[261] T. Wada, A. Inoue, A. L. Greer, Enhancement of room-temperature plasticity in a bulk metallic glass by finely dispersed porosity, Appl. Phys. Lett., 86 (2005) 251907. https://doi.org/10.1063/1.1953884

[262] A. Inoue, W. Zhang, T. Tsurui, A. R Yavari, A. L. Greer, Unusual room-temperature compressive plasticity in nanocrystal-toughened bulk copper-zirconium glass, Philos. Mag. Lett., 85 (2005) 221-229. https://doi.org/10.1080/09500830500197724

[263] D. V. Louzguine, H. Kato, A. Inoue, High-strength Cu- based crystal-glassy composite with enhanced ductility, Applied Physics Letters, 84 (2004) 1088-1089. https://doi.org/10.1063/1.1647278

[264] C. C. Hays, C. P Kim, W. L Johnson, Microstructure controlled shear band pattern formation and enhanced plasticity of bulk metallic glasses containing in situ formed ductile phase dendrite dispersions, Phys. Rev. Lett., 84 (2000) 2901–2904. https://doi.org/10.1103/PhysRevLett.84.2901

[265] S. Pauly, G. Liu, G. Wang, J. Eckert, Microstructural heterogeneities governing the deformation of $Cu_{47.5}Zr_{47.5}Al_5$ bulk metallic glass composites, Acta Mater., 57 (2009) 5445–5453. https://doi.org/10.1016/j.actamat.2009.07.042

[266] P. Gargarella, S. Pauly, K.K. Song, J. Hu, N.S. Barekar, M. Khoshkhoo, A. Teresiak, H. Wendrock, U. Kühn, C. Ruffing, E. Kerscher, J. Eckert, Ti–Cu–Ni shape memory bulk metallic glass composites, Acta Materialia 61 (2013) 151–162. https://doi.org/10.1016/j.actamat.2012.09.042

[267] H.J. Park, S.H. Hong, H.J. Park, W.-M. Wang, K.B. Kim, Development of high strength Ni–Cu–Zr–Ti–Si–Sn in-situ bulk metallic glass composites reinforced by hard B2 phase, Metals and Materials International, 24 (2018) 241-247. https://doi.org/10.1007/s12540-018-0039-1

[268] M.L. Lee, Y. Li, C.A. Schuh, Effect of a controlled volume fraction of dendritic phases on tensile and compressive ductility in La-based metallic glass matrix composites, Acta Mater. 52 (2004) 4121–4131. https://doi.org/10.1016/j.actamat.2004.05.025

[269] D. C. Hofmann, J. Y. Suh, A. Wiest, M. L. Lind, M. D. Demetriou, W. L. Johnson, Development of tough, low-density titanium-based bulk metallic glass matrix composites with tensile ductility, Proc. Natl. Acad. Sci. USA, 105 (2008) 20136–20140. https://doi.org/10.1073/pnas.0809000106

[270] D. C. Hofmann, J. Y. Suh, A. Wiest, G. Duan, M. L. Lind, M. D. Demetriou, W. L. Johnson, Designing metallic glass matrix composites with high toughness and tensile ductility, Nature, 451 (2008) 1085-1089. https://doi.org/10.1038/nature06598

[271] J. W. Qiao, T. Zhang, F. Q. Yang, P. K. Liaw, S. Pauly, B. S. Xu, A tensile deformation model for in-situ dendrite/metallic glass matrix composites, Scientific Reports, 3 (2011) 2816-2820.

[272] Y. Wu, Y. Xiao, G. Chen, C. T. Liu, Z. Lu, Bulk metallic glass composites with transformation-mediated work-hardening and ductility, Adv. Mater., 22 (2010) 2270-2273. https://doi.org/10.1002/adma.201000482

[273] D. C. Hofmann, Shape memory bulk metallic glass composites, Science, 329 (2010) 1294-1295. https://doi.org/10.1126/science.1193522

[274] S. Pauly, S. Gorantla, G. Wang, U. Kühn, J. Eckert, Transformation-mediated ductility in CuZr-based bulk metallic glasses, Nat. Mater., 9 (2010) 473-477. https://doi.org/10.1038/nmat2767

[275] P. Gargarella, S. Pauly, K.K. Songa, J. Hu, N.S. Barekar, M. S. Khoshkhoo, A. Teresiak, H. Wendrock, U. Kühn, C. Ruffing, E. Kerscher, J. Eckert, Ti-Cu-Ni shape memory bulk metallic glass composites, Acta Materialia, 61 (2013) 151–162. https://doi.org/10.1016/j.actamat.2012.09.042

[276] S. Pauly, S. Gorantla, G. Wang, Transformation-mediated ductility in CuZr-based bulk metallic glasses, Nature Mater., 9 (2010) 473–477. https://doi.org/10.1038/nmat2767

[277] J. Das, M. B. Tang, K. B. Kim, R. Theissmann, F. Baier, W. H. Wang, J. Eckert, Work-hardenable ductile bulk metallic glass, Phys. Rev. Letters, 94 (2005) 205501. https://doi.org/10.1103/PhysRevLett.94.205501

[278] F. O. Mear, T. Wada, D. V. Louzguine-Luzgin, A. Inoue, Highly inhomogeneous compressive plasticity in nanocrystals toughened Zr–Cu–Ni–Al bulk metallic glass, Philosophical Magazine Letters, 89 (2009) 276–281. https://doi.org/10.1080/09500830902817861

[279] D. V. Louzguine-Luzgin, A. Vinogradov, G. Xie S. Li, A. Lazarev, S. Hashimoto, A. Inoue, High-strength and ductile glassy-crystal Ni-Cu-Zr-Ti composite exhibiting stress-induced martensitic transformation, Philosophical Magazine, 89 (2009) 2887–2901. https://doi.org/10.1080/14786430903128577

[280] I.S. Golovin, V.Yu. Zadorozhnyy, A.Yu. Churyumov, D.V. Louzguine-Luzgin, Internal friction in a Ni–Ti-based glassy-crystal alloy, Journal of Alloys and Compounds, 579 (2013) 633–637. https://doi.org/10.1016/j.jallcom.2013.07.102

[281] A. Yu. Churyumov, A. I. Bazlov, A. N. Solonin, V. Yu. Zadorozhnyi, G. Q. Xie, S. Li, D. V. Louzguine-Luzgin, Structure and mechanical properties of Ni-Cu-Ti-Zr composite materials with amorphous phase, The Physics of Metals and Metallography, 114 (2013) 773-778. https://doi.org/10.1134/S0031918X13090044

[282] A. A. Tsarkov, A.Y. Churyumov, V.Y. Zadorozhnyy, D.V. Louzguine-Luzgin, High-strength and ductile (Ti-Ni)-(Cu-Zr) crystalline/amorphous composite materials with superelasticity and TRIP effect, Journal of Alloys and Compounds, 658 (2016) 402-407. https://doi.org/10.1016/j.jallcom.2015.10.175

[283] V. Yu. Zadorozhnyy, A. Inoue, D. V. Louzguine-Luzgin, Formation and investigation of the structure and mechanical properties of bulk metallic glassy composite (Ti-Zr)–(Cu-Ni-Co) alloys, Intermetallics, 31 (2012) 173-176. https://doi.org/10.1016/j.intermet.2012.07.008

[284] Y. Jiang, X. Shi, K. Qiu, Numerical study of shear banding evolution in bulk metallic glass composites, Materials & Design, 77 (2015) 15. https://doi.org/10.1016/j.matdes.2015.04.010

[285] G.P. Zheng, Y. Shen, Simulation of shear banding and crack propagation in bulk metallic glass matrix composites, Journal of Alloys and Compounds, 509 (2011) 136-140. https://doi.org/10.1016/j.jallcom.2010.08.131

[286] Y. Shen, G.P. Zheng, Modeling of shear band multiplication and interaction in metallic glass matrix composites, Scripta Materialia, 63 (2010) 181-184. https://doi.org/10.1016/j.scriptamat.2010.03.046

[287] J.S.C. Jang, T.H. Li, P.H. Tsai, J.C. Huang, T.G. Nieh, Critical obstacle size to deflect shear banding in Zr-based bulk metallic glass composites, Intermetallics, 64 (2015) 102-105. https://doi.org/10.1016/j.intermet.2015.05.001

[288] Y. Xue, X. Zhong, L. Wang, Q. Fan, L. Zhu, B. Fan, H. Zhang, H. Fu, Effect of W volume fraction on dynamic mechanical behaviors of W fiber/Zr-based bulk metallic glass composites, Materials Science and Engineering: A, 639 (2015) 417-424. https://doi.org/10.1016/j.msea.2015.05.086

[289] A.Yu. Churyumov, A.I. Bazlov, A.A. Tsarkov, A.N. Solonin, D.V. Louzguine-Luzgin, Microstructure, mechanical properties, and crystallization behavior of Zr-based bulk metallic glasses prepared under a low vacuum, Journal of Alloys and Compounds, 654 (2016) 87–94. https://doi.org/10.1016/j.jallcom.2015.09.003

[290] Y.P. Jiang, X.P. Shi, K. Qiu, Numerical study of shear banding evolution in bulk metallic glass composites, Materials & Design, 77 (2015) 32-40.

[291] A.I. Bazlov, A.Yu. Churyumov, A.A. Tsar'kov and D.M. Khazhina, Studies of the structure and mechanical properties of $Ti_{43.2}Zr_{7.8}Cu_{40.8}Ni_{7.2}Co_1$ alloy containing amorphous and crystalline phases, Phys. Met. Metallogr., 116 (2015) 684–689. https://doi.org/10.1134/S0031918X15050038

[292] L. Guo, H. Guo, S. Gong, H. Xu, Improvement on the phase stability, mechanical properties and thermal insulation of Y_2O_3-stabilized ZrO_2 by Gd_2O_3 and Yb_2O_3 co-doping, Ceram. Int. 39 (2013) 9009-9015. https://doi.org/10.1016/j.ceramint.2013.04.103

[293] V.Yu. Zadorozhnyy, M.V. Gorshenkov, M.N. Churyukanova, M.Yu. Zadorozhnyy, A.A. Stepashkin, D.O. Moskovskikh, S.V. Ketov, L.Kh. Zinnurova, A. Sharma, D.V. Louzguine-Luzgin, S.D. Kaloshkin, Investigation of structure and thermal properties in composite materials based on metallic glasses with small addition of polytetrafluoroethylene, Journal of Alloys and Compounds, 707 (2017) 264–268. https://doi.org/10.1016/j.jallcom.2016.11.359

[294] K. Hajlaoui, A.R. Yavari, A. LeMoulec, W.J. Botta, F.G. Vaughan, J.Das, A.L Greer, A. Kvick, Plasticity induced by nanoparticle dispersions in bulk metallic glasses, Journal of Non-Crystalline Solids, 353 (2007) 327-331. https://doi.org/10.1016/j.jnoncrysol.2006.10.011

[295] A. Inoue, T. Zhang, M. W. Chen, T. Sakurai, J. Saida, M. Matsushita, Ductile quasicrystalline alloys, Appl. Phys. Lett., 76 (2006) 967-969. https://doi.org/10.1063/1.125907

[296] A. Inoue, C. Fan, A.Takeuchi, High-strength bulk nanocrystalline alloys in a Zr-based system containing compound and glassy phases, J. Non-Cryst. Sol., 250-252 (1999) 724. https://doi.org/10.1016/S0022-3093(99)00168-4

[297] J. Eckert, U. Kühn, N. Mattern, A. Reger-Leonhard, M. Heilmaier, Bulk nanostructured Zr-based multiphase alloys with high strength and good ductility, Scripta Mater, 44 (2001) 1587–1590. https://doi.org/10.1016/S1359-6462(01)00779-5

[298] J. Saida, H. Kato, A. D. Setyawan, A. Inoue, Characterization and properties of nanocrystal-forming Zr-based bulk metallic glasses, Rev. Adv. Mater. Sci., 10 (2005) 34-38.

[299] D. V. Louzguine-Luzgin, Y. Zeng, A. D. H. Setyawan, N. Nishiyama, H. Kato, J. Saida, A. Inoue, Deformation behavior of Zr- and Ni-based bulk glassy alloys, J. Mater. Res., 22 (2007) 1087-1090. https://doi.org/10.1557/jmr.2007.0126

[300] M. C. Gao, R. E. Hackenberg, G. J. Shiflet, Deformation-induced nanocrystal precipitation in Al-base metallic glasses, Materials Transactions, 42 (2001) 1741-1747. https://doi.org/10.2320/matertrans.42.1741

[301] W. H. Jiang, M. Atzmon, Plastic flow of a nanocrystalline/amorphous $Al_{90}Fe_5Gd_5$ composite formed by rolling, Intermetallics, 14 (2006) 962-965. https://doi.org/10.1016/j.intermet.2006.01.013

[302] Z.F. Zhang, G. He, H. Zhang, J. Eckert, Rotation mechanism of shear fracture induced by high plasticity in Ti-based nano-structured composites containing ductile dendrites, Scripta Mater. 52 (2005) 945-949. https://doi.org/10.1016/j.scriptamat.2004.12.014

[303] A. M. Glezer, M. R. Plotnikova, R. V. Sundeev, and N. A. Shurygina, Self blocking of shear bands and the delocalization of plastic flows in amorphous alloys upon megaplastic deformation, Bulletin of the Russian Academy of Sciences, Physics, 77 (2013) 1391–1396. https://doi.org/10.3103/S1062873813110129

[304] D. Torre, F. H. Dubach, A. Schallibaum, J. Loffler, Shear striations and deformation kinetics in highly deformed Zr-based bulk metallic glasses, Acta Materialia, 56 (2008) 4635-4646. https://doi.org/10.1016/j.actamat.2008.05.021

[305] A. V. Sergueeva, N. A. Mara, D. J. Branagan, A. K. Mukherjee, Strain rate effect on metallic glass ductility, Scr. Mater., 50 (2004) 1303-1307. https://doi.org/10.1016/j.scriptamat.2004.02.019

[306] Q.P. Cao, J.F. Li, Y.H. Zhou, A. Horsewell, J.Z. Jiang, Effect of rolling deformation on the microstructure of bulk $Cu_{60}Zr_{20}Ti_{20}$ metallic glass and its crystallization, Acta Materialia, 54 (2006) 4373–4383. https://doi.org/10.1016/j.actamat.2006.05.030

[307] G. Kumar, T. Ohkubo, T. Mukai, K. Hono, Plasticity and microstructure of Zr–Cu–Al bulk metallic glasses, Scr. Mater., 57 (2007) 173–176. https://doi.org/10.1016/j.scriptamat.2007.02.013

[308] A.R. Yavari, K. Georgarakis, J. Antonowicz, M. Stoica, N. Nishiyama, G. Vaughan, M. Chen, M. Pons, Crystallization during bending of a Pd-based metallic glass detected by X-ray microscopy, Physical Review Letters, 109 (2012) 085501. https://doi.org/10.1103/PhysRevLett.109.085501

[309] A. Inoue, K Ohtera, A. P. Tsai. T. Masumoto, Glass transition behavior of Al-Y-Ni and Al-Ce-Ni amorphous alloys, Jpn. J. Appl. Phys., 27 (1988) 479. https://doi.org/10.1143/JJAP.27.L1579

[310] G.J. Shiflet, Y. He, S. J. Poon, Mechanical properties of a new class of metallic glasses based on aluminum, J. Appl. Phys., 64 (1988) 6863. https://doi.org/10.1063/1.341978

[311] Y. H. Kim, A. Inoue, T. Masumoto, Ultrahigh tensile strengths of $Al_{88}Y_2Ni_9M_1$ (M=Mn or Fe) amorphous alloys containing finely dispersed fcc-Al particles, Mater. Trans. JIM, 31 (1990) 747-749. https://doi.org/10.2320/matertrans1989.31.747

[312] A. Inoue, H. Kimura, K. Amiya, Developments of aluminum-and magnesium-based nanophase high-strength alloys by use of melt quenching-induced metastable phase, Mater. Trans., 43 (2002) 2006-2016. https://doi.org/10.2320/matertrans.43.2006

[313] Z. C. Zhong, X. Y. Jiang, A. L Greer, Microstructure and hardening of Al-based nanophase composites, Mater. Sci. Eng. A, 226–228 (1997) 531-535. https://doi.org/10.1016/S0921-5093(97)80062-7

[314] D. V. Louzguine, and A. Inoue, Investigation of structure and properties of the Al-Y-Ni-Co-Cu metallic glasses, Journal of Materials Research, 17 (2002) 1014-1018. https://doi.org/10.1557/JMR.2002.0149

[315] V. S. Zolotorevsky, A. I. Bazlov, A. G. Igrevskaya, A. S. Aronin, G. E. Abrosimova, D. V. Louzguine-Luzgin, Significant mechanical softening of an Al-Y-Ni-Co metallic glass on cold and hot rolling, JOM, 71 (2019) 4079-4085. https://doi.org/10.1007/s11837-019-03430-x

[316] T. Nasu, S. Kanazawa, S. Hayashizaki, S.X. Zhao, S. Takahashi, T. Usuki and Y. Kameda, Work softening behavior of zirconium-aluminum-nickel-copper bulk-metallic-glass by rolling, Mater. Trans., 56 (2015) 249. https://doi.org/10.2320/matertrans.M2014342

[317] Y. Zhang, W. H. Wang and A. L. Greer, Making metallic glasses plastic by control of residual stress, Nature Mater., 5 (2006) 857. https://doi.org/10.1038/nmat1758

[318] A. M. Glezer, M. R. Plotnikova, R. V. Sundeev and N. A. Shurygina, Self-blocking of shear bands and the delocalization of plastic flows in amorphous alloys upon megaplastic deformation, Bull. Russ. Acad. Sci. Phys., 77 (2013) 1391–1396. https://doi.org/10.3103/S1062873813110129

CHAPTER 6

Magnetic Properties

Magnetic materials compose another very important and well developed application field of metallic glassy and nanostructured alloys. The connection between nanostructure and magnetic properties is a topic of intensive investigations. Ferromagnetic alloys can exhibit hard or soft magnetism depending on their coercive force (H_c) also called coercivity. Magnetic materials having a coercivity above about 10^4 A/m are considered to be hard while soft magnetic materials have a coercivity below 10^3 A/m. Current Chapter is devoted to this subject.

Contents

6.1   Soft magnetic alloys

Fe-, Ni- and Co-based bulk metallic glassy alloys exhibit good soft magnetic properties [1,2,3] with high saturation magnetization (M_s), high permeability (μ) and a coercivity of about 10 A/m and even lower while Nd-based alloys show hard magnetic properties with a coercivity above 10^4 A/m. Compared to other bulk metallic glasses, (Fe-Co)-based alloys are also particularly attractive for engineering applications due to their combination of ultrahigh strength, good wear resistance, good glass forming ability and rather low cost. A typical magnetization curve of the soft magnetic alloy (also called magnetically soft alloy) is given in Fig. 6.1.

Fe-Zr-B soft magnetic materials also exhibit high magnetization saturation (M_s) of 1.60-1.70 T as well as high effective permeability of 15000 at 1 kHz [4]. In the case of the (Fe, Co, Ni)$_{70}$Zr$_{10}$B$_{20}$ glassy alloys H_c is as low as 3 A/m [5]. The saturation magnetization increases to 0.9 T with increasing Fe content, while the saturated magnetostriction (λ_s) equals zero in the Co-rich composition range and increases monotonously to 15×10^{-6} with increasing Fe content. μ reaches the maximum of about 20000 in the Fe- and Co-rich composition ranges. These glassy alloys exhibit good soft magnetic properties including the saturation magnetization of 0.9 T, H_c of 3–6 A/m, λ_s of $12–15\times10^{-6}$ and μ of 20000 in the Fe-rich range and the saturation magnetization of 0.5 T, H_c of 6 A/m, nearly zero λ_s and μ of 20000 in the Co-rich range.

Large-scale cores are made from metallic glassy powders by sintering thus overcoming limitation connected with low glass-forming ability of some metallic glassy alloys. The addition of the SiC was effective in improving the high frequency magnetic properties [6]. This approach is promising for creation net shape products.

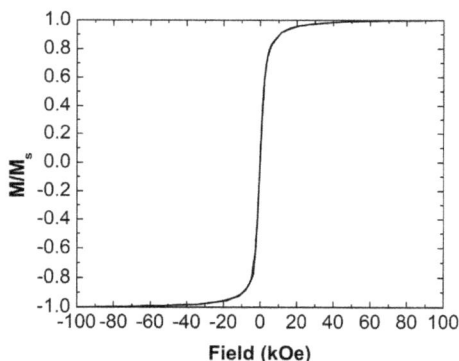

Fig. 6.1. Magnetization curve of a Co-Fe-Ta-B metallic glassy sample.

The coercivity of soft magnetic materials changes with grain size (d) and both coarse grain and nanostructured alloys are known to have low H_c. Fig. 6.2 shows the dependence of coercivity on the grain size. Atomic clusters of sub-nanometer size as the structural units of glassy alloys make them ideal soft magnetic materials. Nanocrystallization is typical for various Fe-based glassy alloys [7] and can be used for further improving their magnetic properties.

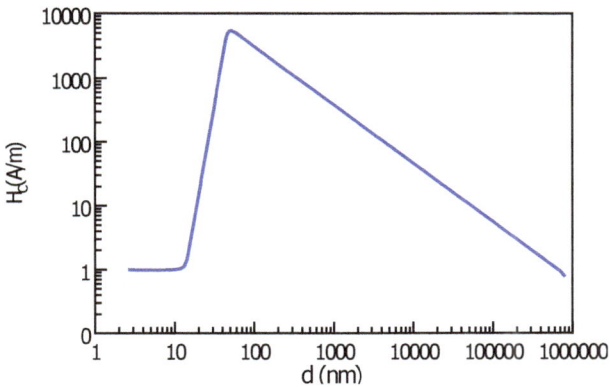

Fig. 6.2. Schematic representation of the changes in H_c of soft magnetic materials as a function of grain size (d).

Well known Fe-based soft magnetic materials with mixed nanocrystalline and amorphous structure are: $Fe_{73.5}Cu_1Nb_3Si_{13.5}B_9$ [8] and $Fe_{84}Zr_{3.5}Nb_{3.5}B_8Cu_1$ [9] alloys. The structure consists of fine BCC Fe nanocrystals dispersed in the amorphous matrix. The $Fe_{84}Zr_{3.5}Nb_{3.5}B_8Cu_1$ alloy shows high μ=100000, high magnetization saturation up to 1.5 T and low hysteresis losses [10]. Cu, Nb and Zr are responsible for BCC Fe grain refinement and formation of a nanostructure in these alloys. 3D atom probe field ion microscopy and high resolution transmission electron microscopy studies showed [11] that Cu formed nano clusters in the $Fe_{73.5}Si_{13.5}B_9Nb_3Cu_1$ amorphous matrix which work as heterogeneous nucleation sites for bcc Fe particles on devitrification. The studies for X-ray absorption fine structure also showed that Cu clusters with nearly FCC structure were present from very early stages of the devitrification process [12]. The density of the clusters is of the order of 10^{-24} m^{-3} while the average cluster size is about 2 nm [13]. This analysis also showed a low Fe content in the grain boundaries between the bcc Fe solid solution nano grains of the nanocrystalline $Fe_{85}Zr_7B_6Cu_2$ alloy. Yavari and Negri [14] discussed nanocrystallization process of soft magnetic Fe-based amorphous alloys using the concentration gradients of the elements that are insoluble in the primary crystalline phase. The origin of the good soft magnetic properties is connected with the formation of the nanoscale BCC Fe structure and a strong magnetic coupling between the BCC grains through the intergranular ferromagnetic amorphous phase.

Nanocrystalline $(Fe_{1-x}Co_x)_{90}Zr_7B_3$ (x=0, 0.1, 0.2, and 0.3) samples have a low coercivity of 9.1A/m and high permeability values of 4500 at 1 kHz along with a high saturation magnetization of 1.74 T [15]. The Fe–M–B (M=Zr, Hf, or Nb) alloys also show low core losses [16]. They also form a nanostructure upon initial crystallization on heating. Nanocrystalline $Fe_{89}(Hf,Nb)_7B_4$ alloys subjected to the optimum annealing exhibit high magnetization saturation above 1.5 T as well as high effective permeability at 1 kHz above 20000 [17].

Nanocrystalline $Fe_{42.5}Co_{42.5}Nb_7B_8$ alloy with a structure consisting of nearly spherical bcc grains with a size ranged from 5 to 10 nm dispersed in the residual amorphous matrix exhibits a high saturation magnetization of 1.9 T and a low coercivity (H_c) [18]. It also exhibits a very high Curie temperature (T_c) exceeding 1173 K.

For long time soft magnetic glassy alloys were limited to marginal glass-formers. However, as soon as a ferromagnetic Fe-(Al,Ga)-metalloid bulk glassy alloy was produced in 1995 [19], Fe-based bulk metallic glasses have attracted significant interest in the scientific community owing to their high strength and good soft magnetic properties [20]. Bulk glassy alloys exhibiting a wide supercooled liquid region before crystallization were also obtained in the Fe–(Co,Ni)–(Zr,Nb,Ta)–(Mo,W)–B system [21]. These alloys have a high T_g of about 870 K and the supercooled liquid region close to 90 K. The critical diameter is 6 mm. The alloys also exhibit high compressive strength of 3800 MPa, high Vickers hardness of HV1360, and high corrosion resistance. These glassy alloys exhibit a large magnetization saturation of 0.74–0.96 T, low coercivity of 1.1–3.2 A/m, high permeability exceeding 12000 at 1 kHz-, and low magnetostriction. Magnetization curves for the as-cast $Fe_{56}Co_7Ni_7Nb_{10}B_{20}$ rod samples of different diameter are presented in [22].

The $(Fe,Ni,Co)_{70}Mo_5P_{10}C_{10}B_5$ bulk metallic glasses with a critical sample diameter exceeding 4 mm showed good thermoplastic formability and good soft magnetic and mechanical properties. These glasses exhibit a wide supercooled liquid region and low viscosity of about 10^7 Pa·s in the supercooled liquid state [23]. The Fe–(Al,Ga)-(P,C,B) and Fe–(Al–Ga)-(P,C,B,Si) bulk glassy alloys also exhibited good soft magnetic properties [24,25]. For example, a ring-shaped sample of the $Fe_{70}Al_5Ga_2P_{9.65}C_{5.75}B_{4.6}Si_3$ glassy alloy with a thickness of 1 mm, an outer diameter of 10 mm and an inner diameter of 6 mm formed by the copper mold casting method exhibits the saturation magnetization of 1.2 T, very low coercivity of 2.2 A/m and rather low λ_s of 21×10^{-6}. The maximum permeability is 110 000. The remarkable improvement of the soft magnetic properties has been demonstrated to result from the significant difference in magnetic domain structure. The domain walls are arranged along the circumference direction for the cast-ring alloy and radial direction for the ribbon ring sheet. The difference in the domain wall structure was reported to originate from the difference in residual stress during the preparations of the ring sample and the melt-spun ribbon. The $Co_{46}Fe_{20}B_{23}Si_5Nb_6$ BMG with a critical diameter of 5 mm exhibited B_s, H_c, μ and fracture strength of ~0.7 T, ~2 A/m, ~20000 and 4400 MPa, respectively [26].

The $Co_{43}Fe_{20}Ta_{5.5}B_{31.5}$ glassy alloy exhibited an extremely low value of the coercive force as low as 0.25 A/m together and a high permeability of 550,000. The Fe–(Co,Ni)–(Zr,Nb,Ta)–(Mo,W)–B system glassy alloys exhibited a large saturation magnetization of 0.74–0.96 T, low coercivity of 1.1–3.2 A/m, high permeability exceeding $1.2 \cdot 10^4$ at 1 kHz, and low magnetostriction of about 12×10^{-6} [27].

The $Fe_{76}Si_9B_{10}P_5$ alloy with a critical diameter of D_{cr} 2.5 mm exhibited saturation magnetization of 1.51 T and H_c below 1 A/m [28]. Minor addition of Cu improved plasticity of this alloy [29]. Fe, C, Si, and P, are the constituent elements in cast-iron produced in a blast furnace. As expensive Fe–Si and Fe–P ferroalloys are in mass production there is no restriction in availability of such materials which offers advantages of lower material cost for industry.

A $Fe_{80}P_{11}C_9$ BMG alloy exhibiting high strength of 3.2 GPa [30] in the partially devitrified state also exhibited good soft-magnetic properties including the saturation magnetization of 1.5 T and coercivity of 4 A/m. The $(Fe_{1-x}Ni_x)_{72}B_{20}Si_4Nb_4$ (x=0.0–0.5) bulk metallic glasses showed the saturation magnetization up to 1.15 T. The maximum Curie temperature was 598 K [31]. The $(Fe_{1-x}Co_x)_{76}Si_9B_{10}P_5$ (x = 0~0.4) ferromagnetic bulk glassy alloys demonstrated excellent combination of high GFA, good soft-magnetic properties as well as high strength [32]. These BMGs exhibited a high saturation magnetization of 1.5 T and low coercive force of 1.2 A/m.

The M_s and μ of the $Fe_{76.5}C_6Si_{3.3}B_{5.5}P_{8.7}$ BMG alloy were enhanced from 1.35 T and 3500 in the as-cast state to 1.57 T and 9890, respectively, upon annealing at 873 K for 30 s [33]. The MRO zones of 1–3 nm size were observed in the glassy matrix and the coercivity was below 20 A/m. The saturation magnetization of the $(Fe,Co,Ni)_{75.5}C_{7.0}Si_{3.3}B_{5.5}P_{8.7}$ bulk metallic glass continuously decreases with increasing Co or Ni content, while the Curie temperature and the permeability increase. The Co-bearing alloys show smaller coercivity and larger permeability than the Ni-bearing alloys [34].

Y addition was confirmed to improve GFA of the $Fe_{72}B_{24}Nb_4$ alloy. Saturation magnetization and coercivity of the as-cast $(Fe_{0.72}B_{0.24}Nb_{0.04})_{95.5}Y_{4.5}$ bulk metallic glassy ring were found to be 0.8 T and 0.8 A/m, respectively, in the relaxed state after annealing at 821 K [35]. The $Fe_{76+x}Si_{9-x}B_{10}P_5$ bulk metallic glasses showed the saturation magnetization above 1.6 T, low coercive force of ~2 A/m and high effective permeability of 17000 [36]. Soft magnetic powder core with a higher saturation magnetization of 1.3 T were developed in the Fe–Nb–B–Si and Fe–Nb–Cr–P–B–Si systems [37]. The Fe–Ni–Cr–Mo–B–Si glassy alloy powders produced by water atomization have also been commercialized with the commercial name AMO-beads owing to the high glass-forming ability for the Fe-based alloy.

Ferromagnetic Co-based BMGs [38] possess low H_c, high μ, nearly zero λ_s, low high-frequency loss, high fracture strength 4 GPa, high glass transition and crystallization temperatures, large supercooled liquid region, high corrosion resistance, and high wear resistance. The $(Co_{0.942}Fe_{0.058})_{67}Nb_5B_{22.4}Si_{5.6}$ alloy exhibits high GFA (critical diameter 4 mm) and good combination properties, i.e., 3600 MPa for fracture strength, 0.43 T for B_s, 0.2 A/m for H_c and 47165 for μ_e at 1 kHz and 1 A/m [39]. Furthermore, the addition of lanthanide element (Er, Tb, Y or Dy) enables a significant increase of the supercooled liquid region and the largest interval reaches 130 K for $(Co_{0.5}Fe_{0.5})_{62}Nb_6Dy_2B_{30}$. In the supercooled liquid region, the alloy exhibits good superplastic formability as well as good nanoscale imprintability [40]. The properties of some alloys are summarized in Table 6.1.

Table 6.1. Composition, critical diameter (D_{cr}) and magnetic properties (B_s, H_c) of typical ferromagnetic BMGs.

Nr	Alloy	D_{cr} (mm)	B_s (T)	H_c (A/m)	Ref.
1	$Fe_{76}Si_9B_{10}P_5$	2.5	1.51	0.8	28
2	$(Fe_{0.75}Si_{0.1}B_{0.15})_{96}Nb_4$	1.5	1.47	2.9	28
3	$Fe_{73}Al_5Ga_2P_{11}C_5B_4$	1	1.29	6.3	28
4	$Fe_{30}Co_{30}Ni_{15}Si_8B_{17}$	1.2	0.92	3.4	28
5	$Fe_{70}Mo_5P_{10}C_{10}B_5$	3	0.93	2.36	41
6	$Fe_{74}Nb_6Y_3B_{17}$	2	0.81	15	42
7	$Fe_{65}Ni_5Mo_5P_{10}C_{10}B_5$	3	0.88	3.2	23
8	$Fe_{55}Co_{10}Ni_5Mo_5P_{10}C_{10}B_5$	4	0.85	2.56	23
10	$(Fe_{0.75}B_{0.15}Si_{0.1})_{96}Nb_4$	1.5	1.2	3.7	43
11	$[(Fe_{0.8}Co_{0.1}Ni_{0.1})_{0.75}B_{0.2}Si_{0.05}]_{96}Nb_4$	2.5	1.1	3	43
12	$Co_{43}Fe_{20}Ta_{5.5}B_{31.5}$	3	0.49	0.25	44

These values are plotted in the diagram in Fig. 6.3.

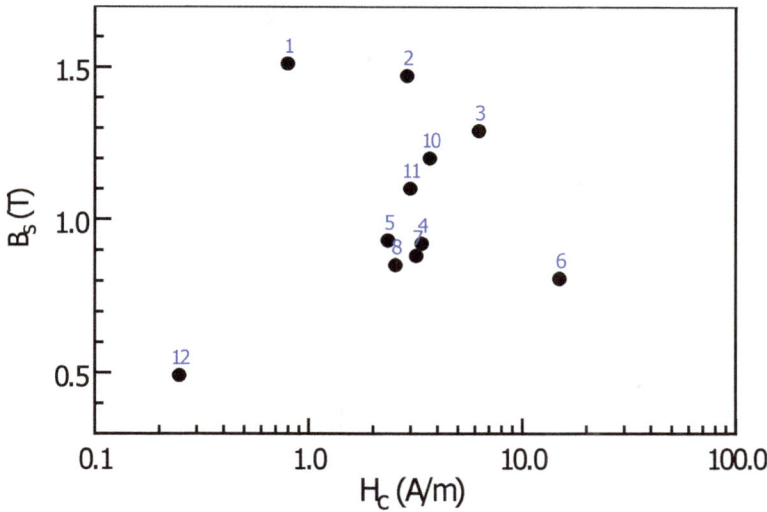

Fig. 6.3. Magnetic properties of the BMGs listed in Table 6.1.

In addition to the massive glassy samples metallic glassy micro- [45] and nano-wires [46] and nanoglasses attract increasing attention. Here there is an area for further application of magnetic materials. Arrays of magnetic nanowires attract interest for perpendicular magnetic recording [47]. The nanowires are potentially capable of producing recording densities in excess of several tens of Gbit per square inch. It has been also reported [48] that the crystalline magnetic nanowires of Fe, Co and Ni show significantly enhanced magnetic coercivity than that of their bulk counterparts. Such an approach can be applied to metallic glassy nanowires as well.

Excellent magnetic softness with coercivity of about 5 A/m and high saturation magnetization of about 1.87 T was found in the partly crystalline $(Fe_{0.7}Co_{0.3})_{83.7}Si_4B_8P_{3.6}Cu_{0.7}$ alloy ribbons and microwires (Fig. 6.4) [49]. Glass–coated $(Fe_{0.7}Co_{0.3})_{83.7}Si_4B_8P_{3.6}Cu_{0.7}$ microwires with a mixed amorphous/nanocrystalline structure exhibit perfectly rectangular hysteresis loops and unusually high domain wall mobility that must be associated to elevated saturation magnetization values. The partially crystalline $(Fe_{0.7}Co_{0.3})_{83.7}Si_4B_8P_{3.6}Cu_{0.7}$ microwires have higher values of $H_c = 480$ A/m and lower saturation magnetization of about 1.6 T. The values of H_c are quite similar to that exhibited by partially nanocrystalline Hitperm-like $Fe_{38.5}Co_{38.5}B_{18}Mo_4Cu_1$ microwires with similar average grain size [50].

Unusually high domain wall mobility observed in $(Fe_{0.7}Co_{0.3})_{83.7}Si_4B_8P_{3.6}Cu_{0.7}$ microwire has been associated to elevated saturation magnetization values.

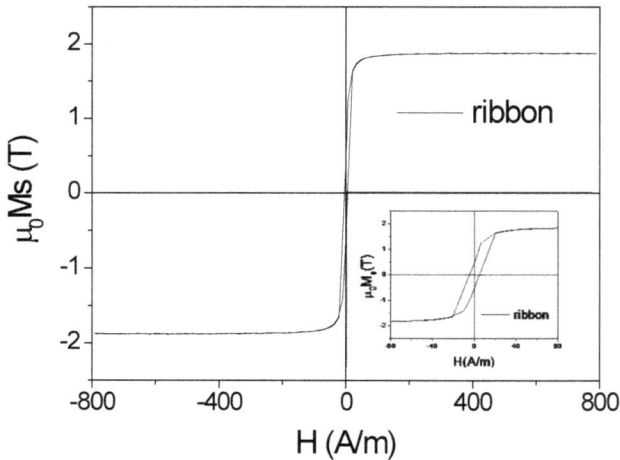

Fig. 6.4. Hysteresis loop of the studied $(Fe_{0.7}Co_{0.3})_{83.7}Si_4B_8P_{3.6}Cu_{0.7}$ partly crystalline sample and a close-up in the insert. Reproduced from [49] with permission of Elsevier.

The $Fe_{84}B_8P_{3.5}Si_{1.5}Mo_2Cu_1$ samples exhibited M_s and H_c of 1.4 T and 4.4 A/m, respectively, for the amorphous and 1.8 T and 2.5 A/m, respectively, for the nanocrystalline BCC-Fe + amorphous phases. The average grain size of bcc-Fe phase in the optimum nanocrystalline state was about 10 nm [51].

The irreversible changes of the magnetic properties and the Curie temperature in a Co-Cr-Zr metallic glass were found after thermal cycling between room and liquid nitrogen temperature [52]. After thermal cycling the coercive force, increased by 35%, magnetization increased by 6%, and remanent magnetization increased by 3%. Aging of thermally cycled samples at 300 K for 10 h led to the further increasing of these values by 60%, 16%, and 10%, respectively, compared to the as-cast state. Subsequent thermal cycling did not change the properties. The observed dependencies were explained as a sequence of phase transitions in the amorphous state.

6.2   Hard magnetic alloys

Hard magnetic alloys (also called magnetically hard alloys) have sufficiently high coercive force as a resistance to demagnetizing fields exceeding 10 kA/m. High magnetic induction is retained because of a strong resistance to demagnetization, for example, as a result of high anisotropy. Permanent magnets must have a large coercive force, high saturation magnetization and high energy product $(BH)_{max}$. These alloys can be used as permanent magnet materials with high magnetic induction.

Ferromagnetic $Nd_{90-x}Fe_xAl_{10}$ bulk amorphous alloys with high coercive force at room temperature were obtained by a copper mold casting method. The maximum diameter of the cylindrical amorphous samples with a length of 50 mm was about 7 mm. Neither glass transition nor supercooled liquid region was observed in these alloys in the temperature range before crystallization, which makes them different from classical bulk glassy alloys exhibiting a wide supercooled liquid region before crystallization [53]. The bulk amorphous $Nd_{70}Fe_{20}Al_{10}$ alloy exhibits ferromagnetism with T_c of about 600 K which is much higher than the highest T_c (about 480 K) for the Nd-Fe binary amorphous alloy ribbons. The remanence (B_r) and H_c for the bulk $Nd_{60}Fe_{30}Al_{10}$ alloy are 0.122 T and 277 kA/m, respectively, in the as-cast state and 0.128 T and 277 kA/m, respectively, after annealing for 600 s at 600 K [54]. The B_r and H_c decrease to 0.045 T and 265 kA/m, respectively, for the crystallized sample. The hard magnetic properties for the bulk amorphous alloys are presumably due to the homogeneous development of ferromagnetic clusters with large random magnetic anisotropy.

Hard magnetic alloys can be produced by crystallization of the glassy phase. For example, a hard magnetic material consisting of the Fe_3B and $Nd_2Fe_{14}B$ phases was obtained by annealing the $Nd_{4.5}Fe_{77}B_{18.5}$ rapidly solidified alloy [55]. The microstructure is composed of the magnetically soft Fe_3B phase nanoscale grains and the magnetically hard $Nd_2Fe_{14}B$ phase. High $B_r = 0.8$ T is obtained due to the permanence enhancement effect of exchange-coupled magnetic grains, the maximum energy product $(BH)_{max}$ is 97 kJ/m^3 while its coercivity H_c=240 kJ/m. The influence of heating rate on the microstructure of $Fe_3B/Nd_2Fe_{14}B$ nanocomposite magnets was also studied [56]. High coercivity values exceeding 300 kA/m were obtained in the amorphous $Nd_5Fe_{72}Cr_5B_{18}$ alloy crystallized into the $Fe_3B/Nd_2Fe_{14}B$ state [57]. Fe-Nd-B amorphous alloys containing 88-90 at% Fe at 923-1023 K form a nanostructure consisting of BCC-Fe, $Fe_{14}Nd_2B$ and the residual amorphous phase. The $Fe_{89}Nd_7B_4$ glassy alloy exhibits good hard magnetic properties, i.e., B_r of 1.28 T, H_c of 252 kA/m and $(BH)_{max}$ of 146 kJ/m^3 for [58].

The α-Fe/$Nd_2Fe_{14}B$ nanomaterials have higher coercivity than those for $Fe_3B/Nd_2Fe_{14}B$ magnets. H_c of 480 kA/m and $(BH)_{max}$ of 160 kJ/m^3 are reported. The structure consists of two phases: magnetically hard $Nd_2Fe_{14}B$ with nanoparticles of α-Fe phase on grain boundaries. Grain size of the $Nd_2Fe_{14}B$ phase is below 30 nm and particle size of the α-Fe phase is below 10 nm [59]. α-Fe/$Nd_2Fe_{14}B$ nanostructured magnet of $Fe_{89}Nd_7B_4$ composition contains residual amorphous phase and shows B_r=1.22 T, H_c=240 kA/m and $(BH)_{max}$=130 kJ/m³. This alloy has high iron and low boron concentration [60].

The rapidly solidified $(Fe_{0.65}Pt_{0.35})_{83}B_{17}$ alloy possesses higher coercivity in the annealed state compared to the binary Fe-Pt alloys [61]. B_r, M_r/M_s, H_c, and $(BH)_{max}$ of the rapidly-solidified $Fe_{80-x}Pt_xB_{20}$ (x = 20,22,24) ribbons in the annealed state are in the range of 0.93–1.05 T, 0.79–0.82, 375–487 kA/m, and 118–127 kJ/m³, respectively (Fig. 6.5). The good hard magnetic properties result from the exchange magnetic coupling between the nanoscale magnetically hard γ_1 tP4 FePt and magnetically soft γ cF4 Fe(Pt) solid solution as well as Fe_2B phase [62]. Fe-Pt-P rapidly solidified alloys were also found to possess good magnetic properties [63]. Although these are rapidly solidified samples, they can be compacted by SPS or hot pressing. The $(Fe_{0.75}Pt_{0.25})_{75-70}B_{25-30}$ alloys were also found to possess good hard magnetic properties including high intrinsic coercivity values up to 400 kA/m in the nano-crystallized state [64]. They are promising candidates for nanocomposite permanent magnets. The structure of the rapidly solidified $(Fe_{0.75}Pt_{0.25})_{75}B_{25}$ alloy contains a limited volume fraction of the nanoscale cubic cF4 Fe(Pt) solid solution particles of about 4 nm in size embedded in the amorphous matrix. The nanoparticles of cF4 Fe(Pt) phase start growing at the elevated temperature and then undergo a transformation forming of the tP4 FePt compound of about 15 nm in size which is followed by the formation of the tI12 Fe_2B phase from the residual amorphous matrix [65].

6.3   The magnetocaloric effect

There is also an interest in the magnetic cooling. The magnetocaloric effect is related to the change in temperature of a material as a result of the alignment of the magnetic domains on exposure to an external magnetic field. Magnetic refrigeration could be more efficient and environmentally friendly than conventional refrigeration systems. RE-based BMGs, depending on the alloying elements, can be used for magnetic refrigeration at temperatures in the 50–200 K. Typical values for the magnetic entropy change (ΔS^{pk}_M) refrigerant capacity (RC) at the field of 1.5 T are S^{pk}_M~3.9 J/kg·K and RC~150 J/kg, respectively [66]. Magnetization of the Gd-based BMGs exhibits a sharp magnetization change at the ordering temperature, though the transition temperatures for the RE-based BMGs up to 100 K are quite low [67]. At large enough external field Fe-based metallic glasses exhibit ΔS^{pk}_M up to ~2 J/kg·K and RC of ~170 J/kg. Under the field of 1.5 T a $Fe_{80}P_{13}C_7$ bulk glassy sample exhibits ΔS^{pk}_M and RC values of 2.20 J/kg·K and 125 J/kg [68]. Reasonable cost, high mechanical strength, relatively high electrical resistivity, and good corrosion resistance make Fe-based BMGs possible magnetic refrigerants.

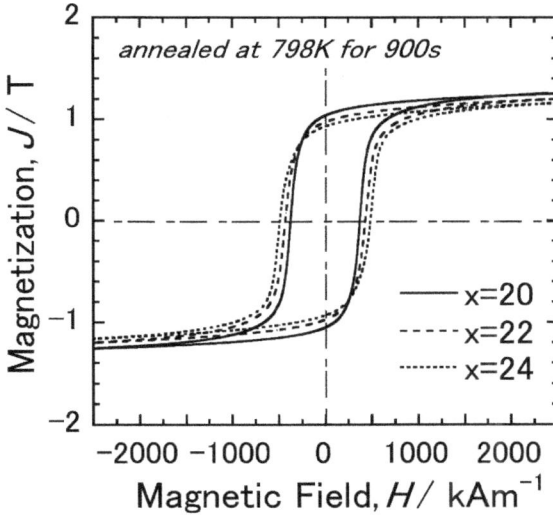

Fig. 6.5. Hysteresis loops of the $Fe_{80-x}Pt_xB_{20}$ (x=20, 22, and 24) metallic glassy samples annealed at 798 K for 900 s. Reproduced from [62] with permission of the American Institute of Physics.

References

[1] A. Makino, A. Inoue, T. Masumoto, Nanocrystalline soft magnetic Fe–M–B (M=Zr, Hf, Nb) alloys produced by crystallization of amorphous phase, Mater. Trans. JIM, 36 (1995) 924. https://doi.org/10.2320/matertrans1989.36.924

[2] K. Amiya, A. Urata, N. Nishiyama, A. Inoue, Magnetic properties of Co–Fe–B–Si–Nb bulk glassy alloy with zero magnetostriction, J. Appl. Phys, 101 (2007) 112. https://doi.org/10.1063/1.2718353

[3] A. Inoue and, B.L. Shen, New Fe-based bulk glassy alloys with high saturated magnetic flux density of 1.4–1.5 T, Mater. Sci. Eng. A, 375–377 (2004) 302. https://doi.org/10.1016/j.msea.2003.10.058

[4] K. Suzuki, A. Makino, The role of boron in nanocrystalline Fe-Zr-B soft magnetic alloys, Materials Science and Engineering: A, 179–180 (1994) 501-505. https://doi.org/10.1016/0921-5093(94)90255-0

[5] A. Inoue, T. Zhang and, T. Itoi, New Fe-Co-Ni-Zr-B amorphous alloys with wide supercooled liquid regions and good soft magnetic properties, Mater. Trans. JIM, 38 (1997) 359. https://doi.org/10.2320/matertrans1989.38.359

[6] G. Xie, H.M. Kimura, D.V. Louzguine-Luzgin, H. Men, A. Inoue, SiC dispersed Fe-based glassy composite cores produced by spark plasma sintering and their high frequency magnetic properties, Intermetallics, 20 (2012) 76. https://doi.org/10.1016/j.intermet.2011.08.023

[7] G. Abrosimova, A. Aronin, D. Matveev, E. Pershina, Nanocrystal formation, structure and magnetic properties of Fe–Si–B amorphous alloy after deformation, Materials Letters, 97 (2013) 15–17. https://doi.org/10.1016/j.matlet.2013.01.092

[8] H. Yoshizawa, K. Yamauchi, T. Yamane and H. Sugihara, Common mode choke cores using the new Fe - based alloys composed of ultrafine grain structure, J. Appl. Phys., 64 (1988) 6047. https://doi.org/10.1063/1.342150

[9] K. Suzuki, N. Kataoka, A. Inoue, A. Makino and T. Masumoto, High saturation magnetization and soft magnetic properties of bcc Fe-Zr-B alloys with ultrafine grain structure, Mater. Trans. JIM, 31 (1990) 743. https://doi.org/10.2320/matertrans1989.31.743

[10] A. Makino, K. Suzuki, A. Inoue and T. Masumoto, Common mode choke cores using the new Fe - based alloys composed of ultrafine grain structure, Mater. Trans. JIM, 32 (1991) 551.

[11] K. Hono, K. Hiraga, Q. Wang, A. Inoue, T. Sakurai, The microstructure evolution of a $Fe_{73.5}Si_{13.5}B_9Nb_3Cu_1$ nanocrystalline soft magnetic material, Acta Metall. et Mater., 40 (1992) 2137. https://doi.org/10.1016/0956-7151(92)90131-W

[12] J. D. Ayers, V. G. Harris, J. A. Sprague, W. T. Elam and H. N. Jones, On the formation of nanocrystals in the soft magnetic alloy $Fe_{73.5}Nb_3Cu_1Si_{13.5}B_9$, Acta Materialia, 46 (1998) 1861. https://doi.org/10.1016/S1359-6454(97)00436-9

[13] M. Ohnuma, K. Hono, H. Onodera, J. S Pedersen and S. Linderoth, Cu clustering stage before the crystallization in Fe–Si–B–Nb–Cu amorphous alloys, Nanostr. Mater., 12 (1999) 693. https://doi.org/10.1016/S0965-9773(99)00219-6

[14] A. R. Yavari and D. Negri, Effect of concentration gradients on nanostructure development during primary crystallization of soft-magnetic iron-based amorphous alloys and its modelling, Nanostruct. Mater., 8 (1997) 969. https://doi.org/10.1016/S0965-9773(98)00047-6

[15] K. Suzuki, N. Ito, J. S. Garitaonandia and J. D. Cashion, High saturation magnetization and soft magnetic properties of nanocrystalline $(Fe,Co)_{90}Zr_7B_3$ alloys annealed under a rotating magnetic field, Journal of Applied Physics, 99 (2006) 08F114. https://doi.org/10.1063/1.2169503

[16] K. Suzuki, A. Makino, A. Inoue and T. Masumoto, Low core losses of nanocrystalline Fe–M–B (M=Zr, Hf, or Nb) alloys, Journal of Applied Physics, 74 (1993) 3316. https://doi.org/10.1063/1.354555

[17] A. Makino, A. Inoue, T. Masumoto, Nanocrystalline soft magnetic Fe–M–B (M=Zr, Hf, Nb) alloys produced by crystallization of amorphous phase, Materials Transactions JIM, 36 (1995) 924-938. https://doi.org/10.2320/matertrans1989.36.924

[18] B. Shen, H. Kimura and A. Inoue, Structure and magnetic properties of $Fe_{42.5}Co_{42.5}Nb_7B_8$ nanocrystalline alloy, Materials Transactions, 43 (2002) 589. https://doi.org/10.2320/matertrans.43.589

[19] A. Inoue, Y. Shinohara, J.S. Gook, Thermal and magnetic properties of bulk Fe-based glassy alloys prepared by copper mold casting, Mater. Trans. JIM, 36 (1995) 1427. https://doi.org/10.2320/matertrans1989.36.1427

[20] A. Inoue, A. Makino, T. Mizushima, Ferromagnetic bulk glassy alloys, Journal of Magnetism and Magnetic Materials, 215-216 (2000) 246. https://doi.org/10.1016/S0304-8853(00)00127-X

[21] A. Inoue, T. Zhang, and A. Takeuchi, Bulk amorphous alloys with high mechanical strength and good soft magnetic properties in Fe–TM–B (TM=IV–VIII group transition metal) system, Appl. Phys. Lett., 71 (1997) 464. https://doi.org/10.1063/1.119580

[22] H. Chiriac and N. Lupu, New FeNbB-based bulk amorphous and nanocomposite soft magnetic alloys, IEEE Transactions on Magnetics, 41 (2005) 3289-3291. https://doi.org/10.1109/TMAG.2005.854721

[23] W. Zhang, C.F. Fang, Y.H. Li, Ferromagnetic Fe-based bulk metallic glasses with high thermoplastic formability, Scripta Materialia, 69 (2013) 77–80. https://doi.org/10.1016/j.scriptamat.2013.03.003

[24] T. Mizushima, K. Ikarashi, M. Yoshida, A. Makino and A. Inoue, Soft magnetic properties of ring shape bulk glassy Fe-Al-Ga-P-C-B-Si alloy prepared by copper mold casting, Mater. Trans. JIM, 40 (1999) 1019.
https://doi.org/10.2320/matertrans1989.40.1019

[25] T. Bitoh, A. Makino and, A. Inoue, Magnetization process and coercivity of Fe-(Al, Ga)-(P, C, B, Si) soft magnetic glassy alloys, Materials Transactions, 45 (2004) 1219.
https://doi.org/10.2320/matertrans.45.1219

[26] Q. K. Man, H.J. Sun, Y.Q. Dong, B.L. Shen, H. Kimura, A. Makino, A. Inoue, Enhancement of glass-forming ability of CoFeBSiNb bulk glassy alloys with excellent soft-magnetic properties and superhigh strength, Intermetallics, 18 (2010) 1876–1879.
https://doi.org/10.1016/j.intermet.2010.02.047

[27] A. Inoue, T. Zhang and A.Takeuchi, Bulk amorphous alloys with high mechanical strength and good soft magnetic properties in Fe–TM–B (TM=IV–VIII group transition metal) system, Appl. Phys. Lett., 71 (1997) 464. https://doi.org/10.1063/1.119580

[28] A. Makino, T. Kubota, C. Chang, M. Makabe and, A. Inoue, Fe-metalloids bulk glassy alloys with high Fe content and high glass-forming ability, J. Mater. Res., 23 (2008) 1339.
https://doi.org/10.1557/JMR.2008.0180

[29] A. Makino, X. Li, K. Yubuta, C. Chang, T. Kubota, A. Inoue, The effect of Cu on the plasticity of Fe–Si–B–P-based bulk metallic glass, Scripta Materialia, 60 (2009) 277–280.
https://doi.org/10.1016/j.scriptamat.2008.09.008

[30] J. F. Wang, R. Li, N.B. Hua, L. Huang, T. Zhang, Ternary Fe–P–C bulk metallic glass with good soft-magnetic and mechanical properties, Scripta Materialia, 65 (2011) 536–539. https://doi.org/10.1016/j.scriptamat.2011.06.020

[31] W. Yang, H. Liu, L. Xue, J. Li, C. Dun, J. Zhang, Y. Zhao, B. Shen, Magnetic properties of $(Fe_{1-x}Ni_x)_{72}B_{20}Si_4Nb_4$ (x=0.0–0.5) bulk metallic glasses, Journal of Magnetism and Magnetic Materials, 35 (2013) 172–176.
https://doi.org/10.1016/j.jmmm.2013.02.004

[32] X. Li, Y. Zhang, H. Kato, A. Makino, A. Inoue, The effect of Co addition on glassy forming ability and soft magnetic properties of Fe-Si-B-P bulk metallic glass, Key Engineering Materials, 508 (2012) 112-116.
https://doi.org/10.4028/www.scientific.net/KEM.508.112

[33] H. Y. Jung, S. Yi, Nanocrystallization and soft magnetic properties of $Fe_{23}M_6$ (M: C or B) phase in Fe-based bulk metallic glass, Intermetallics, 49 (2014) 18–22. https://doi.org/10.1016/j.intermet.2014.01.005

[34] H. Y. Jung, M. Stoica, S. Yi, D.H. Kim, J. Eckert, Electrical and magnetic properties of Fe-based bulk metallic glass with minor Co and Ni addition, Journal of Magnetism and Magnetic Materials, 364 (2014) 80–84. https://doi.org/10.1016/j.jmmm.2014.04.028

[35] S. Lee, H. Kato, T. Kubota, K. Yubuta, A. Makino and A. Inoue, Excellent thermal stability and bulk glass forming ability of Fe-B-Nb-Y soft magnetic metallic glass, Materials Transactions, 49 (2008) 506. https://doi.org/10.2320/matertrans.MBW200732

[36] J.H. Zhang, C.T. Chang, A. Wang, B.L. Shen, Development of quaternary Fe-based bulk metallic glasses with high saturation magnetization above 1.6 T, Journal of Non-Crystalline Solids, 358 (2012) 1443–1446. https://doi.org/10.1016/j.jnoncrysol.2012.03.023

[37] H. Matsumoto, A. Urata, Y. Yamada and A. Inoue, Novel $Fe_{(97-x-y)}P_xB_yNb_2Cr_1$ glassy alloys with high magnetization and low loss characteristics for inductor core materials, IEEE Trans. Magn., 46 (2010) 373. https://doi.org/10.1109/TMAG.2009.2033708

[38] L. Bie, Q. Li, D. Cao, H.X. Li, J. Zhang, C.T. Chang, Y.F. Sun, Preparation and properties of quaternary CoMoPB bulk metallic glasses, Intermetallics, 71 (2016) 7-11. https://doi.org/10.1016/j.intermet.2015.12.008

[39] H.J. Sun, Q.K. Man, Y.Q. Dong, B.L. Shen, H. Kimura, A. Makino, A. Inoue, Effect of Nb addition on the glass-forming ability, mechanical and soft-magnetic properties in $(Co_{0.942}Fe_{0.058})_{72-x}Nb_xB_{22.4}Si_{5.6}$ bulk glassy alloys, J. Alloys Comp., 504 (2010) S31–S33. https://doi.org/10.1016/j.jallcom.2010.03.044

[40] Q.K. Man, A. Inoue, Y.Q. Dong, J. Qiang, C.L. Zhao, B.L. Shen, A new CoFe-based bulk metallic glasses with high thermoplastic forming ability, Scr. Mater., 69 (2013) 553–556. https://doi.org/10.1016/j.scriptamat.2013.07.002

[41] W. Zhang, C.F. Fang, Y.H. Li, Ferromagnetic Fe-based bulk metallic glasses with high thermoplastic formability, Scripta Materialia, 69 (2013) 77–80. https://doi.org/10.1016/j.scriptamat.2013.03.003

[42] D.S. Song, J.H. Kim, E. Fleury, W.T. Kim, D.H. Kim, Synthesis of ferromagnetic Fe-based bulk glassy alloys in the Fe–Nb–B–Y system, J. Alloys Compd., 389 (2005) 159. https://doi.org/10.1016/j.jallcom.2004.08.014

[43] A. Inoue and B.L. Shen, Soft magnetic bulk glassy Fe–B–Si–Nb alloys with high saturation magnetization above 1.5 T, Materials Transactions, 43 (2002) 766. https://doi.org/10.2320/matertrans.43.766

[44] A. Inoue, B. Shen, H. Koshiba, H. Kato and. A. R. Yavari, Cobalt-based bulk glassy alloy with ultrahigh strength and soft magnetic properties, Nature Materials, 2 (2003) 661 - 663. https://doi.org/10.1038/nmat982

[45] A. Zhukov, A. Talaat, M. Churyukanova, S. Kaloshkin, V. Semenkova, M. Ipatov, J.M. Blanco, V. Zhukova, Engineering of magnetic properties and GMI effect in Co-rich amorphous microwires, Journal of Alloys and Compounds, 664 (2016) 235-241. https://doi.org/10.1016/j.jallcom.2015.12.224

[46] K. S. Nakayama, Controlled formation and mechanical characterization of metallic glassy nanowires, Adv. Mater., 22 (2010) 872–875. https://doi.org/10.1002/adma.200902295

[47] S.I. Iwasaki, Perpendicular magnetic recording—Its development and realization, Proc. Jp. Acad. Ser. B Phys. Biol. Sci., 85 (2009) 37–54. https://doi.org/10.2183/pjab.85.37

[48] K. Nielsch, R. B. Wehrspohn, S. F. Fischer, H. Kronmiller, J. Kirsehner and U. Gosele, Magnetic properties of 100 nm-period nickel nanowire arrays obtained from ordered porous-alumina templates, Mater. Res. Soc. Symp. Proc., 9 (2001) 636. https://doi.org/10.1557/PROC-636-D1.9.1

[49] V. Zhukova, M. Ipatov, P. Corte-Leon, J.M. Blanco, E. Zanaeva, A.I. Bazlov, J. Jiang, D.V. Louzguine-Luzgin, J. Olivera, A. Zhukov, Excellent magnetic properties of $(Fe_{0.7}Co_{0.3})_{83.7}Si_4B_8P_{3.6}Cu_{0.7}$ ribbons and microwires, Intermetallics, 117 (2020) 106660. https://doi.org/10.1016/j.intermet.2019.106660

[50] M. Kuhnt, X. Xu, M. Amalraj, P. Kozikowski, K.G. Pradeep, T. Ohkubo, M. Marsilius, Th. Strache, Ch. Polak, M. Ohnuma , K. Hono, G. Herzer, The effect of Co addition on magnetic and structural properties of nanocrystalline (Fe,Co)-Si-B-P-Cu alloys, J. Alloys Compd., 766 (2018) 686-693. https://doi.org/10.1016/j.jallcom.2018.07.013

[51] E.N. Zanaeva, A.I. Bazlov, D.A. Milkova, A.Yu. Churyumov, A. Inoue, N.Yu. Tabachkova, F. Wang, F.L. Kong, S.L. Zhu, High-Frequency soft magnetic properties of Fe-Si-B-P-Mo-Cu amorphous and nanocrystalline alloys, Journal of Non-Crystalline Solids, 526 (2019) 119702. https://doi.org/10.1016/j.jnoncrysol.2019.119702

[52] A. P. Zhukov, B. L. Shtangeev, Cooling-induced phase transition in amorphous CoCrZr alloy, J. Appl. Phys., 73 (1993) 5716. https://doi.org/10.1063/1.353601

[53] A. Inoue, T. Zhang, W. Zhang and A. Takeuchi, Bulk Nd-Fe-Al amorphous alloys with hard magnetic properties, Materials Transactions JIM, 37 (1996) 99. https://doi.org/10.2320/matertrans1989.37.99

[54] A. Inoue and T. Zhang, Thermal stability and glass-forming ability of amorphous Nd-Al-TM (TM = Fe, Co, Ni or Cu) alloys, Mater. Sci. Eng. A, 226–228 (1997) 393. https://doi.org/10.1016/S0921-5093(97)80050-0

[55] R. Coehoorn, D. B. Mooij, J. P. W. B. Duchateau and K. J. H. Buschow, Novel permanent magnetic materials made by rapid quenching, J. de Phys. C, 8 (1988) 669. https://doi.org/10.1051/jphyscol:19888304

[56] Q. Y. Wu, D. H. Ping, B. S. Murty, H. Kanekiyo, S. Hirosawa and K. Hono, Influence of heating rate on the microstructure and magnetic properties of $Fe_3B/Nd_2Fe_{14}B$ nanocomposite magnets, Scripta Materialia, 45 (2001) 355. https://doi.org/10.1016/S1359-6462(01)01042-9

[57] S. Hirosawa and H. Kanekiyo, Nanostructure and magnetic properties of chromium-doped Fe_3B-$Nd_2Fe_{14}B$ exchange-coupled permanent magnets, Mater. Sci. Eng. A, 217/218 (1996) 367. https://doi.org/10.1016/S0921-5093(96)10284-7

[58] A. Inoue, A. Takeuchi, A. Makino and T. Masumoto, Hard magnetic properties of Fe-Nd-B alloys containing intergranular amorphous phase, IEEE Trans. Magn., 31 (1995) 3626. https://doi.org/10.1109/20.489590

[59] A. Manaf, R.A. Buckley and H. A. Davies, New nanocrystalline high-remanence Nd-Fe-B alloys by rapid solidification, J. Magn. Magn. Mater., 128 (1993) 302. https://doi.org/10.1016/0304-8853(93)90475-H

[60] A. Inoue, A. Takeuchi, A. Makino and T. Masumoto, Hard magnetic properties of nanocrystalline Fe-rich Fe-Nd-B alloys prepared by partial crystallization of amorphous phase, Mater. Trans. JIM, 36 (1995) 962. https://doi.org/10.2320/matertrans1989.36.962

[61] K. Inomata, T. Sawa, S. Hashimoto, Effect of large boron additions to magnetically hard Fe - Pt alloys, J. Appl. Phys. 64 (1988) 2537. https://doi.org/10.1063/1.341638

[62] W. Zhang, D. V. Louzguine, and A. Inoue, Synthesis and magnetic properties of Fe-Pt-B nanocomposite permanent magnets with low Pt concentrations, Applied Physics Letters, 85, (2004) 4998-5000. https://doi.org/10.1063/1.1824172

[63] A. A. Kündig, N. Abe, M. Ohnuma, T. Ohkubo, H. Mamiya, K. Hono, Rapidly solidified $(FePt)_{70}P_{30}$ alloy with high coercivity, Appl. Phys. Lett. 85 (2004) 789. https://doi.org/10.1063/1.1776333

[64] A. Inoue, W. Zhang, T. Tsurui and D. V. Louzguine, Formation, crystallized structure and magnetic properties of Fe-Pt-B amorphous alloys, Materials Transactions, 46 (2005) 891-894. https://doi.org/10.2320/matertrans.46.891

[65] D. V. Louzguine-Luzgin, W. Zhang and A. Inoue, Nanoscale precipitates and phase transformations in a rapidly-solidified Fe-Pt-B amorphous alloy, Journal of Alloys and Compounds, 402 (2005) 78-83. https://doi.org/10.1016/j.jallcom.2005.03.089

[66] W.H. Wang, Bulk metallic glasses with functional physical properties, Adv. Mater. 21 (2009) 4524–44. https://doi.org/10.1002/adma.200901053

[67] Q. Luo, D. Q. Zhao, M. X. Pan, W. H. Wang, Magnetocaloric effect in Gd-based bulk metallic glasses, Appl. Phys. Lett., 90 (2007) 211903. https://doi.org/10.1063/1.2741120

[68] W. Yang, J. Huo, H. Liu, J. Li, L. Song, Q. Li, L. Xue, B. Shen, A. Inoue, Extraordinary magnetocaloric effect of Fe-based bulk glassy rods by combining fluxing treatment and J-quenching technique, J. Alloy. Compd., 684 (2016) 29-33. https://doi.org/10.1016/j.jallcom.2016.05.142

CHAPTER 7

Applications Related to Structural, Functional, Magnetic, Chemical and Biological Properties

Bulk metallic glasses are suitable for different applications owing to their excellent structural, functional, magnetic, chemical and biological properties. This chapter is devoted to this subject.

Contents

7.1   Structural and functional materials

Bulk metallic glasses have important applications owing to high glass-forming ability, good casting ability, good formability in the supercooled liquid region as well as good mechanical and chemical properties. These materials are used in sportive goods [1,2], watches, electromagnetic wave shields, optical devices, power inductors, mini transformers, micro-geared motor parts, pressure sensors [3], Coriolis flow meters, coating materials, for medical instrument, etc… Metallic glasses are also useful as a strong light cord, strong light plate loaded in compression, elastic hinge, springs, diaphragms, knife blades, thermal shock resistant materials [4]. Various types of connection adapters have been tested for applications to advanced medical equipments, including endoscope and micro-pump, precision optics, and micro-machines.

Owing to large elastic limit of about 2 % the Zr–Cu–Al–Ni glassy alloy diaphragms produced by net-shape casting have been applied to pressure sensors with unique features such as high sensitivity, high-pressure endurance, and small size which cannot be achieved for conventional stainless steel diaphragms.

Even alloys with low glass-forming ability can be utilized when produced by using special techniques. A plate of the $Al_{86}Si_{0.5}Ni_{4.06}Co_{2.94}Y_6La_{0.5}$ amorphous matrix nanocomposite with a maximum thickness of 12 mm and high microhardness of 420 HV was prepared by spray-forming [5]. Nano-scale FCC-Al particles are homogeneously dispersed in the amorphous matrix.

Also, recent finding indicates that the long standing problem of the annealing-induced embrittlement in metallic glasses was solved using crystalline W coatings. The excess volume could be preserved in a $Zr_{50.7}Cu_{28}Ni_9Al_{12.3}$ bulk metallic glass upon annealing as it is not able to sink from BMG into crystalline W. Thus, the strength and ductility of the BMG are preserved [6].

Successful applications of metallic glasses in micromechanical devices was demonstrated by Saotome et al. [7,8]. The deformation characteristics of the $Zr_{65}Al_{7.5}Cu_{27.5}$ $Zr_{55}Al_{10}Cu_{30}Ni_5$ and $Pt_{48.75}Pd_{9.75}Cu_{19.5}P_{22}$ metallic glasses were obtained. The alloys exhibited the Newtonian viscous flow in the supercooled liquid state under low stresses and a superior micro-formability was achieved at the microscopic scale.

Microgeared motor parts were made of the Zr–Cu–Al–Ni and Ni-Zr-Ti-Fe-Co-Cu glassy alloys by a net-shape casting technique (Fig. 7.1). By using these parts, a small geared motor with a diameter of 1.5 mm and a length of 9.9 mm was produced and applied. A three-stage micro-geared motor made of the Zr-based glassy alloy gears has high torque which is about

20 times higher than that for a conventional vibration motor with a diameter of 4 mm used in the ordinary cell phones. In addition, the number of rotations to failure is about 34 times larger for the Zr-based glassy alloy gear and 313 times larger for the Ni-based glassy alloy gear than that for conventional tool steel gears [9]. Later, the diameter and length of the micro-geared motor have been reduced to 0.9 mm and 2.0 mm, respectively.

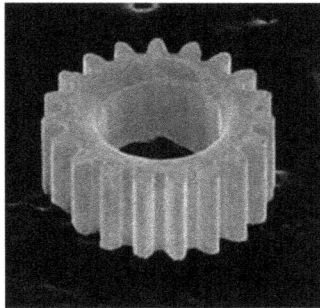

Fig. 7.1. A part of a microgeared motor.

When the size of the mechanical components is reduced to the micrometer and sub-micrometer level, the native surface oxide layer begins to play an important role in the contact mechanical behavior of metallic glasses. The nanoscale tribological properties of the $Ni_{62}Nb_{38}$ [10,11] and Pt-Cu-Ni-P [12] metallic glasses are shown in Fig. 7.2. These materials were found to be suitable as wear resistant materials for micro-electro-mechanical applications. Other results on the subject were shown in Chapter 5.

Metallic glasses have high elastic strain limit and fracture toughness resulting from the amorphous structure without structural defects like dislocations and grain boundaries. These attractive functional properties are required for the micro-electro-mechanical systems. A Fe-base metallic glass with good soft magnetic properties is a promising material for mirror actuation [13]. The scanning mirror was fabricated using a Fe-based metallic glass (Fig. 7.3) in order to achieve high performance and mechanical rigidity. The optical tilting angle exceeding ±110 degrees was obtained by driving with Lorentz force and magnetostatic force at 100 mA. The mirror and the gimbal were actuated at different frequency in two axes (Fig. 7.4) and dual axial scanning was possible. The large scanning angle was possible due to good magnetic properties, large elastic strain limit and fracture toughness of the Fe-based metallic glass.

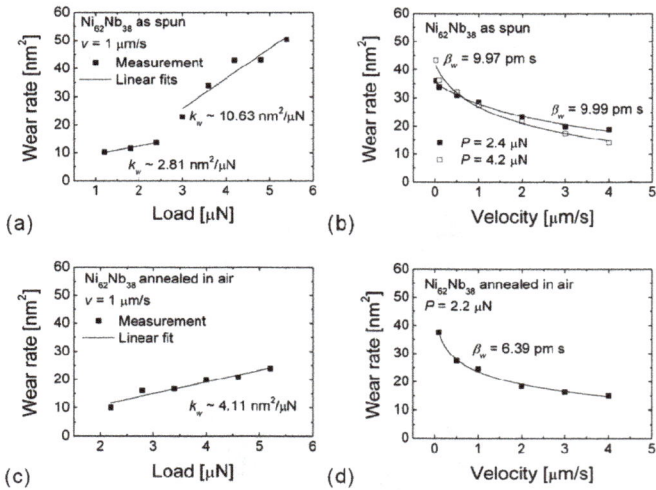

Fig. 7.2. Wear rate as a function of the loads [(a) and (c)] at constant velocity and as a function of the velocity [(b) and (d)] at constant load for the as spun [(a) and (b)] and annealed [(c) and (d)] $Ni_{62}Nb_{38}$ metallic glass samples. Reproduced from [11] with permission of the American Institute of Physics.

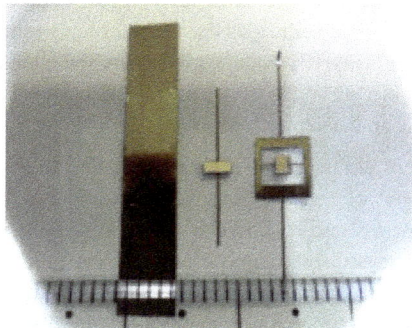

Fig. 7.3. SEM and optical microscope image for the micro-mirror structure. Reproduced from [13] with permission of the Japan Society of Applied Physics.

Metallic Glasses and Their Composites – 2nd Edition   Materials Research Forum LLC
Materials Research Foundations **85** (2021)   https://doi.org/10.21741/9781644901014

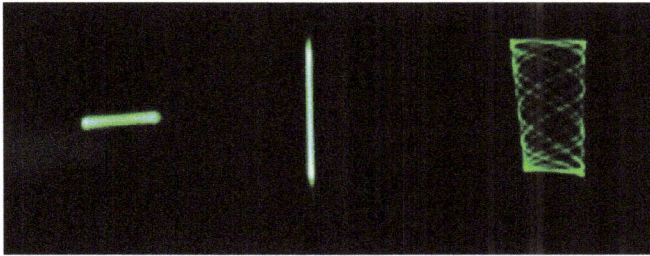

Fig. 7.4. Scanning patterns of the dual axial scanner. (a) Horizontal scanning pattern by mirror movement at 310 Hz. (b) Vertical scanning pattern by movement at 370 Hz. (c) dual axial scanning pattern by the mirror with dual AC at 300 and 360 Hz. Reproduced from [13] with permission of the Japan Society of Applied Physics.

Ti-based glassy alloy pipes of 2 mm in outer diameter, 0.2 mm in thickness, and 300 mm in length produced by net-shape casting technique exhibit high tensile strength and a large elastic strain of 0.02 and high corrosion resistance which are suitable for application as sensing element in a Coriolis flow meter to measure the mass flow of liquid or gas inside the glassy alloy pipe subjected to reinforced oscillation (Fig. 7.5). The sensitivity of the Coriolis flow meter using the Ti-based glassy alloy pipe was reported to be 28–53 times higher than that for conventional SUS316 pipe [14]. Such a Coriolis flow meter can be used in various industrial fields such as chemical, environmental, semiconductor, and medical science fields.

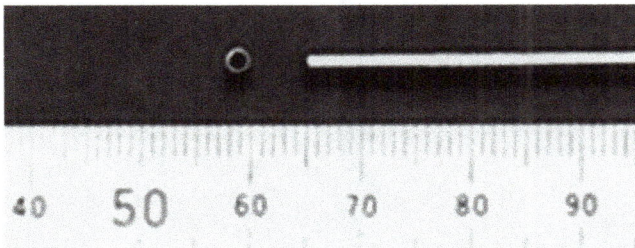

Fig. 7.5. A Ti-based glassy alloy pipe for the Coriolis flow meter.

Glassy alloys exhibit high hydrogen solubility and significant resistance to embrittlement. Such materials were found to be promising for future applications in separators for fuel cells owing to hydrogen permeation characteristics of Zr-Hf-Ni, Ni-Nb-Zr [15] and other glassy membranes. Hydrogen flow through $(Ni_{0.6}Nb_{0.4})_{100-x}Zr_x$ (x=0 to 40 at.%) metallic glassy membranes increased with temperature and the square-root of hydrogen pressures across the membrane. The hydrogen flow was proportional to the square-root of hydrogen pressure difference. The diffusion of hydrogen through the membrane was found to be a rate-controlling factor for hydrogen permeation. The permeability increased with increasing Zr content. The maximum hydrogen permeability for the $(Ni_{0.6}Nb_{0.4})_{70}Zr_{30}$ amorphous alloy was 1.3×10^{-8} (mol/m·s·Pa$^{1/2}$) at 673 K, higher than that of pure Pd. These permeation characteristics indicate the possibility of future practical use of the melt-spun amorphous alloys as a hydrogen permeable membrane.

Flexible, long metallic glassy nanowires were produced by the drawing process based on deformation above the glass transition temperature [16]. The wire with high elasticity was shown to oscillate forming a sine-wave pattern. Metallic glassy nano-wires may have applications ranging from interconnects to sensors.

AMO-beads powders of the Fe–Ni–Cr–Mo–B–Si alloy were produced by water atomization [17] as shot peening balls and fine precise polishing medium. These glassy balls have much longer endurance times compared to the cast steel shots and high speed steel shots, in addition to the important advantage of extremely high resistance to powder explosion reaction in air. Good mechanical properties, such as high Vickers Hardness of HV900, high fracture strength of 3000 MPa and large elastic strain of 0.02 together with high corrosion resistance and a smooth outer surface make AMO-beads a good tool for the shot peening treatment. The AMO-beads are used to make the shot peening balls to high class of automobile high-alloy steel gears, for saving of steel gear weight by about 45 % by improving its fatigue strength by 50-80 % increasing the residual stress on the surface to above 2000 MPa.

The Fe–Cr–Mo–C–B system glassy alloy-coated layer exhibits a better corrosion resistance than SUS304 plates, higher Vickers hardness than hard chromium plates, and better wear resistance than SKD tool steels and FC cast iron. The glassy alloy-coating technique has been applied to surface glassy alloy layers inside solder-melting vessels in continuous solder casting machines. The Fe-Cr-W-C-B glassy alloy coatings are used as heat-resistant materials.

7.2   Magnetic applications

7.2.1   Soft magnetic applications

Fe-based BMGs exhibiting high GFA, low coercive force (H_c), high permeability (μ), high electrical resistivity, low core losses, high mechanical strength, large elastic strain and large supercooled liquid region leading to viscous flow deformation, and the saturation

Metallic Glasses and Their Composites – 2nd Edition   Materials Research Forum LLC
Materials Research Foundations 85 (2021)   https://doi.org/10.21741/9781644901014

magnetization M_s approaching 1.5 T are utilized as magnetic materials [18]. Fe-based BMGs were commercialized under the trademarks of "Liqualloy", "SENNTIX", "GALOA" [19].

Liqualloy powder produced by water atomization can be deformed into flaky shape with a thickness of 2–3 μm to be mixed with a resin. The Liqualloy Fe–Cr–P–C–B–Si system magnetic cores produced by cold consolidation processes of the atomized glassy alloy powders with resin exhibit nearly constant relative permeability in a wide frequency range up to several megahertz, good linear relation between permeability and DC bias field. Low core loss conversion inductors are used in the transformers of notebook PCs. They are also used as power sensing devices with excellent temperature stability. The parts made of Liqualloy are also used as power inductors in CPU and graphics power supply circuits for notebook PC converters. The Liqualloy sheet between metal device part and loop antenna is able to increase transmission distance of magnetic flux line to the inside of loop antenna. A significant increase in the antenna sensitivity was obtained. The Liqualloy cores have better soft magnetic properties than those for the other soft magnetic cores, especially at high frequency. The resulting Liqualloy powder cores have been used as a power inductor in laptop-type personal computers because of higher efficiency and much smaller heat generation.

The Fe–Nb–B–Si–P systems soft magnetic powder cores named SENNTIX were formed and exhibited low core losses [20]. A soft magnetic SENNTIX II alloy of the Fe–P–B–Nb–Cr system shows the high GFA and low H_c [21]. The use of SENNTIX-II powder cores in the Fe–Nb–Cr–B–P and Fe-Nb-Cr-B-P-Si systems reduced by more than 50% core loss compared to the existing metal powder and reduce thermal energy loss in personal computers, resulting in the extension of battery life time of notebook personal computers. Fe-based metallic glasses are also used as magnetoelastic sensors.

The high thermal stability of the structure and the high Curie temperature are required for high temperature application. A Co-based BMG can be applied as a material for the position sensor, antenna for radio-controlled watch, solenoid valve and magnetic sensor. Co–based BMGs exhibit sharp signal, high output voltage and high mechanical strength. However, Co-based BMGs have higher materials cost compared to Fe-based glasses.

A room temperature magnetic semiconductor was recently produced from a ferromagnetic metallic glass [22]. Introduction of oxygen into a ferromagnetic metallic glass allowed producing a room temperature p-type $Co_{28.6}Fe_{12.4}Ta_{4.3}B_{8.7}O_{46}$ magnetic semiconductor with a Curie temperature above 600 K with a unique combination of optical, electrical and ferromagnetic properties. p-n heterojunctions and electric field control of the room-temperature ferromagnetism obtained indicate its potential for various electronic devices applications (Fig. 7.6).

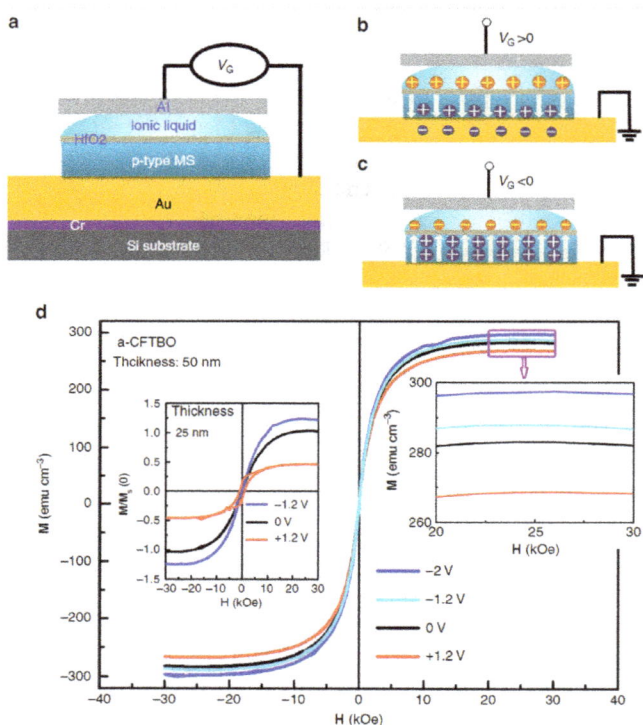

Fig. 7.6. Electric field control of ferromagnetism in the $Co_{28.6}Fe_{12.4}Ta_{4.3}B_{8.7}O_{46}$ magnetic semiconductor. (a) A schematic view of the experimental set-up for applying gate voltage on the thin film through a drop of ionic liquid. The thickness of an insulating HfO_2 layer is about 2 nm. (b) Illustration of electrically induced decrease of carrier concentration at a positive gate voltage. (c) Illustration of electrically induced increase of carrier concentration at a negative gate voltage. (d) Variation of M–H curves with different gate voltages measured at 300 K. The thickness of the a-CFTBO MS is 50 nm. Left inset shows the plot of the normalized $M/M_s(0)$ versus magnetic field measured at different gate voltages for the $Co_{28.6}Fe_{12.4}Ta_{4.3}B_{8.7}O_{46}$ magnetic semiconductor with a thickness of 25 nm, $M_s(0)$ denotes the saturation magnetization without any gate voltage. Right inset is an enlargement of the part indicated by a rectangle. Reproduced from [22] with permission of Nature Publishing Group.

7.2.2 Hard magnetic applications

Hard magnetic alloys with large coercive force exceeding 10 kA/m can be used as permanent magnets. High magnetic induction is retained because of a strong resistance to demagnetization, for example, as a result of high anisotropy. Permanent magnets must have large coercive force, high saturation magnetization and the energy product BH_{max}. Ferromagnetic $Nd_{90-x}Fe_xAl_{10}$ bulk amorphous alloys exhibit high coercive force (of about 280 kA/m) [23]. The bulk amorphous $Nd_{70}Fe_{20}Al_{10}$ alloy shows ferromagnetism with the Curie temperature (T_c) of about 600 K. The remanence (B_r) and coercive force for the bulk $Nd_{60}Fe_{30}Al_{10}$ alloy are 0.122 T and 277 kA/m, respectively. The hard magnetic properties for the bulk amorphous alloys are presumably due to the development of homogeneous ferromagnetic clusters with large random magnetic anisotropy.

7.3 Corrosion resistant alloys

High corrosion resistance of some BMGs for example, Zr-based BMGs [24] and Zr-based BMGs containing Nb [25,26], makes them useful for chemical vessels and other applications. The $(Cu_{0.6}Hf_{0.25}Ti_{0.15})_{90}Nb_{10}$ bulk metallic glass composite prepared by copper mold casting containing ductile dendritic Nb-rich crystalline phase dispersed in the glassy matrix also exhibited much higher corrosion resistance in 1N HCl and 3 mass% NaCl solutions than the Nb-free alloy. Nb forms a highly protective surface film with a higher chemical stability in HCl and NaCl solutions [27]. The compressive fracture strength and compressive plastic strain of the $(Cu_{0.6}Hf_{0.25}Ti_{0.15})_{90}Nb_{10}$ BMG composite are 2625 MPa and 12.5%, respectively. The ductile dendrites block the shear band propagation and promote generation of multiple shear bands.

The $(Ti_{0.45}Zr_{0.1}Pd_{0.1}Cu_{0.31}Sn_{0.4})_{100-x}M_x$ (M: Ta and Nb) BMG alloys with a large supercooled liquid region (ΔT_x) of 53~67K do not contain toxic elements like Ni, Al and Be. These BMG alloys have good mechanical properties, remarkable GFA and large supercooled liquid region. The $Ti_{44.1}Zr_{9.8}Pd_{9.8}Cu_{30.38}Sn_{3.92}Nb_2$ bulk glassy alloy has a high potentiality to be applied in as dental biomaterial. Its compressive strength of 1990 MPa exceeds those values of crystalline Ti and conventional Ti6Al4V alloy. The $Ti_{44.1}Zr_{9.8}Pd_{9.8}Cu_{30.38}Sn_{3.92}Nb_2$ BMG alloy shows high corrosion resistance (Fig. 7.7) with good passivity in a wide range and demonstrates lower passive current densities of approximately $10^{-2}A/m^2$ in 1 mass% lactic acid, $10^{-3}A/m^2$ in PBS (phosphate-buffered saline without calcium and magnesium salts solution containing 8 g/l NaCl, 0.2 g/l KCl, 1.15 g/l Na_2HPO_4, 0.2 g/l KH_2PO_4) and and $10^{-2}A/m^2$ in HBSS (Hank's balance without Ca and Mg or phenol red salts solution containing 8 g/l NaCl, 0.4 g/l KCl, 0.09 g/l $Na_2HPO_4 \cdot 7H2O$, 0.06 g/l KH_2PO_4, 0.35 g/l $NaHCO_3$, 1.0 g/l glucose) aqueous solutions open to air at 310 K compared to those of Ti and the Ti-6Al-4V alloy [28].

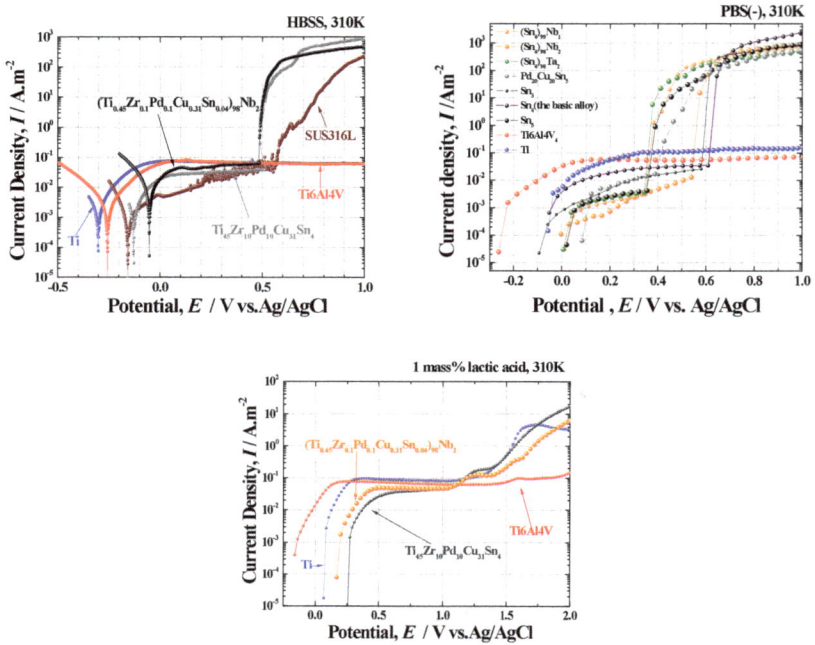

Fig. 7.7. Anodic polarization curves of the $(Ti_{0.45}Zr_{0.1}Pd_{0.1}Cu_{0.31}Sn_{0.4})_{100-x}M_x(at.\%)$ alloys in 1 mass% lactic acid, PBS(-) and HBSS at 310K, as indicated. All samples were rods of 3 mm in diameter (the exposed area was 1 cm². Reproduced from Ref. [28] with permission of Elsevier.

The tribocorrosion degradation mechanisms of the $Zr_{55}Cu_{30}Ni_5Al_{10}$ and $Zr_{65}Cu_{18}Ni_7Al_{10}$ alloys applicable as potential load-bearing implant materials were studied [29]. The $Zr_{65}Cu_{18}Ni_7Al_{10}$ alloy exhibited higher plasticity but relatively reduced wear resistance during the ball-on-disc tests owing to its lower hardness of HV=5.7±1.8 GPa compared to that (HV=6.8±1.6 GPa) of the $Zr_{55}Cu_{30}Ni_5Al_{10}$ alloy. Vickers hardness in the worn of the wear tracks in dry condition was higher than in the unworn areas indicating possible formation of the intermetallic compounds. The spontaneous passivation of these BMGs in Phosphate Buffer Saline solution was mainly attributed to the formation of a passive film. The galvanic coupling was established between the depassivated wear track and the surrounding passive area as a degradation mechanism for the $Zr_{65}Cu_{18}Ni_7Al_{10}$ BMG in the tribological corrosion environment. The $Zr_{55}Cu_{30}Ni_5Al_{10}$ alloy showed a higher stability against tribological corrosion [29].

7.4 Oxidation and surface oxides for electronic devices

Oxidation is usually not desired as it reduces the surface quality and may induce surface crystallization of the glassy matrix. Fine-grained microstructure consisting of ZrO_2 and Cu particles was found on the surface of $Zr_{70}Cu_{30}$ ribbons stored for a few years at room temperature owing to surface oxidation [30]. An amorphous oxide growing on the metal substrate can be stable up to a certain critical thickness [31]. Usually surface oxides on BMGs are a few nanometers thick [32]. A native oxide on the $Cu_{47}Zr_{45}Al_8$ surface is a mixture of the amorphous ZrO_2 and Al_2O_3 with nanoscale inclusions of the Cu_2O crystalline phase [33]. An amorphous Ti(Zr) oxide with Ni rich layer beneath the oxide layer was formed first upon oxidation of a Ti-Zr-Ni-Cu metallic glass [34]. At a higher temperature crystallization of the Ni enriched layer forming Ni_3Ti structure took place, followed by oxide growth with a lamellar structure consisting of Ti(Zr) amorphous oxide and Ni_3Ti.

The local current-voltage characteristics (CVCs) of the Ni-Nb/NiNb oxide/Pt system, carried out in contact mode of an atomic force microscope (AFM) showed n-type conductivity in the case of native oxides, while with the artificial one it exhibited p-type conductivity. X-ray photoelectron spectroscopy (XPS) measurements revealed that the structure of natural Ni-Nb oxide film consists of $Ni-NbO_x$ top layer and nickel enriched bottom layer which provides n-type conductivity. In the artificial oxide film Nb is oxidized completely to Nb_2O_5, Ni atoms migrate into bulk Ni-Nb matrix and the electron depletion layer is formed at the Ni-Nb / Nb_2O_5 interface providing p-type conductivity (Fig. 7.8) [35]. An electronic conducting device can be made using this principle.

7.5 Materials for cleaning water pollutions

The outstanding efficiency of the $Fe_{73}Si_7B_{17}Nb_3$ metallic glass powders in degrading organic water contaminants was reported. The powders completely decompose the $C_{32}H_{20}N_6Na_4O_{14}S_4$ azo dye in an aqueous solution in a short time, about 200 times faster than the conventional Fe powders [36]. MgZn-based metallic glass powders also have excellent functional ability in degrading azo dyes which are typical organic water pollutants [37]. Their azo dye degradation efficiency is about 1000 times higher than that of commercial crystalline Fe powders, and 20 times higher than the Mg-Zn alloy crystalline counterparts. The high Zn content in the amorphous Mg-based alloy enables a greater corrosion resistance in water and higher reaction efficiency with azo dye compared to crystalline Mg. Even under complex environmental conditions, the MgZn-based metallic glass powders retain high reaction efficiency (Fig. 7.9) [37].

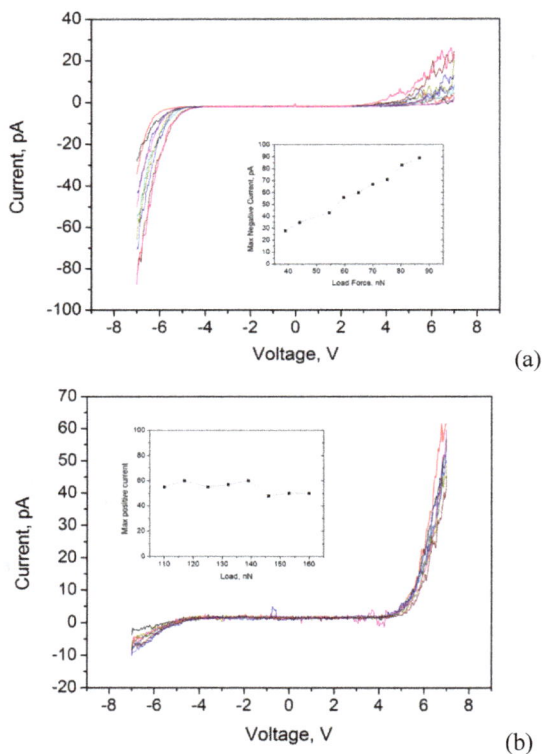

(a)

(b)

Fig. 7.8. A family of CVCs recorded for different cantilever loads for native (a) and artificial (b) Ni-Nb oxide films. Load ranges were from 39 to 87 nN for (a) and from 110 to 160 nN for (b). Insets show dependence of maximum current in conductive CVC region on applied cantilever load. Reproduced from Ref. [35] with permission of the American Institute of Physics.

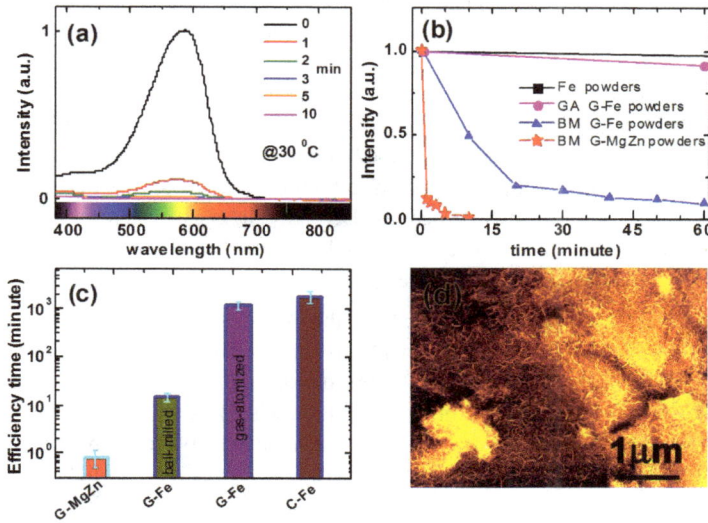

Fig. 7.9. (a) The UV absorption spectrum of the azo dye solutions decolored by G-MgZn powders for different time at 30 ^0C. The absorption peak disappears in 10 minutes. (b) The degradation process by 4 different powders. (c) The comparison of reaction efficiency of the 4 metallic powders. (d) The high-resolution SEM image of the surface morphology of the reacted $Mg_{73}Zn_{21.5}Ca_{5.5}$ glassy powders. The nano-size reaction products distribute homogeneously on the surface of powder particle indicating high reaction activity. Reproduced from Ref. [37] with permission of Nature Publishing Group.

7.6 Materials for medicine

Owing to higher strength, larger elasticity and better corrosion resistance metallic glasses hold promise as a new generation of biomaterials. Both *in vitro* and *in vivo* testing had been done to evaluate their performance as biomaterials. They can be separated to non-biodegradable, typically: Ti-based BMGs, Zr-based BMGs, Fe-based BMGs and biodegradable, typically: Mg-based BMGs, Ca – based BMGs, Zn-based BMGs, Sr-based BMGs [38]. For example, non-biodegradable BMGs belong to Ni-free Ti-Zr-Cu-Pd [39,40] and similar alloys [41,42], Zr-Cu-Fe-Al BMGs [43], Zr-Ti-Cu-Al BMGs [44], these alloys with Nb [45], Cu and Ni-free Zr-based alloys [46]. The $(Ti_{0.45}Zr_{0.1}Pd_{0.1}Cu_{0.31}Sn_{0.4})_{98}Nb_2$ glassy alloy is also a good candidate [28]. The $Fe_{41}Co_7Cr_{15}Mo_{14}C_{15}B_6Y_2$ BMG [47] and similar alloys [48] showed good corrosion resistance in both simulated body fluids.

The $Zr_{56}Al_{16}Co_{28}$ bulk metallic glass suitable for biomedical application exhibited excellent stress corrosion cracking resistance in various simulated body fluids including Hanks', PBS and 0.9% NaCl solutions open to air at 37 °C at a strain rate ranged from 5×10^{-7} to 5×10^{-4} s^{-1} [49]. Good biocompatimility was found for a Mg-based metallic glass-polymer composite. In vivo studies reveal that the composites are biologically compatible [50]. Porous bulk metallic glassy samples can also be used as bio-implants or damping materials owing to their reduced Young's Modulus compared to monolithic BMGs. Multi-component BMGs usually contain toxic elements such as Be, Al, Ni et al., thereby degrading such bioengineering applicability. However, the studies on application of BMGs as biomaterials showed that Be-bearing BMGs are, in general, biocompatible with tissue cells [51].

Biodegradable BMGs were developed in the Mg–Zn–Ca [52 , 53] and Ca-Mg-Zn ($Ca_{65}Mg_{15}Zn_{20}$) [54] systems. Alloying with Yb improved the ductility of Mg-Zn-Ca BMGs [55]. Also in comparison with that of Yb-free control, an *in vitro* cell culture study confirms an improved biocompatibility of these Mg-based BMGs alloyed with Yb. The $Ca_{48}Zn_{30}Mg_{14}Yb_8$ BMG exhibited a low degradation rate without observable hydrogen evolution in Hank's solution [56]. The Ca-Li-Mg-Zn-, glasses are very light and have good thermoplastic formability close to room temperature but degrade rapidly [57]. Zn-based metallic glasses like $Zn_{38}Ca_{32}Mg_{12}Yb_{18}$ can also be applied [58].

The cellular responses to surfaces of biomaterials are significantly influenced by the surface roughness, topography and chemical composition. The ability to tailor surfaces to control cellular growth may pave the way for the design of novel biomaterials that can be integrated in the human body more effectively. A coating of submicro-nano hierarchically structured $Ti_{30.9}Zr_{15.0}Cu_{21.6}Pd_{30.2}$ metallic glass was created [59]. A submicro-nanostructured morphology on the surface of glassy samples was produced without any electrochemical treatment. This submicro-nano hierarchically structured glassy layer promoted cell attachment and proliferation. This indicates that introduction of a thin layer of submicro-nano structured metallic glass composite could be optimal to modify Ti-based BMGs. The $Ti_{30.9}Zr_{15.0}Cu_{21.6}Pd_{30.2}$ glassy alloy shows much finer submicro-scale surface morphology. Spherical particles ranging from 100 nm to 300 nm in size are dispersed in the relatively flat matrix. At a higher magnification, a more detailed structure can be detected. As shown in Fig. 7.10, the spherical particles are further composed of much smaller particles of about 10 nm in . The matrix also consists of similar particles of 10 nm in diameter indicating formation of a hierarchical $Ti_{30.9}Zr_{15.0}Cu_{21.6}Pd_{30.2}$ glassy structure.

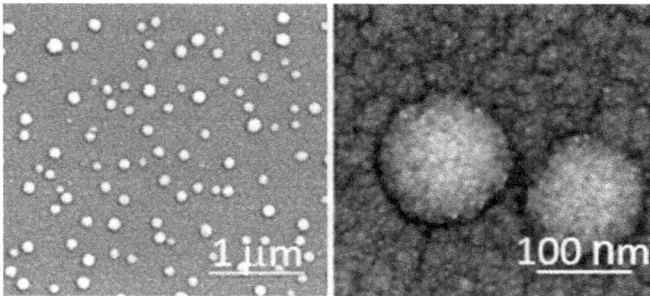

Fig. 7.10. SEM images of the $Ti_{30.9}Zr_{15.0}Cu_{21.6}Pd_{30.2}$ glassy alloy surface morphologies at different magnification. Reproduced from Ref. [59] with permission of the Royal Society of Chemistry.

According to the binding energy states of the elements involved in the $Ti_{30.9}Zr_{15.0}Cu_{21.6}Pd_{30.2}$ Ti2p spectrum was assigned to be TiO_2 and Ti metallic states. Ti is mainly combined in oxide on the surface layer. Zr3d spectrum indicated that Zr is mainly combined in ZrO_2 oxide. Cu and Pd sustain their metallic states. The XRD pattern obtained for the top thin layer indicated that the TiO_2 and ZrO_2 oxides are amorphous. They protect the underlying metallic alloys and enhance the bioactivity.

Biocompatibility of the $Ti_{30.9}Zr_{15.0}Cu_{21.6}Pd_{30.2}$ glassy alloy NGC-1 with a submicron-scale morphology (Fig. 7.11) was compared with another $Ti_{30.9}Zr_{15.0}Cu_{21.6}Pd_{30.2}$ glassy alloy NGC-2 sample consisting of clusters with a size of about 30 nm that are composed of nanometer-sized particles with a diameter of around 5 nm produced at different sputtering conditions. The NGC-2 shows a lower cell activity compared to that of NGC-1. However, the cell bioactivity of NGC-2 is still comparable with that of the Ti controls (Fig. 7.11).

Similar results were obtained for the Zr-Pd metallic nanoglasses [60]. Fine nano-structured $Zr_{62.5}Pd_{37.5}$ metallic glassy alloy exhibits a good thermal stability versus crystallization and good resistance to oxidation in dry air up to 573-673 K. The samples also show a very high corrosion resistance and spontaneous passivation in a simulated body fluid. Spontaneous passivation is achieved and low current density of about 3.5 mA/m^2 maintains up to a very high potential of 1 V. Osteoblast cells cultivation on the nanoglass show its good biocompatibility. The material has potential in biochemistry as a basis for biosensors and artificial tissue engineering.

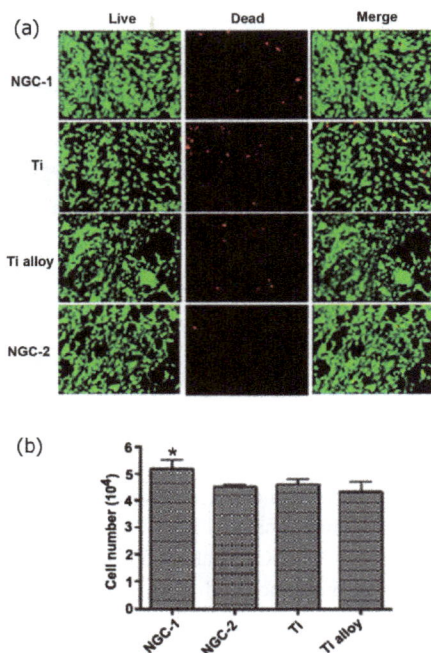

Fig. 7.11. Cell proliferation on NGC-1, NGC-2, pure Ti and Ti alloy surfaces: (a) live/ dead staining; (b) cell number, magnification: 20x. Asterisk (*) indicates statistical significance when compared with NGC-2 and pure Ti, and Ti alloy (n=6 for all the conditions). Significant differences (unpaired Student's t-test): P < 0.05. Reproduced from Ref. [59] with permission of the Royal Society of Chemistry.

A combination of magnetostriction and soft magnetic properties makes BMGs ideal candidates for developing high performance biosensors. The magnetoelastic biosensors are based on Fe–B amorphous metallic glasses. The sensors based on amorphous alloys show a very high elastic–magnetic energy conversion efficiency. The detection of Salmonella in liquid using these phage-based ME biosensors was demonstrated. The ME biosensor exhibited high sensitivity and a detection limit higher than 50 CFU/mL [61].

7.7   Catalysts

Binary and ternary metallic glasses containing Pd and Zr exhibited sufficiently high catalytic activity for several reactions such as reduction and/or oxidation reaction. The derivatives from amorphous Zr_2Pd alloy form nanocrystalline metal/ceramic composites consisting of Pd particles embedded in a ZrO_2 matrix have been shown by oxidation at high temperature [62]. A high catalytic performance for diesel soot combustion was observed for the products of the $Zr_{65}Pd_{35}$ metallic glass odiation at 800 °C [63].

Similar to the nano-structured metallic glasses discussed above Au-based metallic nano-glasses (Fig. 7.12) with large surface area produced from a BMG forming alloy using magnetron sputtering (Fig. 7.13) [64] open up a new application area of such a material as a catalyst. This material showed a good catalytic activity for the following reaction of dimethylphenylsilane ($PhMe_2Si$-H) with water in the presence of Au-based metallic nanoglass. The reaction proceeded at room temperature for 24 h and the desired dimethylphenylsilanol was obtained (Table 7.1).

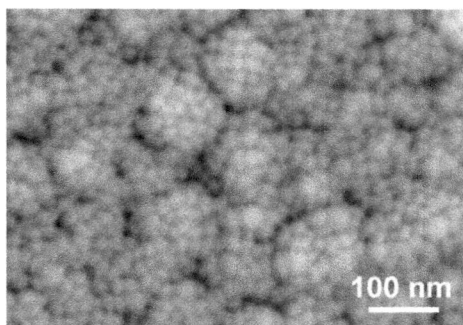

Fig. 7.12. SEM image of the as-deposited Au-based nanostructured metallic glass.

Table 7.1. Material-catalyzed oxidation of dimethylphenylsilane[a]. Reproduced from [64] with permission of Elsevier.

Entry	Route	Yield[b]
1	NA3	93
2	Reuse 1	94
3	Reuse 2	95
4	Reuse 3	100
5	Reuse 4	98

[a] Reactions were carried out using 1 (1.0 mmol), H_2O (0.1 mL), and NA3 (0.87 mol%) in 1.5 mL of acetone at RT for 24 h. [b] Isolated yield.

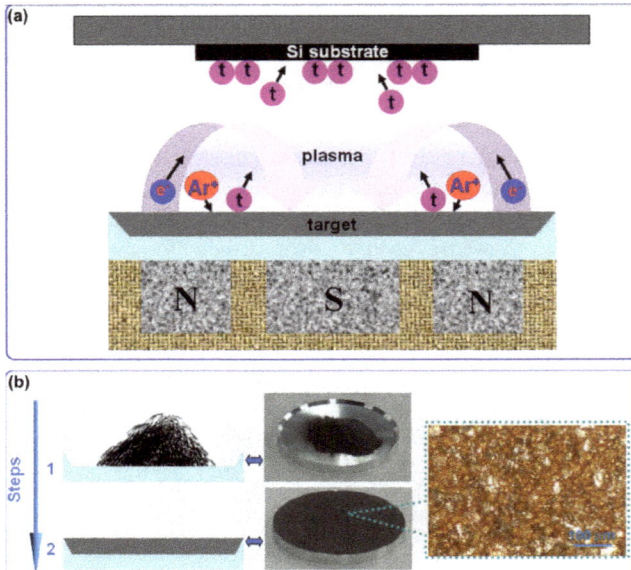

Fig. 7.13. (a) Schematic diagram for synthesis of the Au-based nanostructured metallic glassy samples. (b) The steps for preparation of powder targets. The right inset of (b) is the surface morphology of the powder targets. Reproduced from [64] with permission of Elsevier.

After electrolytic polishing of native surface oxide the Zr-Pd nanoglassy sample with an average grain size of as small as 6 nm (see Fig. 3.12) also exhibit catalytic activity in Suzuki-coupling reaction [60].

$Pt_{57.5}Cu_{14.7}Ni_{5.3}P_{22.5}$ bulk metallic glass nanowires, are applicable in electrochemical devices. High surface area nanowires exhibited good catalysts with high conductivity and activity for methanol and ethanol oxidation. After 1000 cycles, these nanowires maintain 96% of their performance exceeding that of a conventional Pt/C catalyst [65].

7.8   Radiation-steady materials

Radiation steady materials should resist swelling (increase in volume) and embrittlement. Metallic glasses have drawn interest as nuclear damage steady materials [66]. Under neutron irradiation to a total fluence of 10^{19} n/cm^2 the density of the $(Mo_{0.6}Ru_{0.4})_{82}B_{18}$ metallic glass decreased by 1.5% and ductility improved [67]. Some thermally embrittled metallic glasses even improved their ductility being irradiated [68]. In-vessel mirrors are necessary for optical diagnostics of plasmas in fusion devices under the influence of the harsh fusion environment. Optcial characteristics of Zr-based bulk metallic glasses were studied under the impact of deuterium and argon plasma ions of different energy. The mirrors made of BMGs preserve the initial optical quality after long-term irradiation treeatment in the erosion-dominated zone of a fusion reactor [69]. The microstructural changes induced by neutron irradiation were studied in the samples irradiated with neutron fluence of of 1.0×10^{15} n/cm^2 with an average kinetic energy of 1.12 MeV. Structural self-healing effect related to the irradiation-induced vacancy-like defects annihilation is observed [70]. There is a peak split in the first-shell Zr-edge partial PRDF of the as-prepared sample, while there is only one peak in that of the irradiated one. Meanwhile, Zr-Zr and Zr-Cu bond are very similar to those of the irradiated one. In some Zr-base glasses the length difference between Zr-Zr and Zr-Cu pairs probably contributes to a peak split in the first-shell of Zr-edge PRDF [71]. In a neutron-irradiated sample, the peak split in Zr-edge PRDF is weakened, probably due to the rearrangement of neighbor atoms around Zr centers, although the average Zr-Zr and Zr-Cu bond lengths are barely changed.

7.9 Final remarks and future prospects

Due to high cost of the main component palladium for a long time the initially-discovered BMGs were not of much interest to the scientists and engineers until they have been produced in alloys based on the industrial metals such as iron, copper, magnesium and titanium, which has opened many opportunities for their application. Bulk metallic glasses having high strength, hardness, wear resistance, large elastic deformation and high resistance to corrosion are readily applicable as structural and functional materials. At the same time the fundamental, not fully solved scientific problems, are: description of the structure of metallic glasses, processes of their plastic deformation and glass-transition on cooling, while the technical challenge is related to further increase in the tensile plasticity and impact fracture toughness of BMGs and their composites. Based on the full set of the results obtained up to date one can expect further development of the field and significant progress in it in the near future.

References

[1] H. Kakiuchi, A. Inoue, M. Onuki, Y. Takano, T. Yamaguchi, Application of Zr-based bulk glassy alloys to golf clubs, Mater. Trans., 42 (2001) 678-681. https://doi.org/10.2320/matertrans.42.678

[2] W.L. Johnson, Bulk amorphous metal–An emerging engineering material, JOM, 54 (2002), 40-43. https://doi.org/10.1007/BF02822619

[3] A. Inoue, N. Nishiyama, New bulk metallic glasses for applications as magnetic-sensing, chemical, and structural materials, MRS Bull., 32 (2007) 651-658. https://doi.org/10.1557/mrs2007.128

[4] A.I. Salimon, M.F. Ashby, Y. Bréchet, A.L. Greer, Bulk metallic glasses: what are they good for? Materials Science and Engineering: A, 375–377 (2004) 385–388. https://doi.org/10.1016/j.msea.2003.10.167

[5] L. Zhuo, B. Yang, H. Wang, T. Zhang, Spray formed Al-based amorphous matrix nanocomposite plate, J. Alloys Compd., 509 (2011) L169-L173. https://doi.org/10.1016/j.jallcom.2011.02.125

[6] Z.Q. Chen, L. Huang, F. Wang, P. Huang, T.J. Lu, K.W. Xu, Suppression of annealing-induced embrittlement in bulk metallic glass by surface crystalline coating, Materials and Design, 109 (2016) 179-185. https://doi.org/10.1016/j.matdes.2016.07.069

[7] Y. Saotome, S. Miwa, T. Zhang, A. Inoue, The micro-formability of Zr-based amorphous alloys in the supercooled liquid state and their application to micro-dies, Journal of Materials Processing Technology, 113 (2001) 64-69. https://doi.org/10.1016/S0924-0136(01)00605-7

[8] Y. Saotome, K. Imai, S. Shioda, S. Shimizu, T. Zhang, A. Inoue, The micro-nanoformability of Pt-based metallic glass and the nanoforming of three-dimensional structures, Intermetallics, 10 (2002) 1241-1247. https://doi.org/10.1016/S0966-9795(02)00135-8

[9] A. Inoue, B. L. Shen and A. Takeuchi, Developments and applications of bulk glassy alloys in late transition metal base system, Mater. Trans., 47 (2006) 1275-1285. https://doi.org/10.2320/matertrans.47.1275

[10] A. Caron, C. L. Qin, L. Gu, S. Gonzalez, A. Shluger, H.-J. Fecht, D. V. Louzguine-Luzgin and A. Inoue, Structure and nano-mechanical characteristics of surface oxide layers on a metallic glass, Nanotechnology, 22 (2011) 095704. https://doi.org/10.1088/0957-4484/22/9/095704

[11] A. Caron, P. Sharma, A. Shluger, H.-J. Fecht, D. V. Louzguine-Luzgin, and A. Inoue, Effect of surface oxidation on the nm-scale wear behavior of a metallic glass, Journal of Applied Physics, 109 (2011) 083515. https://doi.org/10.1063/1.3573778

[12] A. Caron, D. V. Louzguine-Luzgin, and R. Bennewitz, Structure vs chemistry: friction and wear of Pt-based metallic surfaces, ACS Appl. Mater. Interfaces, 5 (2013) 11341–11347. https://doi.org/10.1021/am403564a

[13] J.-W. Lee, Y.-C. Lin, N. Chen, D. V. Louzguine, M. Esashi, and T. Gessner, Development of the large scanning mirror using Fe-based metallic glass ribbon, Japanese Journal of Applied Physics, 50 (2011) 087301. https://doi.org/10.7567/JJAP.50.087301

[14] C. Ma, N. Nishiyama and A. Inoue, Fabrication and high performance characteristics of metallic glassy alloy tubes, J. Metastab. Nanocryst., 24-25 (2005) 515-518. https://doi.org/10.4028/www.scientific.net/JMNM.24-25.515

[15] S. Yamaura, Y. Shimpo, H. Okouchi, M. Nishida, O. Kajita, H.M. Kimura, A. Inoue, Hydrogen permeation characteristics of melt-spun Ni-Nb-Zr amorphous alloy membranes, Materials Transactions, 44 (2003) 1885. https://doi.org/10.2320/matertrans.44.1885

[16] K. S. Nakayama, Y. Yokoyama, T. Ono, M. W. Chen, K. Akiyama, T. Sakurai, A. Inoue, Controlled formation and mechanical characterization of metallic glassy nanowires, Advanced Materials, 22 (2010) 863. https://doi.org/10.1002/adma.200902295

[17] A. Inoue, I. Yoshii, H. Kimura, K. Okumura and J. Kurosaki, Enhanced shot peening effect for steels by using Fe-based glassy alloy shots, Mater. Trans., 44 (2003) 2391-2395. https://doi.org/10.2320/matertrans.44.2391

[18] A. Inoue, F.L. Kong, Q.K. Man, B.L. Shen, R.W. Li, F. Al-Marzouki, Development and applications of Fe- and Co-based bulk glassy alloys and their prospects, Journal of Alloys and Compounds, 615 (2014) S2–S8. https://doi.org/10.1016/j.jallcom.2013.11.122

[19] H. Matsumoto, A. Urata, Y. Yamada, A. Inoue, Novel $Fe_{(97-x-y)}P_xB_yNb_2Cr_1$ glassy alloys with high magnetization and low loss characteristics for inductor core materials, IEEE Trans. Magn., 46 (2010) 373–376. https://doi.org/10.1109/TMAG.2009.2033708

[20] A. Inoue, F.L. Kong, Y. Han, S.L. Zhua, A.Churyumov, E. Shalaan, F. Al-Marzouki, Development and application of Fe-based soft magnetic bulk metallic glassy inductors, Journal of Alloys and Compounds, 731 (2018) 1303-1309. https://doi.org/10.1016/j.jallcom.2017.08.240

[21] H. Matsumoto, A. Urata, Y. Yamada, A. Inoue, FePBNbCr soft-magnetic glassy alloys with low loss characteristics for inductor cores, J. Alloys Comp., 504 (2010) S139–S141. https://doi.org/10.1016/j.jallcom.2010.03.029

[22] W.J. Liu, H.X. Zhang, J.A. Shi, Z.C. Wang, C. Song, X.G. Wang, S. Lu, X.G. Zhou, L. Gu, D. V. Louzguine-Luzgin, M.W. Chen, K.F. Yao and N. Chen, A room-temperature

magnetic semiconductor from a ferromagnetic metallic glass, Nature Communications, 7 (2016) 13497. https://doi.org/10.1038/ncomms13497

[23] A. Inoue and, T. Zhang, Thermal stability and glass-forming ability of amorphous Nd-Al-TM (TM = Fe, Co, Ni or Cu) alloys, Mater. Sci. Eng. A, 226–228 (1997) 393. https://doi.org/10.1016/S0921-5093(97)80050-0

[24] S. Hiromoto, K. Asami, A. P. Tsai, T. Hanawa, Surface characterization of amorphous Zr-Al-(Ni, Cu) alloys immersed in cell-culture medium, Mater. Trans. JIM, 43 (2002) 261-266. https://doi.org/10.2320/matertrans.43.261

[25] S. Pang, T. Zhang, H. Kimura, K. Asami, A. Inoue, Corrosion behavior of Zr-(Nb-)Al-Ni-Cu glassy alloys, Mater. Trans. JIM, 41 (2000) 1490-1494. https://doi.org/10.2320/matertrans1989.41.1490

[26] V. R. Raju, U. Kühn, U. Wolff, F. Schneider, J. Eckert, R. Reiche, A. Gebert, Corrosion behaviour of Zr-based bulk glass-forming alloys containing Nb or Ti, Mater. Lett., 57 (2002) 173-177. https://doi.org/10.1016/S0167-577X(02)00725-5

[27] C. –L. Qin, W. Zhang, K. Amiya, K. Asami and A. Inoue, Mechanical properties and corrosion behavior of $(Cu_{0.6}Hf_{0.25}Ti_{0.15})_{90}Nb_{10}$ bulk metallic glass composites, Materials Science and Engineering: A, 449–451 (2007) 230-234. https://doi.org/10.1016/j.msea.2006.02.244

[28] J.J. Oak, D. V. Louzguine-Luzgin, A. Inoue, Investigation of glass-forming ability, deformation and corrosion behavior of Ni-free Ti-based BMG alloys designed for application as dental implants, Materials Science and Engineering C, 29 (2009) 322–327. https://doi.org/10.1016/j.msec.2008.07.009

[29] G.-H. Zhao, R. E. Aune, H. Mao, N. Espallargas, Degradation of Zr-based bulk metallic glasses used in load-bearing implants: A tribocorrosion appraisal, Journal of the Mechanical Behavior of Biomedical Materials, 60 (2016) 56–67. https://doi.org/10.1016/j.jmbbm.2015.12.024

[30] U. Koster, L. Jastrow, Oxidation of Zr-based metallic glasses and nanocrystalline alloys, Mater. Sci. Eng. A, 449 (2007) 57-59. https://doi.org/10.1016/j.msea.2006.02.316

[31] F. Reichel, L.P.H. Jeurgens, E.J. Mittemeijer, The thermodynamic stability of amorphous oxide overgrowths on metals, Acta Materialia, 56 (2008) 659. https://doi.org/10.1016/j.actamat.2007.10.023

[32]. S.K. Sharma, T. Strunskus, H. Ladebusch, V. Zaporojtchenko, F. Faupel, XPS study of the initial oxidation of the bulk metallic glass $Zr_{46.75}Ti_{8.25}Cu_{7.5}Ni_{10}Be_{27.5}$, J. Mater. Sci., 43 (2008) 5495. https://doi.org/10.1007/s10853-008-2834-4

[33] D.V. Louzguine-Luzgin, C.L. Chen, L.Y. Lin, Z.C. Wang, S.V. Ketov, M.J. Miyama, A.S. Trifonov, A.V. Lubenchenko and Y. Ikuhara, Bulk metallic glassy surface native oxide: Its atomic structure, growth rate and electrical properties, Acta Materialia, 97 (2015) 282–290. https://doi.org/10.1016/j.actamat.2015.06.039

[34] M.Y. Na, Y.J. Kim, W.T. Kim, D.H. Kim, Oxidation behavior of Ti-Zr-Ni-Cu metallic glass, Corrosion Science, 163 (2020) 108271. https://doi.org/10.1016/j.corsci.2019.108271

[35] A. S. Trifonov, A. V. Lubenchenko, V. I. Polkin, A. B. Pavolotsky, S. V. Ketov and D. V. Louzguine-Luzgin, Difference in charge transport properties of Ni-Nb thin films with native and artificial oxide, Journal of Applied Physics, 117 (2015) 125704. https://doi.org/10.1063/1.4915935

[36] J.-Q. Wang, Y.-H. Liu, M.-W. Chen, G.-Q. Xie, D. V. Louzguine-Luzgin, A. Inoue, and J. H. Perepezko, Rapid degradation of azo dye by Fe-based metallic glass powder, Advanced Functional Materials, 22 (2012) 2567–2570. https://doi.org/10.1002/adfm.201103015

[37] J.-Q. Wang, Y.-H. Liu, M.-W. Chen, D. V. Louzguine-Luzgin, A. Inoue and J. H. Perepezko, Excellent capability in degrading azo dyes by MgZn-based metallic glass powders, Scientific Reports, 2 (2012) 418. https://doi.org/10.1038/srep00418

[38] H.F. Li, Y.F. Zheng, Recent advances in bulk metallic glasses for biomedical applications, Acta Biomaterialia, 36 (2016) 1–20. https://doi.org/10.1016/j.actbio.2016.03.047

[39] S.L. Zhu, X.M. Wang, F.X. Qin, A. Inoue, A new Ti-based bulk glassy alloy with potential for biomedical application, Mater. Sci. Eng. A, 459 (2007) 233–242. https://doi.org/10.1016/j.msea.2007.01.044

[40] J.-J. Oak, D.V. Louzguine-Luzgin, A. Inoue, Fabrication of Ni-free Ti-based bulk-metallic glassy alloy having potential for application as biomaterial, and investigation of its mechanical properties, corrosion, and crystallization behavior, J. Mater. Res., 22 (2007) 1346–1353. https://doi.org/10.1557/jmr.2007.0154

[41] Y.B. Wang, H.F. Li, Y. Cheng, Y.F. Zheng, L.Q. Ruan, In vitro and in vivo studies on Ti-based bulk metallic glass as potential dental implant material, Mater. Sci. Eng. C, 33 (2013) 3489–3497. https://doi.org/10.1016/j.msec.2013.04.038

[42] S.J. Pang, Y. Liu, H.F. Li, L.L. Sun, Y. Li, T. Zhang, New Ti-based Ti-Cu-Zr-Fe-Sn-Si-Ag bulk metallic glass for biomedical applications, J. Alloys Compd., 625 (2015) 323–327. https://doi.org/10.1016/j.jallcom.2014.07.021

[43] L. Huang, C. Pu, R.K. Fisher, D.J.H. Mountain, Y. Gao, P.K. Liaw, A Zr-based bulk metallic glass for future stent applications: materials properties, finite element modeling, and in vitro human vascular cell response, Acta Biomater., 25 (2015) 356–368. https://doi.org/10.1016/j.actbio.2015.07.012

[44] J. Li, L.L. Shi, Z.D. Zhu, Q. He, H.J. Ai, J. Xu, $Zr_{61}Ti_2Cu_{25}Al_{12}$ metallic glass for potential use in dental implants: biocompatibility assessment by in vitro cellular responses, Mater. Sci. Eng. C, 33 (2013) 2113–2121. https://doi.org/10.1016/j.msec.2013.01.033

[45] L. Liu, C.L. Qiu, C.Y. Huang, Y. Yu, H. Huang, S.M. Zhang, Biocompatibility of Ni-free Zr-based bulk metallic glasses, Intermetallics, 17 (2009) 235–240. https://doi.org/10.1016/j.intermet.2008.07.022

[46] Z. Liu, K.C. Chan, L. Liu, Development of Ni- and Cu-Free Zr-based bulk metallic glasses for biomedical applications, Mater. Trans., 52 (2011) 61–67. https://doi.org/10.2320/matertrans.M2010068

[47] Y.B. Wang, H.F. Li, Y. Cheng, S.C. Wei, Y.F. Zheng, Corrosion performances of a Nickel-free Fe-based bulk metallic glass in simulated body fluids, Electrochem. Commun., 11 (2009) 2187–2190. https://doi.org/10.1016/j.elecom.2009.09.027

[48] Y.B. Wang, H.F. Li, Y.F. Zheng, M. Li, Corrosion performances in simulated body fluids and cytotoxicity evaluation of Fe-based bulk metallic glasses, Mater. Sci. Eng. C, 32 (2012) 599–606. https://doi.org/10.1016/j.msec.2011.12.018

[49] A. Kawashima, T. Wada, K. Ohmura, G. Xie, A. Inoue, A Ni- and Cu-free Zr-based bulk metallic glass with excellent resistance to stress corrosion cracking in simulated body fluids, Mater. Sci. Eng. A, 542 (2012) 140–146. https://doi.org/10.1016/j.msea.2012.02.047

[50] A. Sharma, A. Kopylov, M. Zadorozhnyy, A. Stepashkin, V. Kudelkina, J.-Q. Wang, S. Ketov, M. Churyukanova, D. Louzguine-Luzgin, B. Sarac, J. Eckert, S. Kaloshkin, V. Zadorozhnyy, H. Kato, Mg-based metallic glass-polymer composites: investigation of structure, thermal properties, and biocompatibility, Metals, 10 (2020) 867. https://doi.org/10.3390/met10070867

[51] Y. B. Wang, Y. F. Zheng, S. C. Wei, M. Li, In vitro study on Zr-based bulk metallic glasses as potential biomaterials, J. Biomed. Mater. Res. B, 96 (2011) 34. https://doi.org/10.1002/jbm.b.31725

[52] B. Zberg, P.J. Uggowitzer, J.F. Loffler, MgZnCa glasses without clinically observable hydrogen evolution for biodegradable implants, Nature Materials, 8 (2009) 887–891. https://doi.org/10.1038/nmat2542

[53] X.N. Gu, Y.F. Zheng, S.P. Zhong, T.F. Xi, J.Q. Wang, W.H. Wang, Corrosion of, and cellular responses to Mg-Zn-Ca bulk metallic glasses, Biomaterials, 31 (2010) 1093–1103. https://doi.org/10.1016/j.biomaterials.2009.11.015

[54] Y. B. Wang, X. H. Xie, H. F. Li, X. L. Wang, M. Z. Zhao, E. W. Zhang, Y. J. Bai, Y. F. Zheng, L. Qin, Biodegradable CaMgZn bulk metallic glass for potential skeletal application, Acta Biomaterialia, 7 (2011) 3196–3208. https://doi.org/10.1016/j.actbio.2011.04.027

[55] H.-J. Yu, J.-Q. Wang, X.-T. Shi, D.V. Louzguine-Luzgin, H.-K. Wu, J.H. Perepezko, Ductile biodegradable Mg-based metallic glasses with excellent biocompatibility, Advanced Functional Materials, 23 (2013) 4793–4800. Functional Materials, 23 (2013) 4793-4800.

[56] W. Jiao, H.F. Li, K. Zhao, H.Y. Bai, Y.B. Wang, Y.F. Zheng, W.H. Wang, Development of CaZn based glassy alloys as potential biodegradable bone graft substitute, Journal of Non-Crystalline Solids, 357 (2011) 3830–3840. https://doi.org/10.1016/j.jnoncrysol.2011.08.003

[57] J.Fu, L.J.Qiang, W.X.Feng, L.Kun, Z.Bo, Z.H.Yang, B.M.Xiang, M.X. Pan and W.H. Wang, Glassy metallic plastics, Science China Physics, Mechanics and Astronomy, 53, (2010), 409–414. https://doi.org/10.1007/s11433-010-0138-6

[58] W. Jiao, K. Zhao, X.K. Xi, D.Q. Zhao, M.X. Pan, W.H. Wang, Zinc-based bulk metallic glasses, Journal of Non-Crystalline Solids, 356 (2010) 1867–1870. https://doi.org/10.1016/j.jnoncrysol.2010.07.017

[59] N. Chen, X. Shi, R. Witte, K. S. Nakayama, K. Ohmura, H. K. Wu, A. Takeuchi, H. Hahn, M. Esashi, H. Gleiter, A. Inoue and D. V. Louzguine, A novel Ti-based nanoglass composite with submicron–nanometer-sized hierarchical structures to modulate osteoblast behaviors, J. Mater. Chem. B, 1 (2013) 2568–2574. https://doi.org/10.1039/c3tb20153h

[60] S. V. Ketov, X.T. Shi, G.Q. Xie, R. Kumashiro, A. Yu. Churyumov, A. I. Bazlov, N. Chen, Y. Ishikawa, N. Asao, H.K. Wu and D. V. Louzguine-Luzgin, Nanostructured Zr-Pd metallic glass thin film for biochemical applications, Scientific Reports, 5 (2015) 7799. https://doi.org/10.1038/srep07799

[61] S. Li, S. Horikawa, M.K, Park, Y. Chai, V. J. Vodyanoy, B. A. Chin, Amorphous metallic glass biosensors, Intermetallics, 30 (2012) 80-85. https://doi.org/10.1016/j.intermet.2012.03.030

[62] H. Kimura, A. Inoue, T. Masumoto, Production of nanocrystalline ZrO_2 and Pd composites by oxidation of an amorphous $Zr_{65}Pd_{35}$ alloy, Mat. Lett., 14 (1992) 232-236. https://doi.org/10.1016/0167-577X(92)90162-D

[63] M. Ozawa, N. Katsuragawa, A. Masuda, M. Hattori, S. Yamaura, Microstructure and soot-combustion catalysis of oxidized $Zr_{60}Ce_5Pd_{30}Pt_5$ amorphous alloy, J. Jpn. Soc. Powder Powder Metall., 65 (2018) 191-193. https://doi.org/10.2497/jjspm.65.191

[64] N. Chen, R. Frank, N. Asao, D. V. Louzguine-Luzgin, P. Sharma, J. Q. Wang, G. Q. Xie, Y. Ishikawa, N. Hatakeyama, Y. C. Lin, M. Esashi, Y. Yamamoto and A. Inoue, Formation and properties of Au-based nanograined metallic glasses, Acta Materialia, 59 (2011) 6433. https://doi.org/10.1016/j.actamat.2011.07.007

[65] M. Carmo, R. C. Sekol, S. Ding, G. Kumar, J. Schroers, and A. D. Taylor, Bulk metallic glass nanowire architecture for electrochemical applications, ACS Nano, 5 (2011) 2979–2983. https://doi.org/10.1021/nn200033c

[66] J. Carter, E.G. Fu, G. Bassiri, B.M. Dvorak, N.D. Theodore, G.Q. Xie, D.A. Lucca, M. Martin, M. Hollander, X.H. Zhang, L. Shao, Effects of ion irradiation in metallic glasses, Nucl. Instrum. Methods Phys. Res., Sect. B, 267 (2009) 1518–1521. https://doi.org/10.1016/j.nimb.2009.01.081

[67] E. A. Kramer, W. L. Johnson, and C. Cline, The effects of neutron irradiation on a superconducting metallic glass, Appl. Phys. Lett., 35 (1979) 815. https://doi.org/10.1063/1.90947

[68] R. Gerling, F. P. Schimansky and R. Wagner, Restoration of the ductility of thermally embrittled amorphous alloys under neutron-irradiation, Acta Metall., 35 (1987) 1001. https://doi.org/10.1016/0001-6160(87)90047-2

[69] V. S. Voitsenya, A. F. Bardamid, A. I. Belyaeva, V. N. Bondarenko, A. A. Galuza, V. G. Konovalov, I. V. Ryzhkov, A. A. Savchenko, A. N. Shapoval, A. F. Shtan', S. I. Solodovchenko and K. I. Yakimov, Modification of optical characteristics of metallic amorphous mirrors under ion bombardment, Plasma Devices and Operations, 17 (2009) 144-154. https://doi.org/10.1080/10519990902903595

[70] L. Yang, H. Y. Li, P. W. Wang, S. Y. Wu, G. Q. Guo, B. Liao, Q. L. Guo, X. Q. Fan, P. Huang, H. B. Lou, F. M. Guo, Q. S. Zeng, T. Sun, Y. Ren and L. Y. Chen, Structural responses of metallic glasses under neutron irradiation, Scientific Reports, 7 (2017) 16739. https://doi.org/10.1038/s41598-017-17099-2

[71] G. Q. Guo, S. Y. Wu, S. Luo & L. Yang, Detecting structural features in metallic glass via synchrotron radiation experiments combined with simulations, Metals, 5 (2015) 2093–2108. https://doi.org/10.3390/met5042093

About the Author

Dmitri V. LOUZGUINE

Address: WPI Advanced Institute for Materials Research, Tohoku University, 2-1-1 Katahira, Aoba-Ku, Sendai, 980-8577, Japan.

Since 2007 he is the Professor and Principal Investigator at WPI Advanced Institute for Materials Research, Tohoku University, Japan.

RESEARCH ACTIVITIES

Investigation of the formation mechanisms, structure, phase transformations, deformation behavior, mechanical and physical properties of metallic glasses, nanostructured glassy-crystalline/quasicrystalline materials, composite materials having a mixed structure containing nanoparticles embedded in the glassy matrix and high-strength crystalline alloys. Long-term experience in various types of structural analysis techniques, in particular, Transmission Electron Microscopy (TEM) of nanostructured and glassy materials including nanobeam diffraction and high-resolution TEM imaging. X-ray diffraction analysis including creation of the radial distribution functions from Synchrotron-radiation XRD data.

Computer-simulation to materials science (phase transformations, crystallography).

In general, the area of expertise is Physical Metallurgy of metals and alloys which establishes the relationships between the structure of materials and their properties. This subject includes the following topics: solidification, micro- and nano-structure, crystallography, phase transformations, processing, mechanical properties and deformation behavior of metallic materials.